SHAPING SCIENCE AND
TECHNOLOGY POLICY

# SCIENCE AND TECHNOLOGY IN SOCIETY

*Series Editors*

Daniel Lee Kleinman
Jo Handelsman

# Shaping Science and Technology Policy

## *The Next Generation of Research*

Edited by

### DAVID H. GUSTON

and

### DANIEL SAREWITZ

THE UNIVERSITY OF WISCONSIN PRESS

The University of Wisconsin Press
1930 Monroe Street
Madison, Wisconsin 53711

www.wisc.edu/wisconsinpress/

3 Henrietta Street
London WC2E 8LU, England

1     3     5     4     2

Printed in the United States of America

Library of Congress Cataloging-in-Publication Data
Shaping science and technology policy: the next generation of research /
edited by David H. Guston and Daniel Sarewitz.
p.   cm.—(Science and technology in society)
Includes bibliographical references and index.
ISBN 0-299-21910-0 (cloth: alk. paper)
ISBN 0-299-21914-3 (pbk.: alk. paper)
1. Science and state—Decision making.
2. Technology and state—Decision making.
3. Research—International cooperation.
4. Science and state—Citizen participation.
5. Technology and state—Citizen participation.
I. Guston, David H.   II. Sarewitz, Daniel R.   III. Series.
Q125.S5164        2006
338.9′26—dc22        2006008594

# CONTENTS

# FOREWORD

LEWIS M. BRANSCOMB

The best view of the future is through the eyes of younger scholars who are not yet committed to the view that the future is an evolutionary extension of the past. The work of these authors, selected competitively from a large number of candidates, addresses a variety of new issues from original perspectives. Many of the ideas they present will not find quick acceptance in the federal bureaucracy. They are explorations of how science and technology ought to reflect the consensus goals of society.

The new perspective takes a more integrated view of policy for science and science for policy, recognizing that each is a driver of the other. Although the papers are organized into sections for policy, for science, for technology, and for the new genetics, this generation of scholars no longer wastes time fussing over the distinctions among these categories. Their view is as often global as it is domestic. Nor is there even a whiff of technological determinism in this writing. The outcomes of debates about policy are, in the authors' views, clearly shaped by social, cultural, and political forces. But neither do the authors fall into the trap of social construction; both technical facts and gaps in knowledge command attention here.

David Guston and Dan Sarewitz have made an important contribution to the health of science and technology policy studies in identifying and giving visibility to this group of up-and-coming scholars. In their work there are agendas for research that deserve support from both government funding agencies and foundations.

Unfortunately, there are very few places where science and technology policy researchers, especially the younger generation of scholars, can find research support. The unintended consequence of this unfortunate situation is that the younger researchers are often looking to funded problem-solving projects outside academic support, with the result that their tendency to look at both theoretical constructs and practical ideas concurrently is reinforced.

This conference is the second of its kind, to my knowledge, the first having taken place in Hawaii some years ago. It ought to become a periodic event, with both the conference and the publication of its papers supported by one or more foundations or federal agencies. So long as science and technology policy research has no institutional home in the United States, mechanisms to bring the younger investigators together occasionally are particularly important.

# ACKNOWLEDGMENTS

The debt owed to individuals and organizations for the production of any book is substantial. The debt accumulated for this volume may be a little more so, as it is the product of a great many talented hands and minds participating over a number of years in the "Next Gen" project. Derived from a conference sponsored by the National Science Foundation (NSF), the chapters of this volume represent but a fraction of the scholarly and professional effort embodied in the Next Generation project.

We owe Rachelle Hollander and Joan Siebert at NSF, along with anonymous reviewers, many thanks for helping us refine and, ultimately, for approving our proposal for the conference and this volume. To our program committee—Steve Nelson, Lewis Branscomb, Sharon Dunwoody, Diana Hicks, Gene Rochlin, Paula Stephan, Willie Pearson, Mike Quear, and Chuck Weiss—we offer our esteem and gratitude for helping us whittle down the original ninety proposals we received in overwhelming response to our solicitation, to the roughly two dozen younger scholars we invited to speak at the conference. Several members of the program committee also served as discussants at the conference, as did Barry Bozeman, Andrew Reynolds, Helga Rippen, Christopher Hill, David Goldston, and Lee Zwanziger.

The project was a cooperative endeavor of Rutgers, The State University of New Jersey, where one of us then taught, and the Center for Science, Policy, and Outcomes, then of Columbia University, which the other of us directs. Linda Guardabascio and Fran Loeser at Rutgers were remarkably helpful with the grant management side of the equation, and Ellen Oates with the administrative and editorial side. At CSPO, Stephen Feinson and Shep Ryen kept the project on track. We must single out Misty Wing and Lori Hidinger of the renamed Consortium for Science, Policy, and Outcomes (reconstituted at Arizona State University), for their invaluable assistance.

The American Association for the Advancement of Science, through the good works of Steve Nelson and the Directorate for Science and Policy Programs, provided a marvelous venue and logistical support for the conference.

We thank Daniel Lee Kleinman and Jo Handelsman, our series editors at the University of Wisconsin Press, for their encouragement of the project, despite its long time in coming.

Finally, and obviously, the "next generation of leaders in science and technology policy" themselves have earned our thanks, from the more than eighty younger scholars who responded to our request for proposals to the two dozen who presented their research at the conference to the sixteen who appear in this volume. They have given us great enthusiasm for the future of scholarship in science and technology policy, and concrete reasons to anticipate it.

The editors gratefully acknowledge funding from the National Science Foundation under Grant No. SES 0135170, which supported the activities that culminate in the publication of this volume. NSF is not responsible for the findings herein, which are the responsibility of the authors and editors. Any opinions, finding, and conclusions or recommendations expressed in this material are those of the authors and do not necessarily reflect the views of the National Science Foundation.

DAVID H. GUSTON
DANIEL SAREWITZ

SHAPING SCIENCE AND
TECHNOLOGY POLICY

# Introduction

In the 1960s and early 1970s, a survey of the science and technology policy landscape in the United States would have revealed not only an arena of energetic discussion and institutional innovation but also justifiable hopes for a future where theoreticians, empiricists, policy analysts, and policy practitioners would be working together to help shape the nation's research and development enterprise. Yet such hopes have, at best, been partially fulfilled. While science and technology policy research continues to yield important new insights in such areas as the relation between democracy and scientific advance, the role of technological innovation in economic development, and the design of institutions at the interface of science and policy, the quality of public discourse about science policy has changed little over the decades and retains its monomaniacal focus on marginal budgetary increments. While science and technology studies curricula have become regular features of the academic portfolio, few of the university science policy programs initiated in the 1960s and 1970s retain their original energy and scope. The death of the Office of Technology Assessment, the evisceration of the power of the president's science advisor, the "science wars" and the

Sokal hoax, and the transformation of *Technology Review* from provoca-
teur to cheerleader are but a few of the tangible indicators of ambitions
unfulfilled.

Science and technology have meanwhile become, if at all possible,
even more powerful agents of change, increasingly challenging society's
capacity to respond. The obviousness of this fact is rendered only more
conspicuous by the very limited influence that conceptual advances in
science policy have had on real-world practice. More than fifty years
ago, the debate between Vannevar Bush and Harley Kilgore forged the
shape of the nation's R&D enterprise, while in subsequent decades the
formative voices of Harvey Brooks, Alvin Weinberg, Michael Polanyi,
and others helped to set the terms of both the public and intellectual science policy agenda. If those terms now seem outdated or insufficient,
we must nevertheless ask: Where are the voices who will guide science
and technology policy in this new century? This book is our effort to answer the question.

With somewhat modest hopes—hopes that were bolstered by a
grant from the National Science Foundation—we issued a solicitation
for abstracts from young scholars and practitioners in science policy,
and were rewarded with ninety proposals from a gratifyingly broad
range of disciplines and institutions (and even a few countries). With the
assistance of a review committee of established and respected voices
in science policy, we reduced this field to twenty-five. We invited these
people to participate in a conference in Washington, DC, where they
presented their ideas and subjected them to the critique of both academic experts and practitioners in their specific areas of interest. From
those twenty-five, we then chose sixteen papers, the cream of a select
crop, to publish. It is rare that the individual contributions to a multi-
authored volume are subject to so thorough a review.

These sixteen papers thus sample the thinking of some of the best
young minds in science policy—academicians and practitioners both.
The papers are unified by a small number of cross-cutting themes: Who
is making the choices about science and technology policy, and who
stands to win or lose from these choices? What criteria of choice are
being used, and by whom? What standards of governance are being
employed, and by whom? Issues of globalization, of the protean boundary between public and private, of the factors that control how benefits
are distributed, all loom large. More fundamental still, these perspectives are unified by the common recognition that science policy is not so

much about science per se as about the explicit shaping of our world. This shaping occurs in the settings where science itself is administered and governed (Section 1: Shaping Policy); in the processes of knowledge production (Section 2: Shaping Science); in the forces behind innovation (Section 3: Shaping Technology); and, ever more conspicuously, in the drive to manipulate and transform the human species itself (Section 4: Shaping Life).

Yet this coherence of theme and purpose is presented through a refreshing diversity of analytical lenses: historical perspectives on communications technologies and intellectual property; philosophical inquiry into the relation between environmental science and political discourse; case-based illustrations of how the forces of globalization collide with the power of local politics and preferences. From game theory to constructivist studies to econometrics to the tactics of grassroots organizing, this collection's methodological and narrative diversity reveals a vibrant intellectual enterprise with much to offer our complex and troubled world.

This is not to say (heaven forfend!) that the authors all believe, with modern thought, that we can know and control all that we make. Rather, with a remarkable sophistication and ideological openness, the authors are telling us that organized scrutiny can lead to incremental improvement in confronting real problems faced by real decision-makers—in areas as diverse as urban transportation, technical standards-setting, and the governance of human biotechnology.

We are pleased with this volume to announce an important societal resource for the new century: the next generation of science and technology policy research.

# PART I

# SHAPING POLICY

Science and technology policy helps to initiate and shape knowledge-based innovation. But the perception endures that the research produces the innovation, and the policy is secondary or even beside the point of technical success—the evaporation of the government role in the creation of the Internet stands as a prominent example. We tend to sleight attention to those policies as either contributory to the success of innovation or as worthy of attention in and of themselves.

The chapters in this section maintain that well-reasoned, well-articulated, and well-crafted policies are critical for achieving public value from the public good of scientific research. They evaluate policies for science and technology, and then show how they can be shaped across a breadth of scales from the overarching rationales of the public research enterprise and its accountability relationships in a democratic society to more nuts-and-bolts considerations of specific innovation policies.

Mark Brown, one of the few political theorists working on science policy, opens the section with a critique of elite science policy making. His criticism falls both on the overall enterprise of public science, for failing to incorporate the actual public in shaping the research agenda, but also on the enterprise of research ethics, for failing to represent the public in a fully democratic way. Brown compares four approaches to setting the agenda for publicly funded research. The first and most familiar approach grants scientific communities sole authority to determine which questions scientists ought to pursue. A second approach allows substantial autonomy but accepts the role of ethics advisory committees to guide research directions. Yet these approaches neither

7

articulate a role for the politics of science nor admit the public into the decision-making arena. A third approach—epitomized by Philip Kitcher's "well-ordered science"—admits the public but only in a hypothetical way by describing an ideal process that the public would agree to, if they actually did participate. Drawing on contemporary cases in bioethics and a critique of Kitcher from the standpoints of a more pragmatic role for philosophy and a stronger role for democracy, Brown concludes that new and existing institutions can both effectively and democratically represent the public in shaping the research agenda.

In the second chapter, Brian Jackson, a chemist-turned-policy-analyst, reduces a topic of similarly large scope—the concept of the federal research and development (R&D) portfolio—to operational concerns. Over the postwar period, rationales for public R&D spending have emphasized its investment nature, and both common parlance and formal proposals commend to science policy the metaphor of the investment portfolio to inform the management and evaluation of R&D. Because of the portfolio's explicit consideration of balance and investment risk, Jackson argues that the metaphor is particularly relevant for an uncoordinated collection of programs funded by many agencies. But to be more analytically useful in designing the portfolio and evaluating the performance of the investment, the metaphor must be extended to address explicit investment goals, identified and articulated before the investment is made. This strategy has particular relevance given the continued efforts of agencies to apply the Government Performance and Results Act (GPRA) to their R&D activities. Although sometimes partially implied in program design, how particular R&D investments relate to these goals is rarely considered systematically. Jackson concludes that a bottom-up approach based on such explicit investment goals would provide a more suitable framework for the application of other techniques to evaluate R&D and could therefore optimize the shaping of the federal R&D portfolio both within and across government R&D agencies.

In the third chapter, economist Bhaven Sampat is concerned with policies for disseminating the outputs of publicly funded research to provide the highest level of social and economic benefits—a central concern in the wartime debates between Vannevar Bush and Harley Kilgore that has resurfaced as one of the most critical issues facing science and technology policymakers. Political interest has been renewed in large part by concerns that publicly funded research is increasingly being privatized. Sampat enriches the debate over the public and private contours of research by providing a theoretical framework for thinking about the virtues and limitations of the public and private domain, historical context on the Bayh-Dole Act and other changes in patent policy, and empirical evidence on the social welfare consequences of recent changes in policy. He concludes that strong evidence is still lacking for a definitive role of Bayh-Dole in either creating large social welfare benefits or in twisting the mission of research universities to commercialization over all else. Nevertheless, he reasons that this lack of evidence still supports policies that would restrain the expansion of commercial motives and tailor Bayh-Dole to allow the public principal greater oversight of, and potential intervention in, the commercial transactions of research universities.

Grant Black, also an economist, is likewise concerned with the shaping of policies that, in turn, shape the public value of the investment in R&D. Furthermore, he taps another conflict with roots back to the Bush-Kilgore debates, that over the geographic distribution of R&D funds. Research suggests that geography is a crucial variable for innovation, but old data, poor geographic resolution, and imperfect measures of innovation cloud much of the available evidence. Black's research addresses these limitations by examining the role of the local technological infrastructure in small-firm innovation at the metropolitan level. The chapter introduces a novel measure of innovation—Phase II awards from the Small Business Innovation Research (SBIR) Program—to estimate the effect of geographically bounded spillovers from industry and academe on the innovative activity of small high-tech firms. SBIR is the largest federal R&D initiative targeting small business. Black finds that geography, manifest in clustering of resources and the flow of knowledge among individuals, firms, and other institutions comprising a region's technological infrastructure, matters. He also finds no consistent pattern with respect to industry. Black concludes that the effectiveness of SBIR may be compromised by political attempts to direct its geographic impact. Rather, regions should focus on shaping policies to enhance their technological infrastructure and facilitate the interactions within it, with a particular focus on research universities.

These authors show us that there is still much innovation possible in science and technology policy—in the details of individual programs as well as in the overall structure and rationale of the enterprise—and suggest that sustained attention to the shape of that policy can lead more directly to desired societal outcomes from the R&D enterprise.

I

# Ethics, Politics, and the Public

*Shaping the Research Agenda*

MARK B. BROWN

## Introduction

Despite the enormous sums of public money spent on scientific research in the United States—total federal research and development funding of \$127 billion in FY 2004 (AAAS 2004)—there is not enough to fund every worthy project. Establishing an agenda for public research is thus an important and controversial task. This chapter compares four approaches to setting the federal agenda for publicly funded scientific research, focusing on the role of ethics advisory committees and lay participation in each. My primary concern is with the normative justification of each of the approaches, especially with regard to their implications for democratic politics. Three of the approaches have played important roles in the public funding of scientific research, and one points toward a plausible alternative. The first approach grants scientists sole authority to determine what projects they will pursue; the second continues to allow scientists considerable autonomy but imposes ethical constraints on research as determined by expert advisory committees. The third approach allocates federal science funding according to the results of self-interested lobbying by research institutions, universities, and other interested parties, complemented by public policy efforts to stimulate the commercialization of publicly funded research. None of these approaches provides a way to ensure that the public research agenda meets the needs and interests of ordinary citizens.

The fourth approach, in contrast, seeks to integrate scientific and ethical concerns into a democratic political process involving the lay public. Ethics advisory committees, in this approach, do not have the task of establishing ideal ethical standards to be implemented by scientists and policymakers. Their goal is rather to articulate and clarify the ideas and ideals already implicit in contemporary practice. From this perspective, I argue, research ethics can be seen as one component within a system of democratic representation. And because representation in a democracy depends on the participation of the represented, establishing an ethically informed and publicly representative research agenda requires institutions that facilitate public involvement in the politics of science policy.

## The Ideology of Autonomous Science

The leading approach to setting agendas for publicly funded science long granted scientific communities sole authority to determine its direction. During most of the Cold War era, an implicit "social contract for science" gave scientists generous public funding and extensive freedom from political control in exchange for new defense, medical, and consumer technologies. This social contract was always a fragile construction, and scientists' autonomy was never as complete as some nostalgic critics of political efforts to regulate science now claim.[1] Nonetheless, it is widely agreed that until at least the early 1980s, two basic premises governed U.S. science policy: the scientific community is capable of regulating itself; and if it is allowed to regulate itself, science will produce technological benefits for society (Guston 2000, 66).

Within this approach to setting the public research agenda, priorities are established both informally by particular scientific communities and formally by peer review. The use of peer review in the selection of projects, the allocation of grants, and the evaluation and publication of research results forces scientists to justify their projects to other scientists, occasionally to scientists from other disciplines, but generally not to nonscientists (cf. Chubin and Hackett 1990). Ethical considerations, according to this view, ought to affect the research agenda only insofar as individual scientists or particular scientific communities allow such considerations to influence their research priorities. Although some scholars have long argued for extending peer review processes to involve nonscientists (cf. Guston 2003, 35; Fuller 2000, 135–47; Funtowicz and Ravetz 1992), their suggestions have yet to enjoy more than token support.

The image of science as a self-regulating community has been justified in various ways.[2] Perhaps most commonly, scientists have argued that they deserve political autonomy because the scientific community is defined by its distinctively apolitical mode of pursuing truth, and the pursuit of truth is good in itself. Polanyi (1962) thus compared the scientific community to an economic market, in which scientists select problems and methods in order to produce as much truth as possible. Individual scientists adjust their efforts in response to the results achieved by other scientists, producing an aggregate result unanticipated and unattainable by any individual working alone. Political control of science, then, like political control of the market, promises only to disrupt this sublime process of mutual adjustment. Another common defense for the self-regulation of science draws on a perceived threat of practical failure: because lay people do not and cannot understand the nature of scientific inquiry, political involvement in science would inhibit its productivity. As we shall see, the ideology of autonomous science has long given way, in the practice if not the rhetoric of science policymakers, to a view of science as intertwined with social values and political interests.

## Research Ethics as Expert Advice

Even at the height of its influence during the early years of the Cold War, the ideology of autonomous science never completely excluded "external" factors from decisionmaking about scientific research agendas. Ethical and religious concerns, in particular, have always placed constraints on research. But in the years after World War II, in the wake of the atrocities committed by scientists working under the Nazi regime as well as several revelations of abuse of human research subjects in the United States, scientists began to develop a more direct approach to questions of research ethics.[3] During the 1960s, the development of increasingly precise techniques for genetic research lent new importance to long-standing concerns about the ethics of genetic engineering. Interestingly, early debates among scientists and theologians on human genetic engineering explicitly addressed themselves to the public as the ultimate source of decision-making authority for the research agenda. The possibility of manipulating human evolution seemed to pose questions too difficult to be decided by experts alone. Scientists soon retreated from this publicly oriented approach, however, as it became clear that public skepticism toward science in general and gene technology in particular could lead to government regulation of genetic research (Evans 2002, 69–71). Ever since, research ethics has often served an alibi function, allowing scientists to demonstrate concern for particular social values while preempting more intrusive involvement by authorities external to science. The famous 1975 Asilomar conference on the ethics of recombinant DNA research, for example, was explicitly conceived as a way of preserving scientific self-regulation and avoiding government involvement, which the organizers assumed would be overly restrictive (Weiner 2001, 210–14). Similarly, Institutional Review Boards (IRBs) and other mechanisms for ensuring the safety of human and animal research subjects and guarding against conflicts of interest, while certainly providing a check on ethically questionable practices have also been used to keep the governance of science closed to nonscientists. Government oversight and enforcement of IRB procedures has been "negligible to nonexistent," and a 1998 recommendation by the inspector general of the Department of Health and Human Services to include more nonscientists on IRBs—current regulations require one nonscientist—has not been implemented (Greenberg 2001, 360–61).

The alibi function of research ethics acquired institutional sanction and stability with the creation of federal ethics advisory committees. Though neither directly authorized by lay citizens nor soliciting their participation, these committees were conceived as representing the public's best interests. "In place of elected representatives, unelected representatives of the public on advisory commissions—who are much more distantly accountable to the public than elected officials are— would make ethical decisions for the public" (Evans 2002, 36, 200). This understanding of the purpose of ethics committees coincided with a long-standing conception of democratic representation as dependent above all on expert knowledge. Indeed, since the seventeenth century, liberal democratic ideology has tended to emphasize the role of expertise— both technical expertise and the ethical expertise associated with gentlemanly or professional status—in the creation of instrumentally effective policy, and hence, in the representation of citizens' best interests (Ezrahi 1990). I argue below that expert assessments of the public interest are today a necessary but not sufficient condition of democratic representation.

The establishment of the first ethics advisory committees was accompanied by the emergence of a new profession of bioethics. Wanting to be taken seriously by policymakers, bioethicists tailored their approach from the beginning to the needs of government advisory committees (Evans 2002, 37ff.). Since its first edition in 1979, the leading textbook in bioethics has thus promoted the method of "principlism," which focuses on a limited set of midlevel moral principles that provide substantive guidance for policy decisions, while remaining compatible with multiple philosophical foundations (Beauchamp and Childress 2001). By settling on a few simple ethical principles, bioethics helps policymakers present their decisions as the transparent and impersonal application of a fixed set of rules, rather than personal preferences. This approach both facilitates decisionmaking and enhances its perceived legitimacy (Evans 2002, 41f.). As a result, however, public bioethical debate today is narrower and more closed to public involvement than during the 1950s and 1960s. With the exception of clinical bioethicists and others working in practical contexts, most research ethicists have devoted little attention to the concrete contexts and institutional frameworks of the moral dilemmas associated with scientific research. This narrowing of focus has created several important difficulties for public discussion of the agenda for scientific research.

First, whereas early debates among scientists and theologians on eugenics involved consideration of social issues, bioethics today, like research ethics more generally, tends to focus on dilemmas faced by individual scientists (Evans 2002, 20). Much attention has been devoted to whether scientists should pursue particular lines of research, such as those involving recombinant DNA or, more recently, reproductive cloning. Other key issues for research ethics have included the informed consent of human subjects, relations between principal investigators and laboratory staff, the use of animals, access to and falsification of data, patents and intellectual property rights, personal and commercial conflicts of interest, as well as sexism, racism, and other forms of social bias in the lab (Schrader-Frechette 1994). These are certainly important issues, but they have generally been cast in terms of the professional responsibilities of individual researchers to colleagues, research subjects, and the general public. Despite the extensive public discussion since the 1960s of the social impacts of science, research ethicists have often neglected the social norms and incentives fostered by scientific institutions. Put differently, research ethics has usually focused on the results of ethical decisionmaking rather than the process of deciding.

Second, the prevailing individualistic approach to research ethics has been exacerbated by a view of philosophers as moral experts. Despite frequent acknowledgment of contextual factors and the development of case-based and narrative approaches, the task of research ethics is still often framed as providing objective answers to moral dilemmas. The view of research ethics as moral expertise does not depend primarily on the self-conception of research ethicists, many of whom see themselves as merely clarifying the terms of societal deliberation rather than recommending specific policy decisions. Ethicists, however, acquire the aura of expertise by being authorized to act as such through their participation on governmental ethics committees, Institutional Review Boards, and other expert advisory bodies. The social prominence of such bodies, and the misplaced expectations of policymakers and the public, make it likely that the published views of research ethicists will be publicly interpreted and appropriated as expert direction rather than moral clarification (Engelhardt 2002, 81). Put simply, a research ethicist has difficulty to avoid being seen as not merely an authority on the moral dimensions of science but as being in authority to sanction public policy.

Third, research ethics has tended to assume a relatively stark divide between ethics and politics. Ethics is seen as being about principles and politics about interests. Writing in response to the stem cell report of the National Bioethics Advisory Commission (NBAC), for example, one philosopher states, "My own view is that these recommendations may be an acceptable answer to the policy question, but they are certainly a flawed answer to the mirror ethical question" (Baylis 2001, 52). In a similar vein, another author writes, "Whereas members of a bioethics commission have to take into account such practical concerns, it is ethical concerns that should drive their analysis" (Parens 2001, 44). This stark split between ethics and politics hinders efforts to clarify and resolve the moral dilemmas faced by scientists and policymakers. Although individual moral problems may be resolvable in light of ideal standards of behavior, collective problems can only be answered democratically with reference to concrete instances of collective decisionmaking. Focusing on ideal standards thus tends to individualize ethical questions. Put differently, the persuasiveness of a philosopher's argument that a particular policy would be right, good, or just, does not by itself make the policy democratically legitimate. The point is not that ideal ethical standards have no useful place in political decisionmaking. But the process of creating standards—be they philosophical, as with questions of research ethics, or technical, as in environmental regulation—needs to be understood as part of an ongoing political process rather than prior to it.[4] Empirical studies of both science advisory bodies and ethics advisory committees have repeatedly highlighted the distinctly political features of standard setting (Jasanoff 1990; Evans 2002). Although the politics of such bodies has not been very democratic, it has made clear that science and politics are often closely intertwined.

## Science and the Politics of Interests and Economics

The understanding of science that underlies both the ideology of autonomous science and the expert-oriented view of research ethics depends on several assumptions that during the past thirty years have become increasingly implausible: that science has the task of "pursuing truth" as such, rather than those particular truths deemed significant at any given time; that scientific significance can be assessed in isolation from social and political values; that the significance of basic research can be clearly separated from that of applied research and technology; and that lay citizens necessarily lack the interest or competence to make intelligent contributions to science policymaking (Kitcher 2001; Jasanoff et al. 1995). Although one cannot assume that policymakers have absorbed the constructivist arguments of philosophers and sociologists of science, they also frequently express the view that politics necessarily plays a role in scientific agenda setting. Summarizing its review of the process for setting federal research priorities, the National Science Board openly states, "The allocation of funds to national research goals is ultimately a political process that should be informed by the best scientific advice and data available" (NSB 2002, 81; cf. Shapiro 1999). The board says little, however, about what kind of politics is at issue. The two most familiar possibilities appear: first, in the efforts of public officials, interest groups, scientific associations, and research universities to shape the agenda of publicly funded science according to their economic, institutional, or ideological goals; and second, in governmental reform programs designed to promote the efficient and productive use of public research money.

Motivated in part by second thoughts about their association with the atomic bomb and other military technologies developed during World War II, scientists during the Cold War occasionally promoted or opposed particular political causes and their associated research programs, e.g., arms control, cooperative research with Soviet scientists, or the "Star Wars" missile-defense program. While generally avoiding electoral politics, it was not unusual for scientists to actively seek public support for research they deemed to be in the public interest (Greenberg 2001, 330–47). During the past few decades, in contrast, scientists have privatized their politics, so to speak, emphasizing the apolitical nature of their profession on the one hand, while seeking to acquire as much public money as possible for their own particular research programs on the other. An important exception might be seen in the 2004 campaign by scientists to call attention to the Bush administration's efforts to suppress and distort advisory committee recommendations that conflicted with administration policy. But this exception actually proves the rule as scientists framed the campaign as one opposed to any sort of "politicization of science" (Union of Concern Scientists [UCS] 2004). As we shall see, some types of politicization are more compatible with democracy than others.

Officials of the National Science Foundation, the National Institutes of Health, and other federal research institutions, while generally refraining from involvement in controversial science policy debates, effectively function as politicians of science, seeking public funds not only or even primarily with the aim of achieving an intellectually balanced and socially just public research agenda. They aim, rather, to increase their own agency's share of the federal pie. Public funding for the Human Genome Project, for example, was driven not primarily by dispassionate evaluation of the project's social and scientific merit but by the bureaucratic instinct for survival: the National Institutes of Health acquired leadership of the project primarily to prevent the Department of Energy from doing so, and the DOE, for its part, became involved in part out of a need to find new work for laboratories threatened with obsolescence by the end of the Cold War. Similarly, funding for the Los Alamos National Laboratory, the Oak Ridge National Laboratory, and other large federal research facilities often has as much to do with the search for jobs, votes, and political alliances as with the country's genuine research needs (Greenberg 2001, 26–27, 30–31).

None of this would pose a problem for democracy if the competitors for federal research dollars each sponsored the same types of work. But that is clearly not the case. The domination of nondefense federal science funding by NIH, for example, gives priority to biomedical research over other fields. And NIH has traditionally focused on the search for high-tech measures to cure disease, and thus on the needs of those who can afford them, potentially at the expense of cheaper and more widely accessible preventative measures (Greenberg 2001, 193, 420; Wolpe and McGee 2001).

In addition to the efforts of federal research institutions, university officials are also often deeply engaged in the interest group politics of science. Either on their own or with the aid of Washington lobbying firms, university officials frequently seek the "earmarking" of federal funds for buildings, equipment, or entire research facilities (Payne, chapter 7, this volume; Greenberg 2001, 183–204; Savage 1999). Often derided as "academic pork," earmarked funds are awarded without peer-review and often without careful consideration by the hurried members of Congress who approve them. Proponents of earmarking claim that it provides a necessary corrective to a peer review process that tends to favor prestigious universities. Critics argue that earmarking amounts to the politicization of research, and that the prestige of some universities simply reflects the merit of their researchers. Both sides, however, tend to assume a traditional conception of peer review that gives little consideration to the social value or distributional justice of research. The dispute has been over whether lobbyists or scientists should determine how federal research funds are allocated, neither of whom are expected to give priority to the public interest. Although it comprises a small percentage of total public funding for research, earmarking poses a significant threat to the democratic legitimacy of the federal research agenda.

This orientation within the interest group politics of science toward the needs of particular institutions and social groups is complemented by certain features of contemporary federal science policy. Public policy, of course, has long shaped the research agenda through such indirect means as education, tax credits, protocols for the treatment of animal and human subjects, environmental regulations, and patent policy, among others. Since about 1980, however, there has been a partial shift from broad attempts to shape science through macroeconomic controls on funding toward more detailed microlevel efforts to direct research (Guston 2000, chap. 5). The Bayh-Dole Act of 1980, for example, made it possible to patent the results of publicly financed research. Related measures have included federal incentives for partnerships between university and industry researchers and the creation of the Office of Technology Transfer at NIH, which helps scientists with the identification, patenting, licensing, and marketing of commercially promising research. These governmental efforts to shape the scientific agenda have been largely driven by economic considerations, insofar as they measure research productivity in commercial terms.

The commercialization of research raises a host of dilemmas that cannot be fully addressed here (see Sampat, chapter 3, this volume). Although individual researchers may well see the commercialization of research as nothing more than an effective way to pursue their own intellectual goals, the effect is to direct scientific research toward areas that promise technological development and economic growth. The commercialization of research may fulfill ethical requirements and promote social goals, as seems likely with regard to cancer or AIDS research, but it may also conflict with such requirements and goals, as may be the case with research on military technology or genetic testing. Moreover, the proprietary rights associated with commercialization may limit cooperation among scientists or restrict public access to new knowledge. It also inevitably favors the so-called hard sciences over the social sciences and humanities.[5]

To summarize the argument so far: ethics advisory committees have tended to ignore politics, and both science lobbyists and science policymakers have often focused on commercial goals at the expense of the public interest. In the first case, the politics of science policy has been suppressed by philosophy; in the second case, it has been reduced to economics. If the politicization of science means reducing science policy to interest group competition, or stacking expert advisory committees with supporters of a particular policy agenda, it clearly conflicts with democracy's need for science in the public interest. But politicization can serve democracy if it is understood as a matter of calling attention to the politics of science, thus helping ensure that the public research agenda serves societal goals and expert advisory committees include a balance of perspectives.[6] The task is thus to politicize—that is, to make political—both science policy and expert advice in a way that avoids excessive partisanship and accommodates social and ethical concerns. As one commentator puts the challenge, "In the interlocking realms of public science and profit-seeking private science, who represents the public?" (Greenberg 2001, 11).

## Science, Ethics, and Democratic Representation

One provocative proposal for how to represent the public in federal science policy appears in the work of a leading philosopher of science, Philip Kitcher (2001).[7] Kitcher seeks a way of setting the public research agenda that is amenable to both science and democracy. His proposal centers on an ideal standard that he calls "well-ordered science." To articulate the ideal, Kitcher asks us to imagine a group of "ideal deliberators" who proceed through an elaborate process of education and deliberation to transform their initial science policy preferences into "tutored personal preferences" (118). After much discussion and consultation, they develop a scientific research agenda "that best reflects the wishes of the community the ideal deliberators represent" (121). Decisions about the best methods for pursuing the research agenda are left to scientists, with the exception of moral constraints on research methods specified by the deliberators. Once the research has produced outcomes of some kind, the deliberative process is then used again to make decisions about the application and dissemination of research results.

This ideal of well-ordered science offers a provocative challenge to the notion that scientists ought to set the public research agenda either entirely themselves or, if necessary, within constraints determined by ethics advisory bodies. It also helps clarify many failings of current science policy. Nonetheless, it is important to note that Kitcher apparently intends his ideal of well-ordered science to highlight only the substantive outcomes of policy, not the processes through which policy is made. He writes, "[T]here's no thought that well-ordered science must *actually institute* the complicated discussions I've envisaged. The thought is that, however inquiry proceeds, we want it to match the outcomes those complex procedures would achieve at the points I've indicated" (123, original emphasis). Kitcher thus suggests that his primary concern is not with democratic deliberation as such, but with the outcomes and results of deliberation. He seems to want a science policy in accord with what the public *would* support, were it informed and able to express its views, but not one that informed members of the public actually shape themselves.

This apparent rejection of actual public participation in shaping the research agenda is especially problematic in light of Kitcher's suggestion that his ideal deliberators represent the public (123–26). As I argue below, a democratic form of political representation depends on the active participation of the represented. Without mechanisms for asking constituents about their wishes, Kitcher's ideal deliberators—as well as those who would use their hypothetical deliberations as an ideal standard for science policy—must rely on speculation, introspection, or intuition to assess popular preferences. Kitcher thus appears to endorse not the political representation of citizens but a philosophical representation of politics.[8] In this respect, despite initial appearances to the contrary, Kitcher's approach remains in the tradition of expert-oriented research ethics.

Is there a better way to bring philosophical ethics into the politics of science policy? Can one combine ethics and politics without allowing the former to hijack the latter? If philosophical ethics is to find a role in science policy that avoids both the elitism of much bioethics and the utopianism of much political philosophy, it needs to integrate ethical and scientific considerations with social and political concerns. Such an approach would be grounded in a view of research ethics as both interdisciplinary and democratic. Given the centrality of political representation to modern democracy, a democratic theory of research ethics will need to consider how ethics advisory committees can best contribute to democratic representation.

Democratic representation is a complex concept. For present purposes it may suffice to say that democratic representation, as I understand it, depends on the combination of at least three components: substantive knowledge of citizens' best interests (sometimes called the trustee component of representation), public participation in the political process broadly conceived (delegate component), and legal authorization and accountability of those who act on other people's behalf (formal component). Given the misleading dichotomy often drawn between participatory and representative democracy, it is important to emphasize that democratic representation does not conflict with participation but depends on it. Participation is important both to help representatives understand public preferences and to foster a sense of being represented among constituents. Democratic representation must be manifest to citizens as such. "A representative government must not merely be in control, not merely promote the public interest, but must also be responsive to the people. . . . Correspondingly, a representative government requires that there be machinery for the expression of the wishes of the represented, and that government respond to these wishes unless there are good reasons to the contrary" (Pitkin 1967, 232–33). Democratic representatives need not always do exactly what their constituents want, but neither may they consistently ignore the views of the represented. And since citizens often lack articulate views on complex issues of public policy, democratic representation cannot simply echo a preexisting public will. It needs to involve the education and solicitation of the same public views to which it must be responsive (Young 2000, 130f.).

Moreover, democratic representation is best conceived not as an attribute of individual governmental institutions but of entire political systems (Pitkin 1967, 221f.). Consequently, each of the elements of democratic representation need not be present to an equal degree in every political institution. The legislative, judicial, and executive branches of government, for example, each represent citizens in different ways, and the potential for democratic representation resides in how they work together. More broadly, democratic representation emerges from ongoing interactions among legally authorized governmental institutions, informal public discourse, and the various semistructured forms of deliberation and participation in civil society (Mansbridge 2003; Habermas 1996, 483–88).

One might argue that because ethics advisory bodies are authorized only to provide advice and not to make policy decisions, they ought to be expected to fulfill only the trustee element of democratic representation. From this perspective, ethics committees need not include provisions for public participation or popular authorization and accountability. Participation could be relegated to civil society, and popular authorization and accountability to elected officials. There are several reasons, however, for considering how the other components of democratic representation might be incorporated into the work of ethics advisory bodies.[9]

With respect to the formal components of democratic representation, ethics committees are generally authorized by and directly accountable to the president or other executive branch officials who appoint them, rather than the general public. Their grant of popular authority is thus considerably weaker than that of elected officials. Indeed, their authority resides less in being popularly authorized than in their professionally certified expertise. Nonetheless, one might consider whether it would foster democratic representation to strengthen the popular authorization of ethics advisory committees, perhaps by selecting them through public elections. On the one hand, the popular election of ethics advisory committees might conflict with one of the central purposes of expert advice: to facilitate the execution of a political program ostensibly sanctioned by the voters in the preceding election. One can plausibly argue, without endorsing an excessively partisan form of politicized expertise, that publicly elected officials have a right to receive advice conducive to their political goals. On the other hand, however, popular election might enhance the public authority of ethics committees and, hence, their contribution to the political efforts of their sponsors. Moreover, in addition to providing advice, ethics advisory committees are now typically expected to both represent diverse interests and inform and engage the general public, functions that might be well served through popular election. If popular election seems practically unfeasible, one might consider whether appointments to ethics advisory committees, like those of many high-level technical advisors, should at least be subject to confirmation by the Senate (cf. NAS 2005). Although these suggestions for enhancing the formal authorization of ethics advisory committees certainly risk politicizing expertise in the pejorative sense of reducing it to a means of partisan competition, they also have the potential of enhancing the contribution of such committees to democratic representation.

To now consider the other component of formal representation, it is worth thinking about possibilities for improving the formal accountability of ethics advisory committees. For what, to whom, and by whom should ethics advisory committees be held accountable? According to the trustee model of ethicists as experts who represent the public's best interests, ethics advisors are held accountable by other professionals for upholding the rules and norms of their discipline. Professional accountability is exercised primarily through informal mechanisms of reputation and status rather than formal requirements for documenting and assessing performance. Given provisions of the United States Federal Advisory Committee Act (FACA) for the public transparency, accessibility, and documentation of advisory committee meetings, the public accountability of ethics advisory committees already goes significantly beyond the trustee model (cf. Smith 1992). In theory, at least, voters and their elected representatives can use the information provided under

FACA to hold public officials accountable for the performance of the advisory committees appointed under their watch. This would be facilitated by, among other things, a clearer public understanding of what ethics advisory committees are supposed to do. And assuming that ethics advisory committees ought to be held accountable not only to the public but also by the public, enhancing the accountability of such committees might also require improving measures for promoting public participation in their work.

Many science policy scholars and activists have over the past thirty years advocated the inclusion of lay citizens in science policymaking (e.g., Kleinman 2000; Sclove 1995; Petersen 1984; Winner 1977). It has often been argued, for example, that because much research is funded with taxpayer money, ordinary citizens ought to be involved in shaping the research agenda. And because science and technology continually transform the lives of individuals and communities, the norms of democracy require that citizens have an opportunity to participate in shaping these transformations. From this perspective, claims for the political autonomy of science obscure important questions regarding who benefits most from science, who has the resources necessary for taking advantage of science, and who suffers the social costs of science (Guston 2000, 48).

In light of the above conception of democratic representation, participation potentially has both epistemic and political benefits for ethics advisory committees. Epistemically, as suggested above, the ethical issues raised by science are too controversial and multifaceted to be subject to decisive analysis by professional bioethicists and philosophers. Ethics advisory bodies need to include representatives of the full range of disciplinary perspectives relevant to the problems at hand, as well as diverse lay perspectives. This might be understood as a way of fulfilling the FACA provision that requires that advisory committees be "fairly balanced in terms of points of view represented and functions performed" (Jasanoff 1990, 47). Whereas most approaches divorce philosophical ethics from empirical research, research ethics has much to gain by including scholars involved in the empirical study of both politics and science. Research ethics depends on social science to help it consider the empirical manifestations and consequences of its normative claims, and it depends on natural science to help it remain responsive to the actual dilemmas researchers face (Weiner 2001, 217; Haimes 2002).

It makes perfect sense, of course, for philosophers to focus on issues of moral logic and justification, and for social scientists to focus on questions of collective behavior and institutional design. The point is that this division of labor should be undertaken in such a way that scholars in different disciplines can benefit from each other's work. If research ethicists entirely ignore empirical questions, they may well obscure rather than illuminate the problems of social science. Put differently, philosophers need to be just as concerned with how the scientific agenda is established, which requires some consideration of the empirical processes studied by social scientists, as with what the agenda should be. Such an approach might be able to avoid situations such as that which developed with Institutional Review Boards: after being heavily involved in articulating principles for human subjects research and the design of IRB protocols, bioethicists failed for many years to assess the effects of the procedures they had recommended (Arras 2002, 43). According to some recent assessments, the type of contextual, interdisciplinary, process-oriented approach advocated here has long been familiar to bioethicists working in clinical contexts (Arras 2002; Wolf 1994), but it seems to be relatively uncommon in research ethics more generally.

In addition to the epistemic benefits of making ethics committees more interdisciplinary, epistemic benefits can also be expected from the incorporation of lay perspectives. Not only does every profession and academic discipline have its own blind spots, but professionals and academics often share particular assumptions that differ from those of other social and economic groups. The inclusion of lay citizens can help identity such assumptions and introduce new ideas into ethical deliberation. Fulfilling this goal, of course, requires a way of selecting lay participants that is not biased toward those already aligned with one or another of the established positions on any given topic. Consensus conferences, citizen panels, and similar participatory experiments have had a certain amount of success with using random selection to generate a pool of potential participants without fixed positions or ideological agendas. As long as one remembers that random selection does not constitute authorization to act on behalf of others, it may be an effective way to incorporate new perspectives and thus enhance deliberation on ethics advisory committees (Renn et al. 1995). Though not a replacement for traditional forms of civic organization and activism, citizen panels provide new opportunities for philosophers and ethicists to engage lay citizens. They have also begun to receive the attention of U.S. law-makers: on 3 December 2003, President George W. Bush signed the Twenty-first Century Nanotechnology Research and Development Act, which includes a section calling for "the convening of regular and ongoing public discussions, through mechanisms such as citizens panels, consensus conferences, and educational events, as appropriate."

In addition to its epistemic advantages, lay participation on ethics advisory committees also promises political benefits. Although ethics advisory bodies are generally not authorized to make decisions, they sometimes have a powerful influence on both individual policy decisions and the overall decision-making agenda. Public officials often neglect their democratic obligation to consider not only expert recommendations but also the expressed views and interests of their constituents. Incorporating some degree of public input into the work of bioethics committees may help ensure that the participatory element of democratic representation receives at least some consideration, even when decisionmakers find it expedient to give ethics committee recommendations more weight than they should. Put differently, if democratic representation depends on combining expertise and participation, it seems advisable to combine them not only at the top, so to speak, in the deliberations of decisionmakers, but also at the lower level of deliberations on ethics advisory committees.

One might note here, again, that lay participants on ethics advisory committees are not authorized to represent the interests of the social groups to which they purportedly belong. Their primary task is to represent—or rather, make representations of—the experiential perspectives and viewpoints associated with particular social groups (Brown forthcoming). Similarly, although ethics committees might be well advised to use opinion polls, public hearings, focus groups, and other means of assessing of public opinion, decisionmakers are still obliged to make their own judgments about their constituents' wishes. No amount of lay input into the work of ethics advisory committees can absolve elected representatives from their responsibility to solicit the views of their constituents (even if properly representing them sometimes requires acting against those views). Nonetheless, if ethics advisory committees draw on multiple components of democratic representation—expertise, participation, and formal authorization and accountability—they can promote the adoption of a public research agenda that both represents the public and is publicly perceived as such.

# Conclusion

Many ethics advisory committees have already gone to considerable lengths to solicit public involvement and represent the views of diverse social groups. President Bush's executive order creating the President's Council on Bioethics states that the Council will, among other things, "provide a forum for national discussion of bioethical issues" (PCB 2001). The charter of the NIH Secretary's Advisory Committee on Genetics, Health, and Society, like that of many other such committees, specifies that "at least two members shall be specifically selected for their knowledge of consumer issues and concerns and the views and perspectives of the general public" (NIH 2002). The writing of the NBAC stem cell report stimulated a half-million visitors to the NBAC website, enormous amounts of mail, and many public comments at hearings held around the country (Meslin and Shapiro 2002). Nonetheless, there is still considerable need for increasing both the quantity and quality of public participation in shaping the research agenda. The former chair and former executive director of the NBAC thus state that "one must recognize that more—much more—could have been done to engage the public" (Meslin and Shapiro, 99). In particular, they note, the NBAC's procedures for involving the public reached only those with the resources necessary to attend meetings, speak in public, and access the Internet. More generally, the debate on stem cell research has failed to adequately represent the interests of the poor and socially disadvantaged, those living in developing countries, and members of future generations (McLean 2001, 204; Resnik 2001, 198; Chapman, Frankel, Garfinkel 2000, 414f.). Indeed, as suggested above, most biomedical research today focuses on diseases of the rich and the old, and it potentially redirects funds from areas that would benefit broader segments of the population.

Developing a public research agenda that more fully reflects the ideals of representative democracy requires more effective public participation and representation in science policymaking. The appropriate contribution of ethics advisory committees to this goal does not lie in the expert discernment of ideal standards to be implemented by scientists and policymakers. Nor does it involve simply echoing public opinion or providing an arena for rhetorical competition among established interest groups. Instead, ethics committees have the task of integrating scientific, moral, and political concerns into a democratic political process involving the lay public. Research ethics of this sort may require the development of new regulatory institutions, as some have suggested (Fukuyama 2002, 212–15), but it also depends on improving the public responsiveness of existing institutions. By avoiding expert-oriented and interest-driven approaches, and by incorporating a broader range of disciplinary and lay perspectives, ethics advisory committees might become both more effective and more democratically representative in their efforts to shape the public research agenda.

NOTES

Many thanks to Marvin Brown, John Evans, and an anonymous reviewer for helpful comments on earlier versions of this chapter.

1. Guston (2000, 39, 62) describes the social contract for science as "a map of institutional arrangements and their intellectual underpinnings," a "dominant ideology," according to which "the political community agrees to provide resources to the scientific community and to allow the scientific community to retain its decision-making mechanisms and in return expects forthcoming but unspecified technological benefits."

2. For a more complete discussion of four distinct arguments for the political autonomy of science, see Bimber and Guston (1995).

3. Other factors contributing to increased concern with research ethics in the postwar period include the enormous increase in the public funding and societal impact of science and technology; the replacement of the guildlike character of the medical profession with a view of medicine as a trade subject to political regulation; and the declining role of Christianity as a de facto moral guide for medical practice (Engelhardt 2002, 76).

4. One might wonder why I present science and ethics as part of the political process rather than politics as part of either science or ethics. The reason is that citizens must engage in politics to make democratic decisions about science and ethics, whereas neither science nor ethics provides a way to make democratic decisions about politics.

5. Federal funding for research in the humanities has never done justice to its societal significance, even considering its intrinsically lower cost compared to the sciences: grants from the National Endowment for the Humanities totaled a mere $150 million in 1999 (Greenberg 2001, 24). Funding by the National Science Foundation for the social and behavioral sciences has always been somewhat precarious, reaching $204 million of $4.1 billion total NSF R&D funding in FY 2004 (AAAS 2004, 34).

6. This distinction between democratic and partisan forms of politicization is implicit in a recent National Academy of Sciences (2005) report on appointments to science advisory committees. The report first acknowledges that

"societal values, economic costs, and political judgments . . . come together with technical judgments in the process of reaching advisory committee recommendations" (43). The report then asserts that a person's policy perspective "is not a relevant criterion for selecting members whose purpose is to provide scientific and technical expertise" (ibid.). But once members have been selected, the report states, "Disclosing perspectives, relevant experiences, and possible biases . . . provides an opportunity to balance *strong* opinions or perspectives through the appointment of additional committee members" (NAS 2005, 45, original emphasis). The report also advocates increased publicity in the appointment process, among other reforms.

7. The following draws on a more complete discussion of Kitcher's book in Brown (2004).

8. This substitution of philosophical for political representation also appears in Rawls (1993, 25–26), upon which Kitcher models his approach.

9. Among other reasons, restricting public involvement to the electoral authorization of decisionmakers, combined with the reward or sanction provided by subsequent elections, places far too great a burden on the electoral process. Electoral systems are not only subject to various injustices; they are incapable, in principle, of providing representatives with clear guidance on all the issues that may arise during a term in office. Even if a large majority of a legislator's constituents agree with his or her position on biotechnology, for example, citizens might still vote against that person due to his or her position on other issues they deem more important.

REFERENCES

American Association for the Advancement of Science (AAAS). 2004. Congressional Action on Research and Development in the FY 2004 Budget. Available at http://www.aaas.org/spp/rd/pubs.htm.

Arras, J. D. 2002. Pragmatism in Bioethics: Been There, Done That. *Social Philosophy and Policy* 19 (2): 29–58.

Baylis, F. 2001. Human Embryonic Stem Cell Research: Comments on the NBAC Report. In *The Human Embryonic Stem Cell Debate: Science, Ethics, and Public Policy,* ed. S. Holland, K. Lebacqz, and L. Zoloth, 51–60. Cambridge, MA: MIT Press.

Beauchamp, T. L., and J. F. Childress. 2001. *Principles of Biomedical Ethics,* 5th ed. New York: Oxford Univ. Press.

Bimber, B., and D. H. Guston. 1995. Politics by the Same Means: Government and Science in the United States. In *Handbook of Science and Technology Studies,* ed. Jasanoff et al., 554–71. Thousand Oaks, CA: Sage Publications.

Brown, M. B. Forthcoming. Citizen Panels and the Concept of Representation. *Journal of Political Philosophy.*

———. 2004. The Political Philosophy of Science Policy. *Minerva* 42 (1): 77–95.

Chapman, A. R., M. S. Frankel, and M. S. Garfinkel. 2000. Stem Cell Research and Applications: Monitoring the Frontiers of Biomedical Research. In *AAAS Science and Technology Policy Yearbook 2000*, ed. A. H. Teich, S. D. Nelson, C. McEnaney, and S. J. Lita, 405–16. Washington, DC: American Association for the Advancement of Science.

Chubin, D. E., and E. J. Hackett. 1990. *Peerless Science: Peer Review and U.S. Science Policy*. Albany: State Univ. of New York Press.

Engelhardt, H. T., Jr. 2002. The Ordination of Bioethicists as Secular Moral Experts. *Social Philosophy and Policy* 19 (2): 59–82.

Evans, J. H. 2002. *Playing God? Human Genetic Engineering and the Rationalization of Public Bioethical Debate*. Chicago: Univ. of Chicago Press.

Ezrahi, Y. 1990. *The Descent of Icarus: Science and the Transformation of Contemporary Democracy*. Cambridge, MA: Harvard Univ. Press.

Fukuyama, F. 2002. *Our Posthuman Future: Consequences of the Biotechnology Revolution*. New York: Picador.

Fuller, S. 2000. *The Governance of Science: Ideology and the Future of the Open Society*. Buckingham, UK: Open University Press.

Funtowicz, S. O., and J. R. Ravetz. 1992. Three Types of Risk Assessment and the Emergence of Post-Normal Science. In *Social Theories of Risk*, ed. S. Krimsky and D. Golding, 251–73. Westport, CT: Praeger.

Greenberg, D. S. 2001. *Science, Money, and Politics: Political Triumph and Ethical Erosion*. Chicago: Univ. of Chicago Press.

Guston, D. H. 2003. The Expanding Role of Peer Review Processes in the United States. In *Learning from Science and Technology Policy Evaluation: Experiences from the United States and Europe*, ed. P. Shapira and S. Kuhlmann. Northampton, MA: Edward Elgar Publishing.

———. 2000. *Between Politics and Science: Assuring the Integrity and Productivity of Research*. Cambridge: Cambridge Univ. Press.

Habermas, J. 1996. *Between Facts and Norms: Contributions to a Discourse Theory of Law and Democracy*, trans. William Rehg. Cambridge, MA: MIT Press.

Haimes, E. 2002. What Can the Social Sciences Contribute to the Study of Ethics? Theoretical, Empirical, and Substantive Considerations. *Bioethics* 16 (2): 89–113.

Jasanoff, S. 1990. *The Fifth Branch: Scientific Advisors as Policymakers*. Cambridge, MA: Harvard University Press.

Jasanoff, S., G. E. Markle, J. C. Petersen, and T. Pinch, eds. 1995. *Handbook of Science and Technology Studies*. Thousand Oaks, CA: Sage Publications.

Kitcher, P. 2001. *Science, Truth, and Democracy*. New York: Oxford Univ. Press.

Kleinman, D. L., ed. 2000. *Science, Technology, and Democracy*. Albany: State Univ. of New York Press.

Mansbridge, J. 2003. Rethinking Representation. *American Political Science Review* 97 (4): 515–28.

McLean, M. R. 2001. Stem Cells: Shaping the Future in Public Policy. In *The Human Embryonic Stem Cell Debate: Science, Ethics, and Public Policy*, ed. S. Holland, K. Lebacqz, and L. Zoloth, 197–208. Cambridge, MA: MIT Press.

Meslin, E. M., and H. T. Shapiro. 2002. Bioethics Inside the Beltway: Some Initial Reflections on NBAC. *Kennedy Institute of Ethics Journal* 12 (1): 95–102.

National Academy of Sciences (NAS). 2005. *Science and Technology in the National Interest: Ensuring the Best Presidential and Federal Advisory Committee Science and Technology Appointments.* Washington, DC: National Academy Press.

National Science Board (NSB). 2002. Federal Research Resources: A Process for Setting Priorities. In *AAAS Science and Technology Policy Yearbook 2002*, ed. A. H. Teich, S. D. Nelson, and S. J. Lita, 79–87. Washington, DC: American Association for the Advancement of Science.

National Institutes of Health (NIH). 2002. "Charter of the Secretary's Advisory Committee on Genetics, Health, and Society." Available at http://www4.od.nih.gov/oba/sacghs/SACGHS_charter.pdf.

Parens, E. 2001. On the Ethics and Politics of Embryonic Stem Cell Research. In *The Human Embryonic Stem Cell Debate: Science, Ethics, and Public Policy*, ed. S. Holland, K. Lebacqz, and L. Zoloth, 37–50. Cambridge, MA: MIT Press.

President's Council on Bioethics (PCB). 2001. Executive Order 13237, Creation of The President's Council on Bioethics (November 28). Available at http://www.bioethics.gov/reports/executive.html.

Petersen, J. C., ed. 1984. *Citizen Participation in Science Policy.* Amherst: Univ. of Massachusetts Press.

Pitkin, H. F. 1967. *The Concept of Representation.* Berkeley: Univ. of California Press.

Polanyi, M. 1962. The Republic of Science: Its Political and Economic Theory. *Minerva* 1 (1): 54–73.

Rawls, J. 1993. *Political Liberalism.* Cambridge, MA: Harvard University Press.

Renn, O., T. Webler, and P. Wiedeman. 1995. *Fairness and Competence in Citizen Participation: Evaluating Models for Environmental Discourse.* Dordrecht, Neth.: Kluwer Academic.

Resnik, D. 2001. Setting Biomedical Research Priorities: Justice, Science, and Public Participation. *Kennedy Institute of Ethics Journal* 11 (2): 181–204.

Savage, J. D. 1999. *Funding Science in America: Congress, Universities, and the Politics of the Academic Pork Barrel.* Cambridge: Cambridge Univ. Press.

Schrader-Frechette, K. 1994. *The Ethics of Scientific Research.* Lanham, MD: Rowman & Littlefield.

Sclove, R. E. 1995. *Democracy and Technology.* New York: Guilford Press.

Shapiro, H. T. 1999. Reflections on the Interface of Bioethics, Public Policy, and Science. *Kennedy Institute of Ethics Journal* 9 (3): 209–24.

Smith, B. L. R. 1992. *The Advisers: Scientists in the Policy Process.* Washington, DC: Brookings Institution.

Union of Concern Scientists (UCS). 2004. *Scientific Integrity in Policymaking: An Investigation into the Bush Administration's Misuse of Science* (February). Available at http://www.ucsusa.org/global_environment/rsi/index.cfm.

Weiner, C. 2001. Drawing the Line in Genetic Engineering: Self-Regulation and Public Participation. *Perspectives in Biology and Medicine* 44 (2): 208–20.

Winner, L. 1977. *Autonomous Technology: Technics-Out-of-Control as a Theme in Political Thought.* Cambridge, MA: MIT Press.

Wolf, S. 1994. Shifting Paradigms in Bioethics and Health Law: The Rise of a New Pragmatism. *American Journal of Law and Medicine* 20 (4): 395–415.

Wolpe, P. R, and G. McGee. 2001. "Expert Bioethics" as Professional Discourse: The Case of Stem Cells. In *The Human Embryonic Stem Cell Debate: Science, Ethics, and Public Policy*, ed. S. Holland, K. Lebacqz, and L. Zoloth, 185–96. Cambridge, MA: MIT Press.

Young, I. M. 2000. *Inclusion and Democracy.* Oxford: Oxford Univ. Press.

2

# Federal R&D

*Shaping the National Investment Portfolio*

BRIAN A. JACKSON

## Introduction

Like any organization, the U.S. federal government allocates its resources to fulfill its obligations and fund the programs it pursues. As part
of its preparation for the future, it invests a portion of its discretionary
resources rather than spending them all on current activities. Unlike individuals, whose investments might include bank accounts, mutual
funds, or real estate, the investments made by the federal government
fall into somewhat different categories: physical assets, education and
training, and research and development (R&D). Rather than producing
purely monetary returns, these national investments build capabilities
and knowledge applicable to the national economy, defense, health,
and overall quality of life. Although investments in physical assets and
education are important for R&D policy, the central focus of national
science and technology (S&T) policy debate is the shaping of federal expenditures on R&D.

The importance of using public funds responsibly fuels ongoing debate about the ways government appropriates funds and assesses the

results of their expenditure. When focused on R&D, these concerns lead to three perennial questions:

1. Is the government placing its investments well?
2. Is it managing them appropriately?
3. How are the investments performing?

Although the S&T policy community has begun to address these questions for R&D, it has not reached consensus on the best ways to consider them in policy design and implementation. Doing so is made complicated, however, by the lack of a centralized "federal investor" whose activities are being evaluated: the federal R&D portfolio is a post hoc construct made up of the decisions of many independent "investors" within the federal government.

This chapter suggests a "bottom-up" approach to shaping federal R&D investments. Such a method has as its basis the individual goals of each R&D activity. This bottom-up methodology appears to have significant advantages for federal R&D management over more commonly applied approaches that consider groups of federal investments, usually at the program or national level. To ground this goal-based approach, the chapter briefly discusses the federal R&D portfolio using investment management concepts in policy design and implementation. It then introduces the concept of investment goals with respect to individual R&D investments. In addition to showing how investment goals make management concepts such as portfolio risk and balance more meaningful, the chapter also discusses their utility in shaping policy by describing the complex returns of R&D, and in performance assessment for both individual R&D programs and the federal R&D enterprise overall.

## Federal R&D as an Investment Portfolio

In the most concrete description of the investment vision for publicly funded R&D to date, McGeary and Smith (1996, 1484) proposed "that the concept of R&D as an investment be extended by introducing the portfolio concept from financial investment theory into decision-making on the allocation of funds for S&T." Framing federal R&D expenditures as an investment portfolio is useful because it provides a structured way to consider concepts of investment risk and diversification. In financial markets, the potential return of investments is related to their risk. In general, to compensate for a higher probability of loss, riskier investments must reward investors with the possibility for higher returns. Because the risk of individual investments cannot be predicted with certainty, investors build a portfolio of investments to gain some benefit from owning high-risk investments while reducing overall risk exposure. Since the results of R&D are difficult to predict, investors generally judge it to be relatively high risk. As a result, understanding appropriate ways to diversify R&D investments is important to maximizing the portfolio's overall benefit to the nation. In the S&T policy literature, appropriate diversification is generally referred to as "balance" in the R&D portfolio (McGeary and Smith 1996).

Although the value of concepts such as investment risk and portfolio balance is obvious, their utility in designing and managing the federal R&D portfolio ultimately depends on their implementation. In this context, concerns about implementation range from how the concepts are used in national policy debate to how (or if) they are used in program management.

The concept of investment risk has taken on a range of meanings depending on the context and perspective of policy discussion. As OTA (1991, 121) reported, "the definition of risky research changes depending on the agency and field of inquiry, and to some extent, every research project is inherently 'risky.'" Given the public emphasis on the utility of scientific results, risk is often defined in terms of the potential usefulness of a project's results and the time frame needed to realize them. As McGeary and Smith (1996, 1484) argued, "[s]uccess, in the form of useful but unforeseen applications, may not be realized until years later." This concept of risk is the basis for the generalization that basic or fundamental research is inherently more risky than applied research or development. Conversely, research projects that are incremental in nature—taking small steps along a predefined "investigation trajectory"—are generally characterized as less risky than more innovative or creative work.[1] The level of risk inherent in a project has also been linked to the track record of the investigator or firm conducting the R&D; in general, established researchers are considered less risky than younger or recently established investigators. Finally, R&D investment risk has been defined according to the probability of financial payoff from a given research project and the ability of any single individual or organization to appropriate the returns. This definition is implied in the belief that certain types of R&D are too high risk for the private sector and, therefore, should be funded by government.

The concept of balance in the federal R&D portfolio also takes on a range of meanings. The most prominent one is that among the basic research, applied research, and development supported by the federal government (see National Science Board 2000 for definitions). Debate has focused on the importance of sufficient support for basic research and the concern that large investments in development projects, particularly in military systems, could unduly skew the federal R&D portfolio (NAS 1995).[2] A second concept focuses on different fields or disciplines of scientific research. Policy documents often cite the importance of investing in many areas, particularly because of the inability to predict which fields will produce "the next big discovery" (see Committee on Economic Development 1998). Policy analyses have also considered balance among federal mission agencies, although this concern is more of a "second order effect"—since different agencies support more or less basic research or different scientific fields. Other concerns about balance include the fraction of the portfolio devoted to peer reviewed projects, new versus continuing R&D awards, large projects versus small awards, and the geographic distribution of R&D investments.

Having such a range of definitions of balance is less problematic than the lack of a consistent concept of investment risk. Even in a financial portfolio, an investor might seek to maintain balance on a number of different axes simultaneously, e.g., between stocks and bonds and between foreign and domestic investments. The range of disparate definitions of balance does, however, complicate national policymaking. With so many rationales, "[e]valuating balance depends on one's program priorities and research field preferences" (Merrill and McGeary 1999, 1679). As a result, individual segments of the S&T policy community may always believe the portfolio is out of balance along at least one possible axis. Without a method to compare different perspectives systematically, such complaints of imbalance are hard to weigh against competing claims. Furthermore, although many of these balance concepts have a long history in policy debate, it is not clear that they are necessarily the most relevant axes along which the federal portfolio should be evaluated.

Without more broadly applicable definitions, the concepts of risk and balance cannot currently provide a solid foundation for management of the federal R&D portfolio. Returning to the field of financial investing can help by providing one additional concept: defined investment goals. Focusing down on the goal level—articulating the reasons why particular investments are made and their intended outcomes—can loosen some of these conceptual knots.

## Investment Goals and the Federal R&D Portfolio

At its most basic, the statement that the goals of an investment need to be defined before it is made is obvious enough to seem uninteresting. However, like the concepts of risk and balance, investment goals can take on a range of different meanings whose implications with respect to investment management are very different. The different usages, which correspond to the level of aggregation at which investments are discussed, range from a "top-down" viewpoint for an entire portfolio, to an intermediate view for parts of portfolios, to a "bottom-up" approach focused on individual investments.

From the top-down perspective, an investor's goal is simply the overall reason why investments are made at all. In financial investing, such overarching goals might include building wealth or providing for future prosperity. For R&D, goals at this level echo the generalities included in the introductory paragraphs of this chapter: to "prepare the country for the future" or "build national capacity for future challenges." Such goals are not particularly useful for management. "The President's Management Agenda" of the Bush administration (OMB 2001, 43) highlights this concern:

> The ultimate goals of this research need to be clear. For instance, the objective of NASA's space science program is to "chart our destiny in the solar system," and the goal of the U.S. Geological Survey is to "provide science for a changing world." Vague goals lead to perpetual programs achieving poor results.

Such goals express legitimate motivations for making investments but do not provide sufficient information to contribute to investment placement, management, or evaluation.

At the intermediate level of aggregation are goals that define the desired outcomes for groups of investments more specifically—here called "program goals" because they often correspond to the aims of programs within an R&D funding agency. In contrast to "top-down" goals, program goals define the specific problems a set of investments is intended to solve. For example, groups of financial investments could be designed to produce retirement income, accumulate resources for a child's education, or save for a home. Because goals at this level go beyond generalities and identify desirable investment characteristics, they also begin to define criteria for making and managing them.

In the case of R&D investments, program goals could include developing a new fighter aircraft, remediation techniques for environmental problems, capacity in biostatistics research, or more secure methods for e-commerce.[3] Unlike the general goals discussed above, goals at the program level provide some insight into the rationale behind collections of R&D investments and, therefore, begin to be useful for management. However, because of their wide variation across the government, goals at this level still do not provide the consistency needed to support a unified approach to managing the federal R&D portfolio.

What is generally not considered, at least in policy documents available for public consumption, are goals at the highest level of detail: the individual R&D award. In the financial world, individual investments can have a range of goals related to creating or preserving wealth. For example, an investor could aim particular investments at generating resource growth, providing an income stream, minimizing taxes, or providing a hedge against various risks. Different investment types such as individual stocks, mutual funds, partnerships, or derivative instruments would each have a particular goal (or combination of goals). The specific goals of an investment identify both its overall characteristics and intended results and, consequently, define appropriate roles for it to play within particular portfolios' strategies. The central challenge in portfolio design is the selection of investments so their component goals will combine appropriately to achieve the desired portfolio goal. Without clearly articulated goals, it is impossible to assess rationally how individual R&D investments fit as components of program or portfolio goals. As with financial investments, goals for R&D investments should be descriptive of both the investment and its intended returns, while being fundamental enough to categorize investments independent of the goals of the programs or portfolios in which they fall.

## "Bottom-Up" Goals for Federal R&D Investments

Suggestions of goals for R&D activities are many and of long standing, though they often intermingle program-level objectives and general portfolio-level aims with investment-level goals. By focusing downward on relevant goals for individual R&D activities, one can distill from a range of policy analyses and other documents a relatively short list of goals:[4]

1. Expansion of the body of knowledge
2. Monetary returns via economic growth
3. Mission-directed needs for R&D results
4. Workforce development and education
5. Maintenance of national scientific and technical infrastructure and capacity

Such a list of individual investment goals has several advantages for R&D portfolio management. It acknowledges that federal R&D investments are made for a range of reasons that may complement or conflict with one another. The list is tractable for decisionmaking, and yet it seems to capture the full range of rationales for public R&D investments. Furthermore, because the goals are independent of agency portfolios or programs, they are applicable across the federal R&D portfolio.

As is the case for financial investments, the desired outcomes are quite different for each of these five goals.[5] *Discovery of new knowledge* focuses attention on informational outputs and their communication to the larger scientific community and society as a whole. As a result, the desired returns are quality research results that are publicly disseminated. In contrast, the goal of *economic growth* focuses on research paying off financially, and investors whose goal is generating economic growth would not (and should not) consider broad information dissemination a valued output since such activity might undermine the work's profitability.

The third goal, *mission-directed needs for R&D results*, addresses the fact that most federal agencies need R&D-developed capabilities or technologies to carry out their mandates. This goal is most commonly associated with the Departments of Defense and Energy, whose national security programs rely on high-technology systems. However, the missions of agencies guide R&D in pursuit of public health, space exploration, agricultural productivity, worker safety, or environmental quality. Unlike the first two goals, investments guided by a mission rationale are instrumental in nature; the desired return is the technology needed for a new fighter aircraft or the ability to cure a particular disease . As a result, the relevant performance criterion is whether the technology or knowledge developed actually serves the purpose for which it was intended.

The last two goals, *workforce development* and *maintenance of the nation's S&T capabilities,* focus on preserving an ability to apply S&T to future challenges.[6] Beyond maintaining academic training and facilities, however, this category would also include commercial contracts to support the capacity of a firm or, more appropriately, an industry. Maintaining the engineering capabilities of critical defense industries or the capacity to accommodate a surge in demand in case of national emergency is a prime example of such an investment. In both these cases, the desired returns on the investments are measured in the production not of scientific results but of people and capability. The outcomes of an investment in the workforce, for example, might be measured as the number or increase in expertise of the workers it produced. On the other hand, the "return" on an investment aimed at preserving commercial production capacity (not unlike some financial derivatives whose only goal is to hedge particular risks) is simply maintenance of the status quo.

## Advantages of a "Bottom-Up" Goal-Based Approach

These investment goals can provide a route to resolve many issues surrounding investment risk and portfolio balance. By articulating the often very different intended returns for diverse R&D investments, specific investment goals provide the essential ingredient for rigorously defining investment risk. In financial investing, the concept of risk includes both the chance that invested resources will lose value (or be lost) and the chance that the investment will not fulfill its intended goal. Because resources invested in R&D are never "returned" in the way that the money invested in a financial instrument is at sale or maturity, of these two risks the only relevant one is that the R&D investment will not fulfill its intended goal.[7] As a result, articulating the goals of an R&D investment is necessary for the concept of investment risk to have any analytic meaning.

In general, discussions of risk do not tie directly or clearly to what the investments try to accomplish. The distinction between basic and applied research is primarily about whether the goal of the work is new knowledge or technology, not its chances of successfully producing either one. Risk of not producing a financial payoff is only applicable for work where that is the primary intent and, while the track record of an investigator can provide some information on his or her potential productivity, its predictive power is dependent on the nature of the work involved. The explicitly goal-based analysis therefore has the potential to turn upside down the conventional wisdom that fundamental research is inherently more risky than applied work. It is conceivable that a fundamental research project (aimed primarily at producing new knowledge) could be pursued in an area so obviously rich in potential and poorly understood that there is virtually no risk that it will not pay off in new knowledge. In contrast, even if an applied research project involves only proven technologies, certain applications may be uncertain enough that the investment would still be very high risk.[8]

Investment goals may simplify discussion of R&D portfolio balance as well. Although the standard approaches to balance implicitly include some goals of R&D, many combine different goals and, as a result, produce confusion. For example, the most widespread categorization of the portfolio—among basic research, applied research, and development— combines the goals of producing new fundamental knowledge with both economic and mission-directed technology goals. The combination of these separate goals is most clearly demonstrated in *Pasteur's Quadrant* where Stokes (1997) found that resolving the confusion inherent in the basic-applied-development characterization required splitting out the knowledge goal from the technology goal on separate axes of a two-dimensional matrix. Similarly, discussion of balance among scientific fields combines goals of knowledge production, technology and economic concerns, and education and workforce development.

Extending Stokes's twofold breakdown of the goals of fundamental research, this more complete set of goals makes it possible to break out the objectives of R&D across the entire federal R&D portfolio. As a result, these goals provide a way to disaggregate the individual components of traditional concepts of balance. For example, rather than simply "lumping" all defense technology development projects at one end of a basic-applied-development continuum, this treatment focuses attention on how those projects might be aimed at maintaining technical capacities and a viable engineering workforce, in addition to producing particular weapons systems. Similarly, this approach focuses on the underlying reasons that are addressed only by proxy in most other discussions of portfolio balance. For example, while the nation may not have an inherent interest in spreading funds among different fields of S&T for their own sake, there is a national interest in supporting workforce development and research capacity across fields. This approach can also introduce new ways to discuss balance within individual program-level portfolios. Because individual R&D efforts have no guarantee of success, a program portfolio should diversify among different approaches to fulfilling its objectives. Explicitly identifying the goals of each project within a portfolio helps ensure that it contains components whose results are not correlated, therefore enhancing the chance that the overall program will be successful.

Moreover, by providing a higher resolution approach to examining R&D activities, individual goals could make it possible to: (1) better understand the often complex returns of R&D efforts (2) make evaluation of federal R&D programs and agencies more straightforward, and (3) allow greater versatility in assessing the federal portfolio as a whole and the varied agency and program portfolios within it.

## Practical Concerns about R&D Investments

### R&D Investments Can Have Complex Returns

After reading the list of five goals included above, a critical reader might object that a single R&D project almost always contributes to more than one, and sometimes all, of these goals. Examples abound of work, aimed at developing fundamental understanding, that led unexpectedly to new and profitable commercial products. Similarly, the combination within universities of research aimed at producing new knowledge and education is routinely cited as a major strength of the American higher-education system. In fact, the ability of R&D investments to produce results that simultaneously serve science and commercial interests, education, and national needs is a primary reason why they are so attractive.

Instead of detracting from a goal-based approach, this observation in fact emphasizes the benefits of such a methodology. An approach where goals are defined before investments are made allows separation of their intent from their outcomes. As a result, they provide a way to take into account that R&D projects sometimes produce returns that are unexpected and may not have been foreseeable when they were funded. The way that R&D investments are debated and evaluated currently provides no systematic way of addressing such serendipitous discoveries. Instead, discussion often relies on anecdotes—penicillin, the Internet, diagnostic imaging, GPS, and the like—that, while amply demonstrating that such effects can occur, are not helpful for relating the potential occurrence of such effects to current programs and activities.

Like investment risk, serendipity has meaning only in the context of the original intent of the work. For example, if a defense project is intended to develop new radar technologies, the project would have only that single, mission-focused goal. If the experiments done during the project lead to a commercial product, then the project would be credited with a significant economic effect that was outside the initial intent of the program. By recording what was intended, it is possible to determine the effects (both positive and negative) that are unexpected. By providing a structure that can be used for defining goals beforehand and for categorizing returns after the fact, this approach generates a structure that can hold all the results of R&D and, as a result, support a more informed debate.

Although this discussion has generally drawn on examples where a single project had goals that fell into only one of the five classes, nothing prevents or discourages a single project's being aimed at more than one goal. An NSF graduate fellowship program, for example, while primarily intended to support the education of graduate students, might also be intended to develop new knowledge. In the case of projects aimed at more than one goal at the outset, the agency or program would simply have to articulate the intended outcomes for each goal area.

## Performance Assessment Requires Rigorous Investment-Level Goals

The passage of the Government Performance and Results Act (GPRA) in 1993 engendered considerable discussion in the S&T policy community about how to implement the law appropriately for R&D funding agencies, which are quite different from many other government activities, in the time frames required and using the methods stipulated by the act (see, for example, COSEPUP 2001; Kladiva 1997; AAAS 1999).

In its status report on the implementation of GPRA, the Committee on Science, Engineering, and Public Policy (COSEPUP) of the National Academy of Sciences, provides an example of how some of the issues discussed in the previous sections can get in the way of effective R&D evaluation. COSEPUP (2001, 1) states:

> Evaluating federal research programs in response to GPRA is challenging because we do not know how to measure knowledge while it is being generated, and its practical use might not occur until many years

after the research occurs and cannot be predicted. For example, today's global positioning system is the result of research conducted 50 years ago in atomic physics.

COSEPUP's proposal (2001, 10–12) for addressing the complexities of research evaluation is a process in which "experts, supplemented by quantitative methods, can determine whether the knowledge being generated is of high quality, whether it is directed to subjects of potential importance to the mission of the sponsoring agency, and whether it is at the forefront of existing knowledge—and therefore likely to advance the understanding of the field."[9] Like most discussions of R&D evaluation, the COSEPUP report defines one set of common goals for all R&D— production of knowledge and supporting the mission of the agency— rather than separating and looking at multiple goals independently. Later in the report, COSEPUP (2001, 23) indicates that GPRA plans should give explicit emphasis to human resources, akin to goal four (see p. 40). COSEPUP's recommendation that the effect of R&D funding on human resources be specifically highlighted, similar to Stokes's conclusion in *Pasteur's Quadrant* that knowledge production and potential usefulness should be separated, foreshadow the benefits of the disaggregated framework described here.

The central concern in this and similar policy documents is whether applying output and outcome metrics will penalize or misrepresent R&D when the total values of the metrics cannot be determined on timescales relevant to policy or budget cycles, or when they have been designed in such a way that they do not capture all the benefits of R&D. Treating all R&D investments as if they had a single common set of goals is a central source of this problem because it minimizes the differences in their potential outputs and provides no insight on the appropriate timescales over which those outputs should be assessed.

Projects directed at the five goals articulated earlier have intended outputs that are different, measurable, and can be assessed on a variety of timescales. Identifying the goal of a project, which might differ greatly from an "overall" goal articulated for federal R&D in general, defines the output that can reasonably be expected. A clear vision of the desired output defines the appropriate "units" in which to measure it, which then allows the targeting of metrics to determine whether it fell short of, met, or exceeded that expectation. Focusing attention on specific, intended outputs reduces the chance that investments will be judged by inappropriate goals and produce policy-irrelevant (and potentially damaging) conclusions. Making immediate judgments about projects whose main goal is production of new knowledge based on practical use considerations is as inappropriate as judging an income-focused mutual fund by criteria appropriate to growth funds. Although an assessment can be made based on inappropriate criteria, the results can only have an accidental value for future investment design. Furthermore, the focus on the potential long-term, often serendipitous results of R&D also unduly discounts the short-term benefits (in education or new knowledge, for example) that these projects produce. By explicitly accounting for these short-term benefits in a goal-focused model, the R&D projects receive credit for results that do accrue in the short term while a structure to revisit their long-term impact is preserved.

In addition, because it provides a structure to distinguish the expected and unexpected returns of R&D investments, a framework based on investment goals provides a better way to evaluate the performance of R&D funding agencies and the managers who make the individual investments. When assessing the performance of an investment manager, it is unfair to judge the performance of single or groups of investments outside the context in which they were placed. By setting up a structure that captures the intent of the federal investor, a goal-based approach makes it possible to base subsequent evaluation on what could legitimately be expected from the investments.[10] Such an approach will also discourage either too richly rewarding or too harshly punishing R&D managers for serendipitous results that, as COSEPUP suggests, "cannot be predicted." Beyond evaluating the performance of particular programs, investment goals can also help support appropriate comparisons among programs. It would be misleading, for example, to make a direct comparison between financial portfolios based on a metric appropriate to only one of the portfolio's goals. Comparing an R&D program aimed at producing trained scientists to a broad basic research program based on how much new knowledge each produced would yield similarly erroneous results.

## A "Bottom-Up" Approach Preserves Flexibility in Aggregating Portfolios

Evaluation of federal R&D investments is also complicated by the fact that there is no centralized "federal investor" making allocation and priority-setting decisions. Within agencies, decisions are made by individual program managers, though often with extensive input from review or other advisory panels. This has led to an additional concern in the application of GPRA to research: the appropriate level of aggregation to carry out the analysis. COSEPUP (2001, 23–24) states:

> One aspect of GPRA that requires closer consultation between agencies and oversight groups is the clause that permits agencies to "aggregate, disaggregate, or consolidate program activities" in formulating GPRA plans and reports. . . . When the degree of aggregation is high, oversight bodies, potential users, and the public might not be able to see or understand the detailed layers of decision-making and management that underlie the GPRA descriptions.

Looking top down from the agency or total government level can imply a level of centralized decisionmaking that simply doesn't exist. Identifying goals at the project level, whether that identification is made by a program manager or a review panel, provides a strategy to address this problem. Because an analysis based on individual investment goals begins from the bottom up, evaluation at different levels of aggregation becomes more straightforward. If assessments are made at the individual project level, the program performance of any office is simply a roll-up of the performance of its individual projects and an assessment of whether the projects were selected appropriately. Aggregation could conceivably be taken up to the agency or national level, assuming that the goals were applied in similar ways across different agencies, and the databases used to make the assessments within agencies could be integrated. Starting at the project level also makes these performance assessments much more useful for operational management of R&D programs within agencies. Agency-level assessments—that the agency overall is or is not meeting its R&D goals—are less useful for individual offices than the knowledge that their particular projects are either falling short of or exceeding their intended goals.

A central and broadly applicable set of goals also provides a better foundation for looking across the organizational barriers that run through the construction and management of the federal R&D portfolio. The treatment of R&D activities in different agencies is strongly affected by the centrality of R&D to their mission (COSEPUP 2001, 23–24) and how their other mission-related functions are assessed.[11] In addition, the unique, mission-biased perspective that must exist within organizations focused on very different objectives thus shapes the evaluation of R&D. Defining a consistent set of fundamental goals that are independent of mission could help make the intent of the agencies more clear and render differing approaches more obvious to analysts looking across the government as a whole. Simply understanding such differences in intent is a critical first step to a meaningful cross-agency analysis.

## Conclusion

Because of the understandable concern that public funds be wisely used, critical attention to the shape of federal investments in R&D is inevitable. GPRA now requires that all government agencies measure and assess the results of their programs, including their R&D programs. The introductory section of this chapter framed the assessment issue according to three basic questions:

1. Is the government placing its investments well?
2. Is it managing them appropriately?
3. How are the investments performing?

In S&T policy debate, these questions are almost universally addressed in the context of whole R&D portfolios, in contrast to the individual project level approach described in this chapter.

It seems clear, however, that a solid understanding of the goals of individual R&D projects is necessary for managing investment portfolios at the program or agency level. Without defining goals before investments are made, concepts such as risk exposure, portfolio balance, and appropriate investment allocation have no operational leverage. Without clearly defined micro-level goals, the selection and management of R&D investments is like building a financial portfolio without knowing the investment styles of the mutual funds that constitute it. In such a situation, a portfolio that fulfills an investor's overall goals will be constructed only by chance.

Supported by investment-level goals, questions of portfolio design can focus on how individual investments correspond to program or agency's higher-level portfolio goals. Just as a retirement portfolio will be made up of different investments than would a child's college fund, an R&D portfolio striking out into a new technology will differ from one aimed at catalyzing economic growth. As time passes, the appropriate

investment mix will shift as well. As financial investors approach retirement, they shift assets to minimize risk over their contracting time
frame. Similarly, as a new technology program builds its knowledge
base and becomes less and less "unknown," investments should gradually shift away from fundamental knowledge production and toward the
technology-focused goal of the program.

The outlines of such a goal-based approach can be found in recent
policy documents. GPRA documents from R&D agencies include such
elements of this framework as highlighting programs focused on education and workforce development versus those aimed primarily at knowledge production. However, they do not generally appear to build their
evaluations of program and agency performance from the bottom up,
and they certainly do not do so in a way that the performance of the aggregate portfolios can be traced back to the performance of their con-
stituent investments. In addition, by segregating concepts such as education and knowledge production into separate discussions of individual
programs whose primary aim is one or the other, the evaluations cannot
systematically highlight the multiple returns to many R&D investments.
They therefore fall back on discussing whether broad goals like "enabling discovery across the frontier of science and engineering, connected to learning, innovation, and service to society" (NSF, 2002, ii)
were successful or unsuccessful. In addition to being more applicable to
the management of R&D programs, the micro-level approach discussed
here could also enable a more informed and compelling discussion of
the successes of R&D in such evaluation documents.

Elements of such a goal-based approach can be also found in recent
documents from the Office of Management and Budget (OMB). One
states, "It is tremendously important that basic research programs are
able to demonstrate responsible management of their inputs, in addition to clearly articulating and demonstrating progress towards expected outputs" (OMB 2002a, 2). Similarly, the document indicates that
its "focus [is] on improving the management of basic research programs, not on predicting the unpredictable" (OMB 2002a, 1). Where
this analysis and the OMB document diverge is in the selection of
the goals for fundamental R&D. OMB adopts the principles of quality,
relevance, and leadership as metrics for basic research programs. Although these three principles are appropriate measures for the goal of
knowledge production, focusing on them alone overlooks the potential
benefits of fundamental R&D in the other four areas included in this

analysis. OMB's applied research document (2002b) includes a set of goals that relate, but are not directly analogous, to the goal categories proposed in this chapter. As is the case with GPRA documents from agencies, both OMB documents are also primarily focused at the level of agency or organizational portfolios rather than building up portfolio management and evaluation from individual R&D awards.

Because of the practical concerns of managing any portfolio of R&D projects, it is almost certain that many agencies or programs already practice some form of individual goal-based approaches. Since managers must structure their activities based on some picture of what they intend to accomplish and guide their future plans based on an assessment of whether they were successful or not, it is likely that every federal R&D effort is managed through, at the minimum, an implicit version of the model described here. However, the true strength of this approach is realized only when it is applied consistently across different R&D programs within an agency and, ideally, across the many agency activities that constitute the federal R&D portfolio. As a result, determining to what extent similar models are already utilized and extracting best practices from these ongoing policy activities could be a first step to determining an effective way to apply the approach to larger pieces of the federal R&D enterprise. Pilot efforts at applying such goal-based approaches across different programs or agency portfolios would be an important second step to validate such an approach. If appropriate models are not available, a consistent set of concepts, such as those put forward here, should be developed to provide the needed government-wide structure. Once developed and validated, such individual level goals should be integrated into the R&D funding and management process—beginning with assignment of a project's goals when it is funded and ending when its final outputs are assessed.

Like the concepts of portfolio management and investment risk, individual investment goals could inform a model of evaluation that would benefit the analysis of federal investments in R&D. By building a solid foundation for portfolio-level analyses, defined goals for R&D activities provide a framework to consider the diverse potential returns of R&D projects and account for their often large serendipitous effects. The fact that a consistent set of goals can provide a structure to judge the appropriateness of a portfolio's design and both systematically and fairly evaluate its merit is particularly relevant as the implementation of GPRA continues.

While the concept of applying individual investment goals to federal R&D activities is relatively simple, shaping the portfolio in this way involves activities that are anything but. This analysis has focused entirely on identifying the goals of R&D and has not, beyond citing a few illustrative examples, dealt with the process of evaluating how the results of projects meet or do not meet them. An extensive literature exists on the evaluation of outcomes of public R&D that is key to the successful use of the framework described here (see Kostoff 1997, and references therein). Some of the goal areas tie directly to areas where significant work has already been done, such as estimation of the economic returns of R&D. Other areas are not as well studied, and significant questions still exist in the field of public R&D evaluation (NSB 2001). This chapter ends where that literature begins. It does not address the unresolved problems in that field as much as it reframes some and suggests others. If the approach described here proves sound, it could provide a structure in which the results of various approaches to evaluation could be applied more systematically and become even more powerful tools to shape the federal investment portfolio.

NOTES

The author gratefully acknowledges the input of Anduin Touw, Emile Ettedgui, and Henry Willis to earlier versions of the chapter.

1. This concept of risk is implied in discussions of the pressures that stringent performance evaluation and program review could put on science funding agencies. With increased pressure to substantiate results, the concern has been that peer reviewers or program managers will favor projects that are more likely to produce measurable results within the time frame of the project review cycle.

2. See, among others, NSB (2000); U.S. House (1998); and Committee on Economic Development (1998).

3. The current discussion of R&D portfolios to some extent artificially restricts discussion. Government could have a portfolio of expenditures, of which R&D is only a part, that has a portfolio goal of "improving fuel economy of automobiles" or "reducing the incidence of HIV in the U.S. population." In these cases, the appropriate role of R&D as part of that portfolio would have to be determined with respect to the other possible ways government could seek to accomplish the goal. Consideration of these possible substitutes for R&D in program portfolios is considered explicitly in OMB's 2002 draft-applied research investment criteria for DOE (OMB 2002b). Deciding among such different

alternatives is particularly difficult in policy discussions, adding an additional complication to effectively shaping science policy (Toulmin 1968).

4. Another analysis of federal investment management can be found in Bozeman and Rogers (2001). Rather than seeking one set of overarching goals, they identify three types of desired information outputs (basic research, technology development and transfer, and software and algorithms) and S&T human capital as the targets of agency research. Although less inclusive than the goals described here, the insights on portfolio balance included in their work would apply equally to a portfolio categorized using these five more general goals.

5. Federal investors' time frames for achieving these goals may not be immediate. R&D activities may be aimed at laying the groundwork for potentially new mission capabilities at a future time. In those cases, the current expenditures on R&D have been characterized as "options" on future capabilities not unlike similar options used in financial investing (see Vonortas and Hertzfeld 1998).

6. The goal of preserving the nation's S&T infrastructure would reasonably include the portion of the federal facilities and equipment investments that are aimed at science and technology capabilities. As a result, the focus of this chapter on R&D investments alone is not entirely appropriate for this portion of the investment portfolio. Some R&D investments do, in fact, support these types of infrastructures because they can pay for basic laboratory equipment, and the overhead dollars associated with awards thus support physical infrastructures. The separation is simply a reflection of another limitation of the current categorization of federal investments (or likely any categorization with a relatively small number of investment classes).

7. Because investments are placed based on assumptions about future needs and conditions, there is also a risk that both financial or R&D investments will not appropriately match future requirements. For example, the conditions an investment was intended to address could never materialize (e.g., money invested for a need that does not actually arise or research done in an area that proves to be unnecessary). Such misplacement can produce significant opportunity costs. Although such misplacement risk is important, this discussion focuses on the risks involved after an investment is placed that affect its potential return to the investor.

8. One example of an explicitly goal-based risk definition occurs in an NSF (2003) grant description: "A project will be considered 'risky' if the data may not be obtainable in spite of all reasonable preparation on the researcher's part." Because the goal of the project is new knowledge, risk is not expressed in a language of utility or return but simply the ability to gather the needed data for the project itself.

9. This chapter does not directly consider mechanisms of evaluation and the assessment of research programs would almost certainly involve expert

reviews like those described in the report. However, this particular portion of the National Academies analysis is highlighted to illustrate how the points cited in the previous section can get in the way of effective evaluation in the context of an effort like GPRA implementation.

10. The next level of evaluation would include whether the goals of the projects link appropriately to the program and higher goals, and whether the investments (even if they achieved their goals) were made cost-effectively. The question of cost-effectiveness is central for investments that, like R&D, do not result in the "return of the investment at maturity." These questions are beyond the scope of this chapter but one that is of particular relevance in the evaluation of the R&D managers that manage the national portfolio (see Link 1996 for cost-effectiveness analysis).

11. Differences in goals for work among agencies will also have significant impacts on how research is funded and administered. For example, some agencies focus their R&D funding on grant mechanisms while others focus on contracts. Understanding the impact of these different mechanisms and how they relate to both agency and national goals for the R&D portfolio is important.

## REFERENCES

American Association for the Advancement of Science (AAAS). 1999. *Science and Technology Policy Yearbook*. Washington, DC: American Association for the Advancement of Science. http://www.aaas.org/spp/yearbook/contents.htm (11 June 2003).

Bozeman, B., and J. Rogers. 2001. Strategic Management of Government-Sponsored R&D Portfolios: Lessons from Office of Basic Energy Sciences Projects. *Environment and Planning C: Government and Policy*. http://rvm.pp.gatech.edu/papers/strtmng05-05.pdf (17 August 2002).

Committee on Economic Development. 1998. America's Basic Research: Prosperity Through Discovery. http://www.ced.org/docs/report/report_basic.pdf (13 July 2003).

Committee on Science, Engineering, and Public Policy, National Academy of Sciences (COSEPUP). 2001. *Implementing the Government Performance and Results Act for Research: A Status Report*. Washington, DC: National Academy Press.

U.S. House, Committee on Science. 1998. Unlocking Our Future: Toward a New National Science Policy. 24 September. http://www.house.gov/science/science_policy_report.htm (4 August 2002).

Kladiva, S. 1997. *Results Act: Observations on Federal Science Agencies*. General Accounting Office Document GAO/T-RCED-97-220, 30 July, Washington, DC.

Kostoff, R. N. 1997. *The Handbook of Research Impact Assessment*. 7th ed. http://www.onr.navy.mil/sci_tech/special/technowatch/reseval.htm (11 July 2003).

Link, A. N. 1996. *Evaluating Public Sector Research and Development.* Westport, CT: Praeger.

McGeary, M., and P. M. Smith. 1996. The R&D Portfolio: A Concept for Allocating Science and Technology Funds. *Science* 274:1484–85.

Merrill, S. A., and M. McGeary. 1999. Who's Balancing the Federal Research Portfolio and How? *Science* 285:1679–80.

National Academy of Sciences (NAS), Committee on Criteria for Federal Support of Research and Development. 1995. *Allocating Federal Funds for Science and Technology.* Washington, DC: National Academy Press.

National Science Board (NSB). 2000. *Science and Engineering Indicators 2000.* Washington, DC: National Science Foundation.

———. 2001. *The Scientific Allocation of Scientific Resources.* National Science Foundation Document NSB 01–39. Washington, DC, 28 March. http://www.nsf.gov/nsb/documents/2001/nsb0139/ (25 September 2002).

National Science Foundation (NSF). "Physical Anthropology: High Risk Research in Anthropology." http://www.nsf.gov/sbe/bcs/physical/highrisk.htm (11 July 2003).

———. "FY 2003 GPRA Final Performance Plan," February 4. http://www.nsf.gov/od/gpra/perfplan/fy2003/Final%20Plan.pdf (11 July 2003).

Office of Management and Budget (OMB). 2001. "The President's Management Agenda," http://www.whitehouse.gov/omb/budget/fy2002/mgmt.pdf (11 July 2003).

———. 2002a. "OMB Preliminary Investment Criteria for Basic Research," Discussion Draft. http://www7.nationalacademies.org/gpra/index.html (11 July 2003).

———. 2002b. "Applied R&D Investment Criteria for DOE Applied Energy R&D Programs," http://www7.nationalacademies.org/gpra/Applied_Research.html (11 July 2003).

Office of Technology Assessment (OTA). 1991. *Federally Funded Research: Decisions for a Decade.* May, Washington, DC .

Robinson, D. 1994. Show Me the Money: Budgeting in a Complex R&D System. Paper presented at symposium, Science the Endless Frontier 1945–1995: Learning from the Past, Designing for the Future, part 1, 9 December 1994. http://www.cspo.org/products/conferences/bush/Robinson.pdf (4 August 2002).

*Science for Society: Cutting Edge Basic Research in the Service of Public Objectives.* 2001. A Report on the November 2000 Conference on Basic Research in the Service of Public Objectives, May 2001.

Stokes, D. E. 1997. *Pasteur's Quadrant: Basic Science and Technological Innovation.* Washington, DC: Brookings Institution Press.

Toulmin, S. 1968. The Complexity of Scientific Choice: A Stocktaking. In *Criteria for Scientific Development: Public Policy and National Goals,* ed. E. Shils, 63–79. Cambridge, MA: MIT Press.

Vonortas, N. S., and H. R. Hertzfeld. 1998. Research and Development Project Selection in the Public Sector. *Journal of Policy Analysis and Management* 17 (4): 621–38.

# 3

# Universities and Intellectual Property

*Shaping a New Patent Policy for*
*Government Funded Academic Research*

BHAVEN N. SAMPAT

## Introduction

Over the past quarter-century, patenting and licensing of publicly funded research by American research universities have grown dramatically. This growth has contributed to some of the highest-profile debates in science and technology policy. Witness, for example, controversies over the high prices of drugs developed from taxpayer-funded academic patents, concerns about the appropriateness of publicly funded researchers "racing" a private firm to sequence (and patent) the human genome, and fears that patents held by a public research university could hinder embryonic stem cell research in the United States.

The shape of the divide between public and private in academic research lies at the heart of each of these debates. Some view the growth of academic patenting and licensing as a new model of university-industry interaction, one that facilitates economic and social returns from publicly funded research. But others see this growth as a socially inefficient "privatization" of academic research and a threat to the ethos of science itself. In this chapter, I put these changes in historical

context, providing a broad overview of changes in universities' patenting policies, procedures, and practices throughout the twentieth century, as well as an assessment of the effects of policy changes—in particular the Bayh-Dole Act of 1980—on economic returns from university
research.

Since much of the policy discussion of these issues centers on economic returns from university research, I begin the chapter by discussing the various channels through which universities contribute to innovation and economic growth. This discussion reveals that patenting
and licensing are only two of many channels through which universities
make economic contributions, channels that, for most industries, are less
important than the contributions universities make by placing scientific
and technological information in the public domain. Indeed, as I discuss
in the second section, fears of compromising the public aspects of academic research—important not only to industry but for the advance of
science itself—led most American universities to avoid active involvement in patenting and licensing throughout much of the twentieth century. Though this reluctance began to fade in the 1970s, the major impetus toward increased university involvement in patenting and licensing
was the Bayh-Dole Act, passed in 1980 to facilitate commercialization of
university inventions. In the third section, I discuss the political history
of Bayh-Dole, showing that the intellectual foundations for this sea
change in federal patent policy were weak. Nevertheless, assessment of
the social welfare effects of Bayh-Dole (and growth of university patenting and licensing more generally) remains an open empirical question.
There is little evidence that increased university patenting and licensing
has facilitated increased technology transfer or any meaningful growth
in the economic contributions of universities. Yet neither is there systematic evidence that this growth is negatively affecting the conduct of
or returns from public science. I conclude with suggestions for shaping
policies for academic patenting and licensing more productively.

## Universities, Innovation, and Economic Growth

Over the past century, American research universities have been extremely important economic institutions. From agriculture to aircraft to
computers to pharmaceuticals, university research and teaching activities have been critical for industrial progress (Rosenberg and Nelson

1994). Most economic historians would agree that the rise of American technological and economic leadership in the postwar era was strongly dependent on the strength of the American university system.

The economically important "outputs" of university research have come in different forms, varying over time and across industries. They include, among others: scientific and technological information (which can increase the efficiency of applied R&D in industry by guiding research toward more fruitful departures), equipment and instrumentation (used by firms in their production processes or their research), skills or human capital (embodied in students and faculty members), networks of scientific and technological capabilities (which facilitate the diffusion of new knowledge), and prototypes for new products and processes.

The relative importance of the different channels through which these outputs diffuse to (or alternatively, "are transferred to") industry also has varied over industry and time. The channels include, among other things, labor markets (hiring students and faculty), consulting relationships between university faculty and firms, publications, presentations at conferences, informal communications with industrial researchers, formation of firms by faculty members, and licensure of patents by universities. Although the recent growth of patenting and licensing by universities has received considerable attention, it is important to keep in mind that patents are one of many channels through which university research contributes to technical change in industry and economic growth.

Indeed, in most industries patents provide a relatively unimportant channel. In a survey of R&D managers of firms in the U.S. manufacturing sector, Cohen, Nelson, and Walsh (2002) asked respondents to rank different channels through which they learn from university research. They found that in most industries, the channels reported to be most important were publications, conferences, and informal information exchange. Patents and licenses ranked near the bottom of the list.[1] A study by Agrawal and Henderson (2002), which focused on two major academic units at the Massachusetts Institute of Technology (MIT), provides corroborating evidence from the "supply" side of academic research. Faculty members reported that a very small fraction of the knowledge transfer from their laboratories to industry (7 percent) occurs via patenting. Other channels—Agrawal and Henderson focus on publications—are more important.

It is noteworthy that the most important channels of university-industry knowledge transfer—publications, conferences, and informal information exchange—are those associated with what the sociologist of science Robert Merton (1973) called the norms of "open science," which create powerful incentives for academics to publish, present at conferences, and share information, thus placing information in the public domain (Dasgupta and David 1994).

Outputs of academic research disseminated via open science are useful not only to industry but also to future academic research. Academic research is a cumulative process that builds upon itself: recall Sir Isaac Newton's aphorism, "if I have seen further, it is by standing on the shoulders of giants." Thus, prominent sociologists of science (Crane 1972; Merton 1973; de Solla Price 1963), philosophers (Polanyi 1962; Kitcher 1993; Ziman 1968), and economists (Dasgupta and David 1994; Nelson 2002) have pointed to the importance of information sharing and communication (via formal and informal channels) for the advance of academic research.

## University Patenting and Licensing before Bayh-Dole

Throughout much of the twentieth century, universities were reluctant to become directly involved in patenting and licensing activities, precisely because of fears that such involvement might compromise, or be seen as compromising, their commitment to open science and their institutional missions to advance and disseminate knowledge. Consequently, many universities avoided patenting and licensing activities altogether, and those that did get involved typically outsourced their patent management operations to third-party operations like the Research Corporation at Columbia University, or set up affiliated but legally separate research foundations like the Wisconsin Alumni Research Foundation, to administer their patents (Mowery and Sampat 2001a; 2001b).

While most universities involved in patenting employed one of these two options in the pre-Bayh-Dole era, there was considerable variance in their formal patent policies, for example, in faculty disclosure policies and sharing rules (Mowery and Sampat 2001b). In the postwar era, many universities had "hands off" policies, refusing to take out patents as institutions but allowing faculty members to patent and retain title if they desired. Thus, before 1980, Columbia's policy left patenting up to the inventor and administration up to the Research Corporation, stating that "it is not deemed within the sphere of the University's scholarly objectives" to hold patents. Some universities required faculty members to report inventions to university administration, which then typically turned the report over to the Research Corporation or university research foundations. Notably, several major universities explicitly forbade the patenting of biomedical research results, evidently based on the belief that restricting the dissemination of health-related inventions was particularly contrary to their missions. At Harvard, Chicago, Yale, Johns Hopkins, Columbia, and Chicago, these prohibitions were not dropped until the 1970s.

Notwithstanding some variance in actual patent policies, over the first three-quarters of the twentieth century American research universities were generally reluctant to become directly involved in patenting and licensing. However, patents did occasionally result from university research efforts—a fact that is not surprising in view of the fact that American universities were never pure "ivory towers" but rather historically active in use-oriented basic and applied research (Rosenberg and Nelson 1994). In such cases, patents were typically held by faculty members or by third-party technology transfer agents, rather than the universities themselves. In most cases, the reason for patenting was explicitly to protect the public interest—to promote commercialization, to protect state taxpayer interests, or to prevent firms from patenting and restricting access to university research (Mowery and Sampat 2001b). Although it is difficult to show conclusively, it is likely that strong norms militating against academic patenting checked any ambitions universities may have had to patent in instances where publication or open dissemination would suffice to achieve these ends.

All of this began to change in the 1970s, and identifying the sources of this shift remains an important topic for future research (see Mowery and Sampat 2001b and Sampat and Nelson 2002 for preliminary attempts). Certainly, one important development was the fruition of commercial applications resulting from the postwar growth of "use-oriented" basic research (Stokes 1997) in fields like molecular biology. At the same time, there was a slowdown in growth of federal funds for academic research (Graham and Diamond 1997), leading some universities to become increasingly interested in patenting as a source of income. Changes in the geographical dispersion of federal R&D funding in the 1970s may also have played a role in inducing "entry" into academic patenting (Mowery and Sampat 2001a). Moreover, changes in government

patent policy during the 1970s—precursors to the Bayh-Dole act—
made it easier for universities to patent publicly funded research.

Whatever the causes, many institutions began to reconsider their
patent policies and procedures during the 1970s. Thus, by the mid-
1970s, Research Corporation's *Annual Report* noted that most major institutions were considering setting up internal technology transfer offices (Mowery and Sampat 2001a), and patenting by universities began
to grow in the decade before Bayh-Dole. Nevertheless, Bayh-Dole mag-
nified and accelerated these changes by providing strong congressional
endorsement for the position that active university involvement in patenting and licensing, far from being ignoble, serves the public interest.

## Patents, Public Funding, and the Bayh-Dole Act

Until the late-1960s, the vast majority of university patents, including
those handled by university-affiliated research foundations, were based
not on federally funded research but rather on research financed by institutional funds, industry, and state and local governments. Before this
time, many federal funding agencies retained title to any patents resulting from research they funded, and others required universities to go
through cumbersome petitioning processes to retain title to publicly
funded patents. Several agencies became more liberal in allowing universities to retain title to patents during the late-1960s and 1970s, but
there was considerable uncertainty about political commitment to these
changes. Bayh-Dole erased this uncertainty, creating a uniform federal
policy allowing universities to retain rights to patents resulting from
publicly funded research.

### Historical Background and Intellectual Foundations
of Patent Policy

Though issues relating to university patenting provided the immediate impetus for the introduction of Bayh-Dole—and its effects on universities were more pronounced than on other contractors—the long-
standing historical concerns about the effects of patenting in academic
environments were ignored during the Bayh-Dole hearings.[2] Instead,
the debates focused primarily on more general issues relating to
whether the governments or private contractors should retain title to

patents resulting from public funds, and the feasibility and desirability of a uniform federal patent policy across funding agencies.

This debate is old: Congress considered the issue of who should retain rights to patents resulting from publicly funded research at least seventy years before the passage of Bayh-Dole (Forman 1957). However, government patent policy first became a prominent issue following the massive expansion of federal R&D during World War II. It was a central point of contention during the debates over the shape of post–World War II science and technology policy between Vannevar Bush, who wanted to expand the wartime policy of allowing contractors to retain patent rights, and Senator Harley Kilgore, who argued that the federal government should retain the rights (see, among others Kevles 1977 and Guston 2000).

The postwar debate highlighted the central issues in controversies over government patent policy for the next three decades. Supporters of the retention of intellectual property rights by government agencies argued that allowing contractors (rather than government agencies) to retain title to patents resulting from federally funded research favored large firms at the expense of small business. Moreover, they asserted, such a policy would harm consumers who would have to pay monopoly prices for the fruits of research they had funded through their taxes. Supporters of allowing contractors to retain title argued that failure to do so would make it difficult to attract qualified firms to perform government research and that absence of title would reduce incentives to invest in commercial development of these inventions.

Another contentious issue in the debates about government patent policy was the desirability of a "uniform" patent policy across all federal agencies. Each of the major federal R&D funding agencies had established its own patent policy following World War II, and the resulting mix of agency-specific policies created ambiguities and uncertainties for contractors and for government employees. Despite numerous congressional hearings on this issue, no legislation was adopted during the 1950–75 period because of the inability of supporters of opposing positions to resolve their differences. The legislative deadlock was reinforced by statements on federal agencies' patent policies issued by Presidents Kennedy in 1963 and Nixon in 1971. Both presidents asserted that agency-specific differences in patent policy were appropriate, in view of the differences in agency missions and R&D programs (US OMB 1963; US OMB 1971).

Very little of the debate in this period focused on universities, which, after all, are minority recipients of federal R&D funds and were themselves historically reluctant to become actively involved in patenting and licensing. Federal policy toward patents resulting from publicly funded university research became a topic of debate only after the release of a report on NIH's Medicinal Chemistry program (GAO 1968) and a second report, for the Federal Council for Science and Technology (FCST), on government patent policy more generally (Harbridge House 1968b). Both reports examined the effects of federal patent policy on research collaboration between U.S. pharmaceutical firms and academic researchers in medicinal chemistry. During the 1940s and 1950s, these pharmaceutical firms had routinely screened compounds developed by NIH-funded university researchers for biological activity at no charge. Depending on the patent policies of particular universities, these pharmaceutical firms might receive exclusive rights to develop and market the compounds. In 1962, the Department of Health, Education, and Welfare (HEW) notified universities that firms screening compounds must formally agree not to obtain patents on any technologies that resulted from NIH funding or that were in the "field of research work" supported by the NIH grant.

The two reports criticized HEW's patent policy, arguing that pharmaceutical firms had stopped screening NIH grantees' compounds because of the firms' concern that HEW policies might compromise their rights to intellectual property resulting from their in-house research (Harbridge House 1968a, II-21; GAO 1968, 11). Both reports recommended that HEW change its patent policy to clarify the circumstances in which rights reverted to the government, and those under which universities could retain title to patents and issue exclusive licenses to firms.

HEW responded to these critical reports in 1968 by establishing Institutional Patent Agreements (IPAs) that gave universities with "approved technology transfer capability" the right to retain title to agency-funded patents. In addition, HEW began to act more quickly on requests from universities and other research performers for title to the intellectual property resulting from federally funded research. Between 1969 and 1974, HEW approved 90 percent of petitions for title, and between 1969 and 1977 the agency granted IPAs to seventy-two universities and nonprofit institutions (Weissman 1990). The National Science Foundation (NSF) instituted a similar IPA program in 1973, and the Department of Defense (DOD) began in the mid-1960s to allow universities with approved patent policies to retain title to inventions resulting from federally funded research.

Thus, by the beginning of the 1970s, U.S. universities were able to patent the results of federally funded research via IPAs (or similar programs at DOD) that were negotiated on an agency-by-agency basis, or via case-by-case petitions. These changes were likely partially responsible for the growth of university patenting during the 1970s.

The Bayh-Dole Act

Universities' concerns over potential restrictions on HEW's IPA program provided the primary thrust for the introduction of the bill that eventually became the Bayh-Dole Act. In August 1977, HEW's Office of the General Counsel expressed concern that university patents and licenses, particularly exclusive licenses, could contribute to higher healthcare costs (Eskridge 1978). HEW ordered a review of its patent policy, including a reconsideration of whether universities' rights to negotiate exclusive licenses should be curtailed. During the ensuing twelve-month review, HEW deferred decisions on thirty petitions for patent rights and three requests for IPAs. HEW's reconsideration of its patent policies followed a similar review at DOD that had led to more restrictive policies toward university patenting (Eisenberg 1996).

Reflecting their increased patenting and licensing activities, U.S. universities expressed concern over these restrictions to Congress (Broad 1979a). In September 1978, Senator Robert Dole (R-Kansas) held a press conference where he criticized HEW for "stonewalling" university patenting, commenting, "rarely have we witnessed a more hideous example of overmanagement by the bureaucracy" and announcing his intention to introduce a bill to remedy the situation (Eskridge 1978, 605). On 13 September 1978, Senator Dole and Senator Birch Bayh (D-Indiana) introduced S. 414, the University and Small Business Patent Act. The act proposed a uniform federal patent policy that gave universities and small businesses blanket rights to any patents resulting from government-funded research.[3] The bill lacked provisions that were typically included in IPAs, including the requirement that in order to receive title universities must have an "approved technology transfer" capability. In contrast to the language of some IPAs, the bill also lacked any language expressing a federal preference for nonexclusive licensing agreements (Henig 1979).

Since at least the earlier Bush-Kilgore debates, there had been strong congressional opposition to any uniform federal policy that granted rights of ownership of patents to research performers or contractors. But Bayh-Dole attracted little opposition, for several reasons. First, as its title suggests, the bill's focus on securing patent rights for only universities and small business weakened the argument (*a la* Kilgore) that such patent-ownership policies would favor big business.[4] Second, the bill included several provisions designed to defuse criticism that it would lead to "profiteering" at the expense of the public interest, including a recoupment provision whereby institutions would have to pay back a share of licensing income or sales to funding agencies, and time limits on exclusive licenses of five years from commercial sale or eight years from the date of the license. Third, and most importantly, the bill was introduced in the midst of debates over U.S. economic competitiveness in the late 1970s. An article in *Science* discussing the debate on the bill observed:

> The critics of such legislation, who in the past have railed about the "giveaway of public funds" have grown unusually quiet. The reason seems clear. Industrial innovation has become a buzzword in bureaucratic circles. . . . [T]he patent transfer people have latched onto this issue. It's about time, they say, to cut the red tape that saps the incentive to be inventive. (Broad 1979b, 479)

Committee hearings on the Bayh-Dole bill in both chambers were dominated by witnesses from small business, various trade associations, and universities active in patenting and licensing. Much of the testimony and commentary during these hearings focused on lagging U.S. productivity growth and innovativeness, suggesting that government patent policy contributed to these woes. In their opening statements in the Senate Judiciary Committee hearings on the bill, Senators Bayh and Dole each pointed to two problems with federal patent policy as of 1979: the "policy" in fact consisted of more than twenty different agency-specific patent policies, and most federal agencies made it difficult for contractors to retain title to patents.

Most of the debate centered on whether publicly funded patents held by private contractors enjoyed higher rates of commercialization than those held by the government itself. Witnesses supporting the Bayh-Dole bill frequently cited results from the Harbridge House study that rates of utilization of government funded patents were higher when contractors, rather than agencies, held title to these patents. Another frequently cited statistic was based on the 1976 FCST report that concluded that fewer than 5 percent of the 28,000 patents owned by the federal government in 1976 were licensed (FCST 1978). Legislators and witnesses used this finding to argue that giving patent rights to contractors would create incentives for development and commercialization that were lacking under the current system. As Eisenberg (1996) points out, however, this inference was invalid. The patents cited in these studies were based primarily on research funded by DOD (83 percent of the patents from the Harbridge House sample and 63 percent of those from the FCST sample), which in fact readily granted patent rights to research performers. Patents derived from DOD-funded research for which contractors elected not to seek ownership rights almost certainly had limited commercial potential, and it is not surprising that they were not licensed.

Moreover, the data in these reports were based primarily on patents resulting from government-funded R&D carried out by private firms. As such, they were of questionable relevance to the debate over the patenting by universities of inventions funded by the federal government. Several representatives from universities did make some points specific to the academic context. A first was that university inventions are "embryonic" when first disclosed, requiring significant additional development before they can be commercially useful, and that firms would not invest in these costly development activities without clear rights to the relevant intellectual property—which, they argued, required university title to the patents and exclusive licenses. Other witnesses suggested that giving title to universities would create incentives for inventors and institutions to become actively involved in the development and commercialization, anticipating arguments recently developed more formally by Jensen and Thursby (2000).

Beyond citing the Harbridge House or FCST statistics—which were of questionable validity and relevance—supporters of Bayh-Dole offered little systematic evidence that university inventions were being "underutilized" because of difficulties universities faced in retaining title or granting exclusive licenses. More importantly, none of the witnesses discussed the potential risks created by university patenting and licensing for the norms of academic science, or any potentially detrimental effects of patenting and licensing for other channels of university-industry technology transfer. A journalist covering the hearings observed that "although the Dole-Bayh bill is receiving nearly unprecedented support, some congressional aides point out that it still leaves unanswered

fundamental questions about patents in general and patents on university campuses in particular" (Henig 1979, 284).

Both the House and Senate passed Bayh-Dole overwhelmingly and
with minimal floor debate in the winter of 1980. The final version of the
bill omitted the recoupment provision and the time limit on exclusive licenses for small firms, although it retained time limits on exclusive licenses for large firms. President Carter signed the Bayh-Dole Act into
law in 1980, and it became effective in 1981.

The act created a uniform federal patent policy for universities and
small businesses, giving them rights to any patents resulting from grants
or contracts funded by any federal agency. Because universities had
begun increasing their involvement in patenting and licensing before
Bayh-Dole, and because the growth of university patenting and licensing probably would have continued in its absence, one of the most important of the law's effects was normative: by endorsing university involvement in patenting and licensing, it assuaged any remaining fears
about the reputational costs of involvement in the "business-side" of
patenting and licensing. University patenting, licensing, and even licensing revenues would no longer be seen as potential sources of political embarrassment—as they had been throughout much of the twentieth century—but rather as indicators of the "entrepreneurialism" and
"economic dynamism" of research universities.

## The Effects of Bayh-Dole

### The Growth of University Patenting and Licensing

In the wake of Bayh-Dole, universities increasingly became directly involved in patenting and licensing, setting up internal technology transfer offices to manage licensure of university patents. Figure 3.1 shows
the distribution of years of "entry" by universities into patenting and
licensing, defined as the year in which the universities first devoted
0.5 full-time equivalent employees to "technology transfer activities"
(AUTM 1998). Consistent with the discussion above, few universities
were involved in patenting and licensing early in the century. Entry
began during the 1970s but accelerated after Bayh-Dole.

University patenting exhibits a similar trend. Figure 3.2 shows the
total number patents issued to research universities over the 1925–2000
period. Here again, growth began during the 1970s but accelerated after

## Figure 3.1. Year of entry into technology transfer activities, 1920-1998

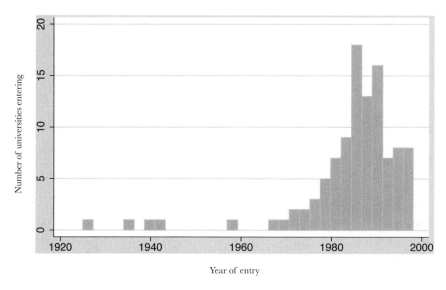

Year of entry

*Source:* AUTM (1998)

## Figure 3.2. Patents issued to research universities by issue year

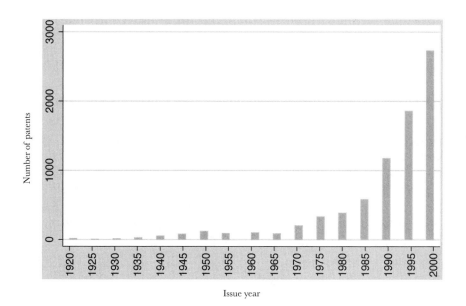

Issue year

Figure 3.3. Distribution of Carnegie University licensing revenues

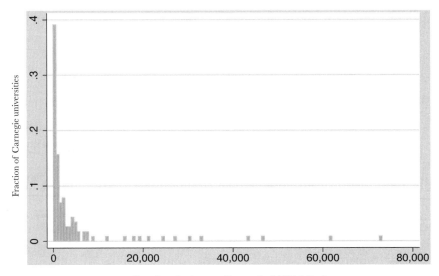

Gross licensing incomes (thousands of 1998 dollars)

1980. Time series on license revenues are more difficult to obtain, as they were not systematically collected until the early 1990s. In 1991, according to a survey by the Association of University Technology Managers (AUTM), universities earned nearly $200 million in gross license revenues. A decade later, this figure increased to over $1 billion (AUTM 2002).

Although the growth rate is impressive, license revenues accounted for less than 5 percent of all research funds at AUTM universities in 2000 (AUTM 2002). Note also that this figure was calculated before subtracting the inventors' share of royalty income (typically 30–50 percent) and before subtracting costs of patent and license management, which can be significant.

Also, a small number of universities account for the lion's share of licensing revenues. Figure 3.3 shows the distribution of university licensing revenues in 1998, the last year for which disaggregated licensing data were available. Note that few universities are making large revenues; in fact, 10% of these universities account for over 60 percent of total licensing revenues. The numbers in figure 3.3 also do not include costs of patent and license management. It is likely that after taking costs into account, the majority of American research universities have been losing money on their patenting and licensing activities (cf. Trune and Goslin 1997).

## Effects on Technology Transfer and the Academic Enterprise

Of course, the primary purpose of the Bayh-Dole act was not to make universities rich but rather to promote "technology transfer."[5] A number of observers in the United States and abroad have looked to such patenting and licensing trends as displayed above and pronounced Bayh-Dole a resounding success (see, e.g., Bremer 2001; *Economist* 2002). Implicit in this interpretation is the assumption that the commercialization and development underlying these trends would not have occurred absent Bayh-Dole, or, more generally, absent university patenting and licensing.

This assumption is bound to be valid in some cases, but certainly not in all. The importance of patents and licensing for development and commercialization of university inventions was not well understood during the Bayh-Dole hearings, and it is not well understood today. Universities can patent any inventions developed by their faculty members, and are known to do so in cases where commercialization would go forward even absent patenting and licensing.[6] For example, the Cohen-Boyer recombinant DNA technique was being used by industry even before the University of California and Stanford began licensure; patenting (and licensing widely) allowed the universities to generate income but did not facilitate technology transfer. In an oral history, Niels Reimers (1998), the manager of the Cohen-Boyer licensing program, made this point explicitly, noting that "whether we licensed it or not, commercialization of recombinant DNA was going forward. As I mentioned, a nonexclusive licensing program, at its heart, is really a tax . . . [b]ut it's always nice to say 'technology transfer.'"

Another invention that fits this bill is Richard Axel's co-transformation process, patented and licensed by Columbia University. In this case, firms were using the technology shortly after it was described in the scientific literature and before a patent was granted. The university compelled firms to license the invention by threatening to sue the firms if they continued to use it without a license.

In these two cases, technology transfer preceded university patenting and licensing activities. These are just two cases, but two important ones: together they account for close to 15 percent of cumulative royalty revenues earned by all research universities between 1983 and 2001.

Here, the university revenues are "taxes" on industry (to use Reimers's language) and ultimately consumers, rather than indicators of the extent of technology transfer.

In cases such as these, where universities are patenting inventions that would have been utilized or developed even absent intellectual property rights, society suffers the standard losses from noncompetitive pricing. Restrictive access to university inventions may also unnecessarily limit further experimentation and development (see Merges and Nelson 1994). The share of these cases and the extent of these costs are unknown because they involve counterfactuals. But a proper evaluation of the effects of Bayh-Dole on technology transfer must take these types of costs into account.

More importantly, a complete assessment of the effects of Bayh-Dole requires analysis of its effects on other channels of knowledge and technology transfer. We have little understanding of whether and how increased academic patenting and licensing affect the other channels through which universities contribute to technical change in industry and economic growth. Yet these issues have been the cause of much concern, and rightly so, in view of the relative importance of these other channels. In particular, the surveys cited above suggest that the most salient economic contribution of universities is via production of information that helps increase the efficiency of applied R&D. Traditionally, this information has been disseminated in publications and conferences, informal communication, and by employment of graduate students. We do not know whether increased patenting and licensing constrict these channels of knowledge transfer, and this topic remains an important one for future research.

A related concern is that academics and universities are limiting disclosure and availability of information and materials that are inputs into scientific research itself. Unfortunately, apart from anecdotal evidence there is little systematic evidence on this front either. But given the importance of communication and information disclosure for the advance of science, this concern too deserves to be taken seriously.

## Conclusion

This chapter shows that the "private parts" of academic science have certainly expanded over the past quarter-century. Universities' historical reluctance to become involved in patenting and licensing has disappeared, replaced with apparent enthusiasm. This change was supported in part by the Bayh-Dole Act, which not only made it easier for universities to patent federally funded research but also provided a political imprimatur and a remote financial incentive for university involvement in patenting and licensing activities.

The political history of Bayh-Dole reveals that it passed based on little solid evidence that the status quo ante resulted in low rates of commercialization of university inventions. The congressional hearings neglected the economic importance of the public aspects of university research; they ignored the possibility of potential negative effects of increased patenting and licensing on open science and on other channels of technology and knowledge transfer.

Nevertheless, the net effects of Bayh-Dole (and the rise of university patenting and licensing activity more generally) on innovation, technology transfer, and economic growth remain unclear, and much more research is necessary on this front. As such, while current efforts to emulate Bayh-Dole type policies in other countries (see OECD 2002) may be premature, we also do not have enough evidence to suggest that major changes to the Bayh-Dole act are necessary in the United States.

Some tinkering, however, could help protect the public domain—and the public interest—especially important in the context of taxpayer-funded research. From a social welfare perspective, we would want universities to patent publicly funded research outputs only when, absent patents, these research outputs would not be effectively utilized. Similarly, we would want universities to issue exclusive licenses on patented inventions only when nonexclusive licensing fails to promote use or commercialization. However, under Bayh-Dole, universities have complete latitude in making decisions about what to patent and how to license, and they typically make patenting and licensing decisions based on their own self-interest rather than public interest, even though the two are often correlated.

This problem can be thought of in principal-agent terms.[7] By way of Bayh-Dole, principals (funding agencies) delegated to agents (universities) the authority to make decisions about patenting and licensing, believing that doing so would facilitate technology transfer and social returns from university research. But the agents make patenting and licensing decisions based on their own self-interest—including concern for licensing revenues—and in some cases may make choices that hinder, rather than facilitate, technology transfer and social returns. Because it is difficult and costly for principals to monitor the decisions of their agents (e.g., because knowledge of the "right" licensing regime requires specialized knowledge about the university, the technology, and the relevant industry), agents can "get away with" making choices contrary to the public interest (Rai and Eisenberg 2002, 161). This principal-agent framework suggests some possible alternatives available to policymakers that could help assure that universities patent and license only when it serves the public interest, consistent with the intent (if not the letter) of Bayh-Dole.

One standard solution to principal-agent problems is to better inform principals about the actions of agents, that is, to improve monitoring. In a sense, policies implemented by NIH and other agencies to improve disclosure and reporting requirements by universities are a step in this direction, although they do not go far enough. It may make sense for funding agencies to require universities to keep more detailed information on what criteria they use in deciding whether to patent federally funded inventions, and how and to whom to license these patents. It would not be feasible for funding agencies to review all of this information, but random (low probability) audits could suffice to curb opportunistic behavior on the part of universities. Moreover, such a system would also remind universities at each step of the process of their responsibility to promote technology transfer rather than their own private interests.

Another solution is to increase the costs that agents face if their actions are discovered to deviate from those desired by principals. Interestingly, throughout much of the twentieth century, universities avoided involvement in patenting and licensing precisely because of fears of political repercussions if they were found to be "profiteering" at public expense. I have argued above that these fears may have disappeared—in part because of Bayh-Dole's endorsement of university involvement in technology transfer—and norms about how a university should behave in these arenas may no longer be checking their financial ambitions. However, in specific instances funding agencies may still be able to induce patenting and licensing consistent with the public interest via hortatory statements making it clear that "in research area X, the public interest is generally served via broad dissemination." NIH followed such an approach with respect to research on single-nucleotide polymorphisms and human genetic DNA sequences (Rai and Eisenberg 2002).

Because such statements clearly define the public interest, they could create political and reputational costs for institutions whose patenting and licensing strategies deviate from it. Along the same lines, Nelson (2002) suggests amending Bayh-Dole to emphasize that in most instances the presumption should be wide dissemination of publicly funded research (via publication or nonexclusive licensing), while recognizing that in specific instances patenting and licensing may be necessary.

Such exhortations may not be enough, and others have suggested strengthening the ability of funding agencies to take corrective actions in instances where patenting and licensing decisions made by universities are contrary to the public interest. Under Bayh-Dole, funding agencies can revoke title to publicly funded patents in "exceptional circumstances," or they can "march in" to compel broader licensure in cases where licensees are not taking appropriate steps to commercialize the invention. These provisions are, however, extremely cumbersome and consequently have not been exercised since the passage of Bayh-Dole. Rai and Eisenberg (2002) have thus proposed amending Bayh-Dole to lower the bar for ex post intervention. An attractive feature of this proposal is that even the threat of such actions by principals could curb opportunistic behavior by agents.

Each of these suggestions for shaping patent policy is based on the recognition that patenting and restrictive of public research licensing facilitates commercialization in some cases but not all—a recognition not embodied in current government patent policy. In addition to these changes, observers (Rai and Eisenberg 2002; Nelson 2002) have further suggested that some potential problems associated with the growth of university patenting and licensing could be ameliorated by more general changes in the patent system, for example, strengthening the utility and nonobviousness requirements, and implementing of a formal research exemption.

Finally, the long-term interests of the universities themselves may point toward the exercise of greater responsibility for managing patenting and licensing with a view to the public interest rather than their own financial interests. Universities are society's best vehicles for advancing fundamental knowledge and disseminating it widely. Over the long run, if they stray too far from this mission and start behaving like profit-maximizing firms, any political will for extensive financial support of academic research will surely diminish. Given that in over twenty years since the passage of Bayh-Dole only a handful of universities are earning significant net licensing revenues, aggressive pursuit of royalty income is probably not worth the effort. Even absent the additional shaping of patent policy by lawmakers and bureaucrats, if American research universities reaffirm their commitment to creating economically and socially useful "public knowledge," we can expect that their spectacular achievements over the last century will be repeated in the new one.

NOTES

I thank Diana Hicks for comments on an earlier version of this chapter.

1. There is considerable interindustry variance. Patents and licenses are considerably more important in pharmaceuticals than in other industries. However, even in pharmaceuticals, the other channels historically have been extremely important (Gambardella 1995).

2. Portions of this section draw from Mowery et al. (2003), chap. 4.

3. Identical legislation (H.R. 2414) was introduced in the House of Representatives by Rep. Peter Rodino (D-New Jersey) in 1979.

4. A contemporary account noted that limiting the bill to universities and small businesses was "a tactical exclusion taken to ensure liberal support" (Henig 1979, 282). A senate aide commented, "We'd like to extend [the policy] to everybody . . . but if we did the bill would never have a chance of passing" (Broad 1979b, 474).

5. See Sampat (2003) for a detailed survey of the empirical evidence on the intended and unintended effects of Bayh-Dole.

6. According to a recent survey of seventy-six major university technology transfer offices, licensing income is the most important criterion by which technology transfer offices measure their own success (Thursby, Jensen, and Thursby 2001).

7. See Guston 2000 for a more general discussion of principal-agent problems in science and technology policy.

REFERENCES

Agrawal, A., and R. Henderson. 2002. Putting patents in context: Exploring knowledge transfer from MIT. *Management Science* 48 (1): 44–60.

Association of University Technology Managers (AUTM). 1998. *The AUTM Licensing Survey 1998, Survey Summary.* Norwalk, CT: AUTM.

———. 2002. *The AUTM Licensing Survey: FY 2001.* Norwalk, CT: AUTM.

Bremer, H. 2001. *The First Two Decades of the Bayh-Dole Act as Public Policy: Presentation to the National Association of State Universities and Land Grant Colleges* [online]. New York: National Association of State Universities and Land Grant Colleges, 11 November 2001 [cited 14 September 2004]. Available at http://www.nasulgc.org/COTT/Bayh-Dohl/Bremer_speech.htm

Broad, W. 1979a. Patent bill returns bright idea to inventor (in News and Comment). *Science* 205: 473–76.

———. 1979b. Whistle blower reinstated at HEW. *Science* 205: 476.

Bush, V. 1943. The Kilgore bill. *Science* 98 (2557): 571–77.

Cohen, W., R. Nelson, and J. Walsh. 2002. Links and impacts: The influence of public research on industrial R&D. *Management Science* 48 (1): 1–23.

Crane, D. 1972. *Invisible Colleges: Diffusion of Knowledge in Scientific Communities.* Chicago: Univ. of Chicago Press.

Dasgupta, P., and P. David. 1994. Towards a new economics of science. *Research Policy* 23 (5): 487–521.

de Solla Price, D. 1963. *Little Science, Big Science.* New York: Columbia Univ. Press.

*The Economist.* 2002. Innovation's Golden Goose. *Economist* 365 (8303): T3.

Eisenberg, R. 1996. Public research and private development: Patents and technology transfer in government-sponsored research. *Virginia Law Review* 82: 1663–727.

Eskridge, N. 1978. Dole blasts HEW for 'stonewalling' patent applications. *Bioscience* 28: 605–6.

Federal Council on Science and Technology. 1978. *Report on Government Patent Policy, 1973–1976.* Washington, DC: U.S. Government Printing Office.

Forman, H. I. 1957. *Patents: Their Ownership and Administration by the United States Government.* New York: Central Book Company.

Gambardella, A. 1995. *Science and Innovation.* New York: Cambridge Univ. Press.

Guston, D. H. 2000. *Between Politics and Science: Assuring the Integrity and Productivity of Research.* New York: Cambridge Univ. Press.

Harbridge House, Inc. 1968a. Effects of government policy on commercial utilization and business competition. In *Government Patent Policy Study, Final Report.* Washington, DC: Federal Council for Science and Technology.

———. 1968b. Effects of patent policy on government R&D programs. In *Government Patent Policy Study, Final Report.* Washington, DC: Federal Council for Science and Technology.

Henig, R. 1979. New patent policy bill gathers Congressional support. *Bioscience* 29 (May): 281–84.

Kevles, D. 1977. The National Science Foundation and the debate over postwar research policy, 1942–45. *Isis* 68: 5–26.

Kitcher, P. 1993. *The Advancement of Science.* New York: Oxford Univ. Press.

Merges, R., and R. Nelson. 1994. On limiting or encouraging rivalry in technical progress: The effect of patent scope decisions. *Journal of Economic Behavior and Organization* 25: 1–24.

Merton, R. K. 1973. *The Sociology of Science: Theoretical and Empirical Investigations.* Chicago: Univ. of Chicago Press.

Mowery, D., and B. Sampat. 2001a. Patenting and licensing university inventions: Lessons from the history of the research corporation. *Industrial and Corporate Change* 10: 317–55.

———. 2001b. University patents, patent policies, and patent policy debates, 1925–1980. *Industrial and Corporate Change* 10: 781–814.

Mowery, D., R. Nelson, B. Sampat, and A. Ziedonis. 2003. *The Ivory Tower and Industrial Innovation: University Patenting Before and After Bayh-Dole.* Palo Alto, CA: Stanford Univ. Press.

Nelson, R. 2002. The market economy and the republic of science. Draft.

Organisation for Economic Co-operation and Development (OECD). 2002. *Benchmarking Science-Industry Relationships.* Paris: OECD.

Polanyi, M. 1962. The republic of science: Its political and economic theory. *Minerva* 1: 54–74.

Rai, A. T., and R. S. Eisenberg. 2002. The public and the private in biopharmaceutical research. Paper presented at the Conference on the Public Domain, at Duke University, Durham, NC.

Reimers, N. 1998. *Stanford's Office of Technology Licensing and the Cohen/Boyer Cloning Patents: An Oral History Conducted in 1997 by Sally Smith Hughes, Ph.D., Regional Oral History Office.* Berkeley, CA: The Bancroft Library.

Rosenberg, N., and R. R. Nelson. 1994. American Universities and Technical Advance in Industry. *Research Policy* 23: 323–48.

Sampat, B., and R. R. Nelson. 2002. The emergence and standardization of university technology transfer offices: A case study of institutional change. *Advances in Strategic Management* 19.

Stokes, D. E. 1997. *Pasteur's Quadrant: Basic Science and Technological Innovation.* Washington, DC: Brookings Institute.

Thursby, J., R. Jensen, and M. Thursby. 2001. Objectives, characteristics and outcomes of university licensing: A survey of major U.S. universities. *Journal of Technology Transfer* 26: 59–72.

Trune, D., and L. Goslin. 1998. University technology transfer programs: A profit/loss analysis. *Technological Forecasting and Social Change* 57: 197–204.

U.S. General Accounting Office. 1968. *Problem Areas Affecting Usefulness of Results of Government-Sponsored Research in Medicinal Chemistry: A Report to the Congress.* Washington, DC: U.S. Government Printing Office.

U.S. Office of Management and Budget. 1963. Memorandum and statement of government patent policy. *Federal Register* 28: 10943–46.

———. 1971. Memorandum and statement of government patent policy. *Federal Register* 36: 16886.

Ziman, J. M. 1968. *Public Knowledge: An Essay Concerning the Social Dimension of Science.* London: Cambridge Univ. Press.

# 4

# Geography and Spillover

*Shaping Innovation Policy through*
*Small Business Research*

GRANT C. BLACK

## Introduction

California's Silicon Valley, Massachusetts' Route 128, and North Carolina's Research Triangle conjure up images of intensely innovative and productive regions at the forefront of economic activity. With these images in mind, politicians—particularly at the state and local levels—have been increasingly interested in growing their own regional hotspots of innovation. Given the importance of small businesses to economic activity in certain industries, an increasingly popular approach is to develop policies that attract and stimulate small high-tech business (Acs and Audretsch 1990; 1993; Acs, Audretsch, and Feldman 1994; Pavitt, Robson, and Townsend 1987; Philips 1991). Indeed, fourteen states in the United States had strategic economic development plans with a bioscience focus by 2000, forty-one states had some type of initiative to support the bioscience industry (Biotechnology Industry Organization 2001; Johnson 2002), and several states have started to fund programs in stem cell research and nanotechnology.

The notion that geographic proximity plays a key role in the innovation process of small firms underlies such policies. Yet understanding of this role is rudimentary. Previous research exploring agglomeration effects and local knowledge spillovers in the innovation process has faced limitations because of data restrictions. The measurement of innovative activity relies almost exclusively on proxies drawn from innovative inputs, such as research and development (R&D) expenditures and employment, or intermediate innovation outputs, such as patents. The link between these measures and innovation, however, is not always straightforward. High levels of R&D expenditures, for example, may not coincide with large numbers of innovations, and patents are a more accurate measure of invention than of innovation. For example, an industry that experiences high patent activity may not always experience high innovative activity. Furthermore, because many innovation measures, such as R&D expenditures, are collected through government-sponsored surveys, data cannot be disaggregated at small units of observation, such as the firm, or at geographic regions smaller than the state, due to federal regulations on data suppression to protect the identity of respondents. For example, a line of studies characterized by Jaffe (1989) and Feldman (1994a; 1994b) began to investigate whether spillover effects are geographically bounded. Many of the earlier studies examined spillovers at the state level due to data constraints. However, the state is generally considered too broad a region to effectively capture the intricacies expected in the spillover process. To correct this drawback, emphasis has shifted toward urban centers of economic activity, such as metropolitan areas (Anselin et al. 1997, 2000; Jaffe et al. 1993). But the problem of limited data has restricted efforts to isolate spillover effects at these smaller units of observation.

Innovation counts—the number of new commercialized products or processes—or citations to innovations are direct measures of innovation that should eliminate such drawbacks. Still, compiling in a systematic manner actual innovation counts or citations is a time-consuming process that few have attempted. Nevertheless, a handful of industry-specific case studies offer insights into innovation for a narrow set of industries. On a larger scale, the Small Business Administration (SBA) carried out a one-time survey of innovations for the United States in 1982. While this effort provided data across numerous industries, it covered only a single year. No systematic collection of innovation counts has taken place in more than twenty years.

To address the shortcomings of traditional measures of innovation, the limited understanding of spillover and agglomeration effects at the metropolitan level, and the sparse research focusing on the innovative activity of small businesses, I propose a novel measure of innovation and use it to examine the role that local technological infrastructure plays in small-firm innovation across U.S. metropolitan areas. This new measure of innovation comes from the Small Business Innovation Research (SBIR) Program, a federal R&D program designed to stimulate commercialized innovation among small firms. Since the 1970s, evidence continues to demonstrate that small firms can substantially contribute to innovation and overall economic growth (Acs 1999; Acs and Audretsch 1990; Acs, Audretsch, and Feldman 1994; Korobow 2002; Phillips 1991). This centralization is in part due to small firms' access to knowledge generated by sources outside the firm, including universities and large, established firms. Therefore, small firms may be particularly affected by the geographic boundaries of spillovers. Identifying the links between knowledge sources and innovating small firms is important, particularly as the expansion of the high-tech sector has spawned the growth of small firms. The resulting evidence indicates the importance of knowledge spillovers, particularly from universities, and to some extent of agglomeration, to both the likelihood and rate of small-firm innovation in a metropolitan area. Moreover, unlike most previous work that focused on a one-year period, this study examines innovative activity aggregated across six years. This evidence sheds light on how to effectively shape innovation and economic development policies targeting the small business sector.

In this chapter I first describe the SBIR Program, present the strengths of using the SBIR Phase II award as a measure of small-firm innovation, and examine the geographically skewed nature of SBIR activity in the United States. I then explore the role geographic proximity plays in small-firm innovation and describe the empirical methodology. Next, I present the empirical findings on the role that local technological infrastructure plays in small-firm innovation before ending with a discussion of the policy implications of my results.

## Small Business Innovation Research Program

Congress created the Small Business Innovation Research Program under the Small Business Act of 1982. The legislative goals of SBIR are in part "to stimulate technological innovation . . . [and] to increase private sector commercialization innovations derived from Federal research and development" among small, domestic businesses (P.L. 97–219). SBIR arose in part from the growing literature citing the significant contribution of small firms to economic growth through innovation and job creation (Birch 1981; Scheirer 1977) and the lack of federal R&D funds captured by the small business sector (Zerbe 1976).

SBIR is the largest federal R&D program for small business. Since 1998, funds designated for it have exceeded $1 billion annually. Participation in SBIR is mandatory for the ten federal agencies with external R&D budgets in excess of $100 million. Participating agencies must set aside 2.5 percent of their R&D budgets for the program. Five agencies (Department of Defense, the National Institutes of Health, the National Aeronautics and Space Administration, Department of Energy, and the National Science Foundation) account for approximately 96 percent of SBIR funds.

The SBIR Program consists of three phases. Phase I is a competitive awarding of limited federal funds for the short-term investigation of the scientific and technical merit and feasibility of a research idea. Phase I currently caps funding at $100,000 per award. Phase II is a competitive awarding of additional federal funds up to $750,000 to develop the research performed in Phase I. Phase II selection is restricted to Phase I awardees, with an emphasis on proposals with strong commercial potential. Approximately one-half of Phase I award recipients receive Phase II funding. Phase III, for which no SBIR funds are awarded, focuses on private commercialization of Phase II projects, for which firms must acquire private or non-SBIR public funding.

The SBIR Phase II award offers distinct advantages over other measures of innovative activity for small firms. First, the Phase I review process is, in effect, an evaluation procedure that helps ensure that Phase II awards go to feasible research projects with the specific goal of commercialization. Phase II awards are similar to patents in this regard, in that they are an intermediate step toward a commercialized innovation. Yet, Phase II awards differ substantially from patents because they more closely approximate a final innovation by their strong relation to commercialization. A sizeable portion of Phase II projects reach commercialization, which is not true of patents. Nearly 30 percent of 834 sampled Phase II projects from early in the program achieved, or had plans to likely achieve, commercialization within four years of receiving the Phase II award (SBA 1995). In addition, more than 34 percent of the surveyed firms indicated intellectual property protection was not needed

Table 4.1. Top five metropolitan areas receiving Phase II awards by industry, 1990–95 (number of awards)

| Chemicals & allied products | Industrial machinery | Electronics | Instruments | Research services |
|---|---|---|---|---|
| San Francisco (51) | New York (46) | Boston (132) | Boston (138) | Boston (371) |
| Boston (47) | Boston (28) | San Francisco (95) | Los Angeles (136) | Washington, DC (212) |
| New York (44) | Seattle (27) | New York (76) | San Francisco (78) | Los Angeles (164) |
| Denver (29) | San Francisco (24) | Los Angeles (71) | Washington, DC (71) | San Francisco (142) |
| Washington, DC (29) | Lancaster (16) | Washington, DC (47) | New York (65) | New York (125) |
| Percent of all Phase II awards received by the top five metropolitan areas | | | | |
| 52.6 | 49.3 | 57.6 | 51.0 | 56.8 |

for their product, further suggesting that patents as a measure of innovation miss a sizeable portion of innovative activity.

Second, the Phase II award offers a unique measure for examining the innovation mechanism of small, high-tech firms. SBIR is mandated to target firms having five hundred or fewer employees and solicits projects in high technology areas. Phase II firms are typically young and small. The Small Business Administration (1995) found that over 41 percent of surveyed firms were less than five years old at the time of their Phase I award, and nearly 70 percent had thirty or fewer employees. SBIR firms also concentrate most of their efforts on R&D; over half the firms in the survey devoted at least 90 percent of their efforts to R&D.

Third, annual data on SBIR are available since 1983, when firms were first awarded SBIR funding. This large sample of firms, spanning twenty years, allows for both time series and longitudinal analysis since an individual firm's participation in the program can be tracked over time.

The distribution of SBIR awards is highly skewed geographically. Firms in a handful of states and metropolitan areas receive the vast majority of SBIR financing. In general, one-third of the states receive approximately 85 percent of all SBIR awards (Tibbetts 1998). Table 4.1

lists the top five metropolitan areas in five broad industries by number of Phase II awards received in 1990–95. The distribution of SBIR awards follows a pattern much like other measures of R&D activity.[1] Innovative activity is concentrated on the East and West Coasts. Boston is ranked first or second across all five industries, with San Francisco and New York among the top five in every industry. The two exceptions are Denver, ranked fourth in chemicals, and Lancaster, Pennsylvania, ranked fifth in machinery.

## The Role of Geographic Proximity in Small Firm Innovation

Local technological infrastructure is typically thought to comprise the institutions, organizations, firms, and individuals that interact and, through this interaction, influence innovative activity (Carlsson and Stankiewicz 1991). This infrastructure includes academic and research institutions, creative firms, skilled labor, and other sources of inputs necessary to the innovation process. Much research has focused on particular elements of the technological infrastructure (such as concentrations of labor or R&D), while far less research has attempted to focus on the broader infrastructure itself.[2] The literature that has explored the infrastructure as a whole generally describes its status in innovative areas, such as Silicon Valley, in an effort to hypothesize about the relationship between the technological infrastructure and innovative activity (Dorfman 1983; Saxenian 1985; 1996; Scott 1988; Smilor, Kozmetsky, and Gibson 1988).

The relationship between SBIR Phase II activity and the local technological infrastructure in a metropolitan area can be described as a model of production in which the output being produced is based on knowledge-related inputs.[3] In this analysis, the "knowledge production function" defines Phase II activity (the measure of the knowledge output) as a function of measurable components of the local technological infrastructure (the knowledge inputs). These components cover a range of knowledge and agglomeration sources that typify the local technological infrastructure. They include private and public research institutions, the concentration of relevant labor, the prevalence of business services, and an indicator of potential informal networking and area size. The technological infrastructure in a metropolitan area is measured using five variables to capture the breadth of that infrastructure. These variables, in effect, collectively measure the role of knowledge spillovers and agglomeration effects transmitted through the local technological infrastructure on Phase II activity in a metropolitan area. The variables used to estimate the impact of the local technological infrastructure on SBIR activity across metropolitan areas are described below and shown in appendix table 4.A1.

The number of R&D labs within a metropolitan area is used as a proxy for knowledge generated by industrial R&D.[4] The more R&D labs located in a metropolitan area, the more likely that innovative activity will increase due to the expected rise in useful knowledge emanating from them and spilling into the public domain. The number of R&D labs is collected from the annual *Directory of American Research and Technology*, the only source of metropolitan R&D.

Two variables are constructed to capture the academic sector: industry-related academic R&D expenditures and whether or not at least one research university is located in the metropolitan area.[5] The focus on research universities rather than other types of academic institutions is important as their substantial performance of research makes them the predominant source of academic knowledge within a region (NSB 2000). The variable indicating the presence of at least one research university within a metropolitan area captures the ease with which knowledge from the academic sector is available in a local area. The stronger the presence of research universities, the greater ease with which knowledge can likely be transferred to small firms in the same metropolitan area, due to a higher probability of firms being aware of the research being performed at local institutions and of increased interaction between university and private-sector researchers.

The other academic variable expands this indication of access, focusing on the level of knowledge produced at research universities as measured by academic R&D expenditures at local research universities matched to an industry. Knowledge contributed by other types of academic institutions likely plays a much smaller role in the knowledge spillover process due to their lower contributions to cutting-edge research. Moreover, research universities generate the highly trained science and engineering workforce through graduate programs, a vital source of tacit knowledge for firms hiring their graduates. Given the high correlation between R&D expenditures and conferred degrees in science and engineering fields, these institutions' R&D expenditures in science and engineering fields proxy the knowledge embodied in the human capital of graduates as well as in research.

The National Science Foundation's WebCASPAR database provides institutional level data on academic R&D expenditures by department. Field-specific academic R&D expenditures are linked to a relevant industry and aggregated based on academic field classifications from the NSF's *Survey of Research and Development Expenditures at Universities and Colleges*.

The importance of a skilled labor force is captured by the concentration of employment within a metropolitan area in high-tech related industries. The intensity of employment concentration is based on the concentration of employment for an industry in a metropolitan area relative to the concentration of employment for that industry across the nation. The higher the concentration of employment in a metropolitan area, the greater the potential for knowledge transfers between firms and workers in the same industry. Industrial employment data come from the Bureau of the Census's *County Business Patterns*.

The level of employment in business services measures the prevalence of services available within a metropolitan area that contribute to successful innovation. Business services include advertising, computer programming, data processing, personnel services, and patent brokerage. As services used in the innovation process become more prevalent, innovation is expected to increase due to the lower cost of producing a successful innovation. Employment data come from *County Business Patterns*.

Population density serves as an indicator of area size and the potential for informal networking in a metropolitan area. For example, the greater the density, the more likely are individuals engaged in innovative activity to encounter other individuals with useful knowledge and to appropriate that knowledge through personal relationships. Many communities have based the development of technology parks and implementation of urban revitalization projects on this hypothesis. Population density at the metropolitan level comes from the Bureau of Economic Analyses' *Regional Economic Information Systems*.

SBIR Phase II activity is measured in two ways. To examine the likelihood of small-firm innovative activity occurring in a metropolitan area, a variable is constructed that measures whether any Phase II awards were received by firms in a given metropolitan area during the sample period. The number of Phase II awards received by firms in a metropolitan area is used to examine the rate of small-firm innovation. Annual SBIR award data come from the Small Business Administration.

To estimate empirically the effect of knowledge spillovers and agglomeration from the local technological infrastructure on small-firm innovation measured by SBIR Phase II activity, this analysis examines 273 metropolitan areas in the United States over the period 1990–95. To control for interindustry differences in the effect of the local technological infrastructure on innovation, five industries are examined: chemicals and allied products, industrial machinery, electronics, instruments, and research services. These broad industrial classifications allow for reliable data collection at the metropolitan level and comparability to other research exploring the spatial variation of innovative activity in high-tech sectors (Anselin, Varga, and Acs 1997; 2000; Ó hUallacháin 1999).

This analysis examines the effect of the local technological infrastructure on innovative activity in two ways.[6] It first estimates the impact of the infrastructure on the likelihood of the occurrence of innovative activity in a metropolitan area, indicated by whether any Phase II awards were received there. This method explores the role of knowledge spillovers and agglomeration on the mere presence of innovative activity, which can suggest whether the size and composition of the technological infrastructure matter for small-firm innovation. It then estimates the impact of the infrastructure on the rate of innovation, measured by the total number of Phase II awards received by small businesses in a metropolitan area. This method examines the importance of spillovers and agglomeration effects from the technological infrastructure to the magnitude of small-firm innovative activity in areas where it occurs.

## Empirical Results

This section provides a discussion of the empirical results and draws heavily from Black (2003), in which the data and econometric results are more fully presented. Table 4.A2 in the appendix summarizes the empirically estimated relationship between components of the technological infrastructure and SBIR activity for the five industries examined.

## The Likelihood of SBIR Phase II Activity

Knowledge spillovers, far more than agglomeration, influence the likelihood of Phase II activity. Spillovers from the academic and industrial
sectors have the most consistent impact across industries on whether a
metropolitan area experiences SBIR activity. The presence of research
universities has a positive and highly significant effect on whether small
businesses in a metropolitan area receive Phase II awards across all five
industries. In other words, the likelihood of Phase II activity is higher in
areas having research-oriented academic institutions. The number of
R&D labs is also significantly related to the probability of a metropolitan area having Phase II activity. Metropolitan areas with more R&D
labs are more likely to experience Phase II activity. These results coincide with previous evidence of small firms' reliance on external knowledge flows in the innovation process. For instance, over 70 percent of
papers cited in U.S. industrial patents come from public science (Narin,
Hamilton, and Olivastro 1997).

Proximity to related industry has mixed effects in its impact on a metropolitan area's Phase II activity. The concentration of employment in
chemicals, machinery, and electronics has no significant impact on the
likelihood of firms within a metropolitan area receiving Phase II awards.
In effect, Phase II activity does not benefit from relatively high concentrations of labor in these industries. The prevalence of large firms in these
industries, which can lead to high levels of employment in a metropolitan area, may drive this result. It may suggest that knowledge flows less
easily between firms in these industries, particularly from large to small
companies. Costs associated with agglomeration in these industries may
also offset any benefits from the clustering of labor. For instruments and
research services, however, a higher concentration of industry-specific
employment leads to a significant increase in the likelihood of Phase II
activity. This positive effect may arise from the prevalence in these industries of small firms that rely more heavily on external knowledge. This result is particularly true of the research services industry, which by its nature relies on firms with related activities to generate business and draws
on a wide range of knowledge due to the broad scope of the industry.

The response in the likelihood of Phase II activity to the prevalence
of business services also varies across industries. In electronics and research services, increased employment in business services has no significant impact on Phase II activity in a metropolitan area. The lack of a

relationship in research services may be expected given the nature of the industry. These firms typically provide contractual services to other firms and may not seek the same level of business services as firms engaged in innovative activity predominately for themselves. A strong, positive effect exists in instruments, and a less significant but positive impact is found in machinery. Small-firm innovation in these industries benefits from the clustering of business services in a metropolitan area. The innovation process in these industries likely relies more heavily on services that small firms do not provide internally. An unexpected negative and significant relationship exists in chemicals between business services employment and the likelihood of Phase II activity. This result implies that a higher concentration of business services reduces the likelihood of Phase II activity in chemicals, suggesting that the costs associated with the clustering of business services outweigh the benefits. The result is perhaps an artifact of peculiarities in the biotechnology sector, in which extremely high costs emerge from the commercialization of some new innovations, such as new pharmaceuticals, and the low probability of long-run success for small firms in this industry. It could imply that metropolitan areas with high levels of business services coincide with areas in which large firms having a greater probability of success dominate the sector, reducing the likelihood of innovative activity among small competitors.

Population density has virtually no discernable effect on the likelihood of Phase II activity in four of the five industries. Agglomeration effects due to the size of an area and the potential for networking play an insignificant role in most industries in determining whether a metropolitan area experiences Phase II activity. This result does not necessarily imply that these agglomeration effects do not matter for all types of innovation, but that they matter little for the SBIR activity of small firms. In chemicals, however, a denser population leads to a significant increase in the probability that small businesses in a metropolitan area receive Phase II awards, suggesting that tacit knowledge or face-to-face interaction plays a greater role in the innovation process in chemicals than in other industries.

## The Rate of SBIR Phase II Activity

What stands out across industries is that, while the local technological infrastructure has a strong impact on whether an area is likely to receive

Phase II awards, there is a much weaker relationship between that infrastructure and the actual number of awards received. Industrial R&D activity does not play a significant role in determining the rate of Phase II activity in a metropolitan area, in contrast to its role in determining the likelihood of Phase II activity. Increasing the number of R&D labs does not lead to a significant change in the number of Phase II awards received by firms in a metropolitan area, regardless of industry, suggesting that either knowledge is not easily spilling over from private-sector R&D efforts to small businesses participating in SBIR, or these firms are relying on other sources for knowledge that influences the magnitude of their innovative activity.

The answer seems to be that these small firms turn to universities instead of industry. The rate of innovation among small firms depends most strongly on the level of R&D activity performed by research-oriented universities compared to the other components of the technological infrastructure. The positive spillovers emanating from local research universities that influence the likelihood of Phase II activity also play a dominant role in determining the rate of that activity in all industries but machinery. Greater R&D activity in the local academic sector contributes to more Phase II awards for small firms. In electronics, academic R&D expenditures are the only significant factor of the technological infrastructure. These results support previous evidence that small firms appropriate knowledge generated by local universities, and that this knowledge is a key determinant of these firms' innovative activity (Mowery and Rosenberg 1998). While universities perform only about 10 percent of all R&D in the United States, the cutting-edge and often exploratory nature of academic R&D make it an attractive source of knowledge for small businesses, particularly since they are frequently incapable of performing substantial R&D efforts internally. Moreover, the public nature of academic R&D provides a mechanism in which knowledge from universities is more easily transferred (see Sampat, chapter 3, this volume). Knowledge can flow from universities through publications, presentations, informal discussion between faculty and industrial scientists, and students taking industrial jobs (Stephan et al. 2004).

The concentration of industry employment is significant only in chemicals and research services, implying that the positive spillovers associated with proximity to similar firms matter less in determining the number of Phase II awards than merely the presence of Phase II activity.

In other words, the transfer of knowledge through networking and informal interactions between employees of different firms in the same industry seems to play an inconsistent role in the rate of Phase II activity across industries. Surprisingly, the presence of a relatively high concentration of chemical workers has a negative effect on the number of Phase II awards received in a metropolitan area—more concentrated employment in the chemical industry results in less small-firm innovation. This result may reflect industry scale effects driven by the large chemical and pharmaceutical manufacturers that dominate the industry. These firms may dictate the high concentrations of employment in metropolitan areas, while the clustering of small firms in the industry may not be driven by proximity to these large players. The SBIR firms in this industry predominately fall in the biotechnology arena, which can be concentrated in quite different areas than those where large chemical manufacturers are located. Therefore, the negative effect is at least in part the result of differences in the geographic distribution of firms in the chemical industry.

The number of Phase II awards that firms within a metropolitan area receive does not depend on the level of business services in that area. Regardless of industry, a higher level of employment in business services does not increase the rate of Phase II activity. This result may stem from the early-stage nature of SBIR research, so that firms engaged in Phase II research may not be at the appropriate stage of the innovation process to significantly utilize external business services.

Interestingly, population density is a more decisive factor in the rate of Phase II activity than in the likelihood of such activity. Moreover, the direction of its impact varies across industries. A denser population in a metropolitan area leads to a significantly lower number of Phase II awards in the chemical and instrument industries but a larger number in machinery, suggesting negative effects of agglomeration outweigh any benefits in chemicals and instruments. For these industries, costs due to increased competition for resources needed to innovate successfully may dampen the rate of innovation for small firms in more densely populated areas. The density of the population plays no significant role in the rate of Phase II activity in electronics and research services, indicating that area size and the potential for networking are not important factors in determining the number of Phase II awards received by small firms in these industries.

## Conclusion

Geography matters in the innovation process of small businesses. The local technological infrastructure influences both the likelihood and rate of innovative activity among small businesses in a metropolitan area. The infrastructure's role emanates from the clustering of its resources and the flow of knowledge among individuals, firms, and institutions within it. These agglomeration effects and knowledge spillovers play a clearer role in determining whether innovative activity occurs and a lesser role in determining the rate of innovation. The innovative activity of small businesses benefits most from knowledge spillovers, particularly those from the local academic research community.

Research universities are the key component of the local technological infrastructure's impact on small-firm innovation. Knowledge spillovers, indicated by the presence of, and R&D activity at, local research universities, contribute to a greater likelihood of innovative activity occurring across all five industries studied and a higher rate of innovative activity in four of the five industries. This reliance by small firms on knowledge from nearby universities suggests these firms tend to appropriate external sources of knowledge—particularly public sources—given the internal resource constraints common to small companies, such as limited capital, labor, and space. Moreover, it indicates that knowledge coming from the local academic research community can significantly help small businesses in the same metropolitan area successfully innovate.

Whereas universities consistently provide stimuli for small-firm innovation across industries, the evidence on the impact of spillovers from industry tells a more complicated story. A rising number of R&D labs within a metropolitan area increases the probability of innovative activity among small businesses. In stark contrast, there is no evidence of a significant impact of increasing the number of R&D labs on the rate of innovation. Knowledge spillovers from industrial R&D may help small companies develop the capacity to innovate, but they do little to influence the propensity of these firms to use this capacity.

Agglomeration effects from the local technological infrastructure vary widely across industries, resulting in no dominant pattern of influence on small-firm innovation at the metropolitan level. The concentration of industry employment shows no consistent impact on either the likelihood or rate of innovation. A higher concentration of employment in instruments and research services significantly increases the likelihood of innovative activity, while a higher concentration only in research services leads to a greater level of innovation. Across industries, the prevalence of business services and population density more clearly influence the likelihood of innovative activity than the rate of innovation within a metropolitan area. Even so, these agglomeration effects show no consistent pattern across the five industries. These findings highlight the importance of recognizing differences in the innovation process across industries, which should not be ignored in the policy arena. A one-size-fits-all policy based on presumed agglomeration effects will likely not work.

Two lines of policy implications emerge from this research. One relates specifically to the SBIR Program, and the other to economic development policies in general. Congress created SBIR to stimulate innovation among small businesses by providing a mechanism to better incorporate small firms in the federal R&D enterprise. Measuring the success of the program has largely focused on the returns to the public investment in the firms receiving SBIR funding. Concerns about the geographic distribution of SBIR funding, however, have escalated among SBIR administrators and legislators, with some questioning whether the highly skewed distribution of SBIR awards accords with the goals of the program. Recent recommendations urge a more equitable distribution of awards at the state level and increased involvement of state and local governments. Indeed, several states now have local SBIR outreach and assistance offices.

Previous research (Tibbetts 1998) indicates that innovation is clustered in regions, such as California or the Northeast, and that these same areas experience greater levels of SBIR activity. This research finds that agglomeration and knowledge spillovers contribute to this clustering of innovative activity at the metropolitan level, which likely drives the skewed distribution of SBIR awards. Inefficient outcomes from the awards-selection process may arise if policies designed to distribute SBIR awards more equitably lead to choosing unsuccessful firms over successful ones. Innovative activity may diminish if these unsuccessful firms contribute less to economic activity than the successful firms they replace. Attempts to create a more equitable distribution, therefore, potentially may reduce the rate at which the SBIR Program funds successful projects. Moreover, SBIR funding may not be enough to stimulate innovation if firms receiving awards reside in areas with weak technological infrastructures. Increasing innovative activity in these areas may require substantial investment in resources that foster beneficial agglomeration effects and knowledge spillovers. Such a policy, however, could well prove too costly to be effective.

On the other hand, SBIR funding may provide financial capital to potentially successful small firms in areas with weak infrastructures, allowing them to pursue innovative activity that would otherwise be too costly in the absence of SBIR assistance. This result would particularly be true if these firms must seek knowledge from sources at increasing distances and, presumably, at increasing cost. A better understanding of the potential outcomes is needed before implementing a goal of geographic equity in SBIR.

In a broader vein, this research offers direction in shaping economic development policy. Concentrated efforts by state and local policymakers to create regional hotspots of innovation are not imaginary (Keefe 2001; Southern Growth Policies Board 2001) and demand that policymakers be well informed about the drivers of innovative activity. My results emphasize the role played by the local technological infrastructure—particularly through the local academic research community—in stimulating small-firm innovation. This result is relevant for policymakers interested in stimulating economic activity at the metropolitan level. The stronger the technological infrastructure, the more likely a metropolitan area will experience innovative activity.

Still, many development policies strive to stimulate economic growth simply by attracting existing firms or promoting the emergence of new ones. Incentives commonly used in this regard include R&D tax credits, corporate tax reductions, targeted funding for education, and government programs to aid small and new firms in the innovation process. Incentives to attract or birth innovative firms may fall short if an infrastructure providing adequate agglomeration benefits and useful knowledge spillovers is unavailable. Perhaps early policy efforts would be better focused on building a suitable technological infrastructure, such as establishing private and public research facilities. If an area focuses on the small-business sector for development, this research implies that policies to improve the research capabilities of local universities would likely lead to increased innovative activity. Areas with no substantial academic research activity may promote policies that reduce the cost of acquiring knowledge from distant academic research communities and foster interaction with these communities to enhance the flow of knowledge. Once an adequate technological infrastructure is in place, subsequent policies could then be geared to stimulate the flow of knowledge between individuals and institutions in the local economy. Such policies would increase the likelihood of discovering useful knowledge and therefore increase the realized benefits of knowledge spillovers to innovation and, subsequently, overall economic activity. For example, many states sponsor regional conferences bringing together individuals and institutions involved in similar activities, provide services to link small businesses with potential partners to improve the success of innovative activity, and publicly sponsor collaborative efforts in targeted technologies as well as firm incubators located at universities.

Economic development policies must take into account the state of the current technological infrastructure. It is not enough to know simply that the technological infrastructure can stimulate innovation. For policy to be most effective, it must draw on the strengths of the current infrastructure, address the infrastructure's weaknesses, and target specific types of innovative activity complementary to the area's industrial composition. Policymakers must refrain from blindly pursuing one-size-fits-all policies or policy fads targeting the latest hot industry or technology.

## APPENDIX

Table 4.A1. Definition of variables

| | |
|---|---|
| Population density | Average number of persons per square kilometer in a metro area, 1990–95 |
| R&D Labs | Average number of R&D labs located within a metro area, 1990–95 |
| Business services employment | Average employment level in business services (SIC 73) within a metro area, 1990–95 |
| Industrial employment concentration | Average location quotient for employment in industry $i$ within a metro area, 1990–95 |
| Presence of research universities | Dummy variable indicating whether (=1) or not (=0) any Research I/II or Doctorate I/II universities were located in a metro area, 1990–95 |
| Academic R&D expenditures | Total level of academic R&D expenditures by Research I/II or Doctorate I/II institutions in fields corresponding to industry $i$ within a metro area, 1990–95 (in thousands of 1992 dollars) |
| Indicator of Phase II activity | Dummy variable indicating whether (=1) or not (=0) a metro area had at least one firm receive any Phase II SBIR awards in industry $i$, 1990–95 |
| Number of Phase II awards | Total number of Phase II SBIR awards received by firms in industry $i$ within a metro area, 1990–95 |

Table 4.A2. Importance of local technological infrastructure to SBIR activity by industry

| Technological infrastructure components | Industry | | | | |
|---|---|---|---|---|---|
| | Chemicals & allied products | Industrial machinery | Electronics | Instruments | Research services |
| *Likelihood of SBIR activity* | | | | | |
| Presence of a research university | + | + | + | + | + |
| Number of R&D labs | + | + | + | + | + |
| Concentration of industrial employment | | | | + | + |
| Employment in business services | − | + | | + | |
| Population density | + | | | | |
| *Rate of SBIR activity* | | | | | |
| University R&D | + | | + | + | + |
| Number of R&D labs | | | | | |
| Concentration of industrial employment | − | | | | + |
| Employment in business services | | | | | |
| Population density | − | + | | − | |

*Note:* +/- indicates a statistically significant positive/negative relationship between the component of the technological infrastructure and the likelihood or rate of SBIR activity.

## NOTES

1. See Feldman (1994b) for a detailed breakdown of the distribution of selected innovation measures by state.

2. For example, see Markusen, Hall, and Glasmeier (1986) and Jaffe (1989).

3. See Griliches (1979), Jaffe (1989), and Feldman (1994b) for the derivation and application of the knowledge production function model. Following Feldman (1994b), I employ a knowledge production function to estimate the relationship between SBIR activity and the local technological infrastructure. I define this knowledge production function as:

$$SBIR_{is} = f(R\&DLABS_{is}, \ UNIV_{is}, \ EMPCON_{is}, \ EMPSIC73_s, \ POPDEN_s)$$

where SBIR is a measure of SBIR Phase II activity; R&DLABS measures industrial R&D activity; UNIV measures industry-related academic knowledge; EMPCON is industrial concentration; EMPSIC73 is the concentration of relevant business services; POPDEN is population density; $i$ indexes industry; and $s$ indexes the spatial unit of observation (metropolitan areas).

4. Industrial R&D expenditures, commonly used as a measure of R&D activity, are unavailable at the metropolitan level due to data suppression.

5. A research university is defined as an institution rated as a Research I/II or Doctorate I/II institution in the Carnegie classification system.

6. Empirical estimation uses a negative binomial hurdle model for count data. See Cameron and Trivedi (1998), Mullahy (1986), and Pohlmeier and Ulrich (1992) for details on this technique. Hausman, Hall, and Griliches (1984) provide the first application of using a count model to examine innovative activity when investigating the effects of industrial R&D on firms' patenting behavior.

## REFERENCES

Acs, Z. 1999. *Are Small Firms Important? Their Role and Impact.* Boston: Kluwer Academic Publishers.

Acs, Z., and D. Audretsch. 1990. *Innovation and Small Firms.* Cambridge, MA: MIT Press.

———. 1993. Has the role of small firms changed in the United States? In *Small Firms and Entrepreneurship: An East-West Perspective,* ed. Z. Acs and D. Audretsch. Cambridge: Cambridge Univ. Press.

Acs, Z., D. Audretsch, and M. Feldman. 1994. R&D spillovers and recipient firm size. *Review of Economics and Statistics* 100 (2): 336–40.

Anselin, L., A. Varga, and Z. Acs. 1997. Local geographic spillovers between university research and high technology innovations. *Journal of Urban Economics* 42:422–48.

———. 2000. Geographic and sectoral characteristics of academic knowledge externalities. *Papers in Regional Science* 79 (4): 435–43.

Biotechnology Industry Organization. 2001. *State Government Initiatives in Biotechnology 2001.* Washington, DC: Biotechnology Industry Organization.

Birch, D. 1981. Who cares about jobs? *Public Interest* 65: 3–14.

Black, G. 2003. *The Geography of Small Firm Innovation.* Boston, MA: Kluwer Academic Publishers.

Cameron, A., and P. Trivedi. 1998. *Regression Analysis of Count Data.* Cambridge: Cambridge Univ. Press.

Carlsson, B., and R. Stankiewicz. 1991. On the nature, function, and composition of technological systems. *Journal of Evolutionary Economics* 1 (2): 93–118.

Dorfman, N. 1983. Route 128: The development of a regional high technology economy. *Research Policy* 12: 299–316.

Feldman, M. 1994a. Knowledge complementarity and innovation. *Small Business Economics* 6 (5): 363–72.

———. 1994b. *The geography of innovation*. Dordrecht: Kluwer Academic Publishers.

Griliches, Z. 1979. Issues in assessing the contribution of research and development to productivity growth. *Bell Journal of Economics* 10 (1): 92–116.

Hausman, J., B. Hall, and Z. Griliches. 1984. Econometric models for count data with an application to the patents-R&D relationship. *Econometrica* 52 (4): 909–38.

Jaffe, A. 1989. Real effects of academic research. *American Economic Review* 79: 957–70.

Jaffe, A., M. Trajtenberg, and R. Henderson. 1993. Geographic localization of knowledge spillovers as evidenced by patent citations. *Quarterly Journal of Economics* 108 (3): 577–98.

Johnson, L. 2002. States fight to keep high-tech industries. *NANDO Times*, 11 October.

Keefe, B. 2002. Biotech industry's newest darling. *Atlanta Journal-Constitution*, 27 June: E1, E3.

Korobow, A. 2002. *Entrepreneurial Wage Dynamics in the Knowledge Economy*. Dordrecht: Kluwer Academic Publishers.

Markusen, A., P. Hall, and A. Glasmeier. 1986. *High Tech America: The What, How, Where, and Why of the Sunrise Industries*. Boston: Allen & Unwin.

Mullahy, J. 1986. Specification and testing in some modified count data models. *Journal of Econometrics* 33: 341–65.

Mowery, D., and N. Rosenberg. 1998. *Paths of innovation: Technological change in 20th-century America*. Cambridge: Cambridge Univ. Press.

Narin, F., K. Hamilton, and D. Olivastro. 1997. The increasing linkage between U.S. technology and public science. *Research Policy* 26 (3): 317–30.

Ó hUallacháin, B. 1999. Patent places: Size matters. *Journal of Regional Science* 39 (4): 613–36.

Pavitt, K., M. Robson, and J. Townsend. 1987. The size distribution of innovating firms in the UK: 1945–1983. *Journal of Industrial Economics* 55: 291–316.

Philips, B. D. 1991. The increasing role of small firms in the high-technology sector: Evidence from the 1980s. *Business Economics* 26 (1): 40–47.

Pohlmeier, W., and V. Ulrich. 1992. Contact decisions and frequency decision: An econometric model of the demand for ambulatory services. ZEW Discussion Paper 92–09.

Saxenian, A. 1985. Silicon Valley and Route 128: Regional prototypes or historical exceptions? In *High-technology, Space and Society*, ed. M. Castells. Beverly Hills, CA: Sage Publications.

———. 1996. *Regional Advantage: Culture and Competition in Silicon Valley and Route 128*. Cambridge, MA: Harvard Univ. Press.

Scheirer, W. 1977. *Small Firms and Federal R&D*. Prepared for the Office of Management and Budget. Washington, DC: U.S. Government Printing Office.

Scott, F., ed. 1988. *The New Economic Role of American States: Strategies in a Competitive World Economy*. Oxford: Oxford Univ. Press.

Small Business Administration. 1995. *Results of Three-Year Commercialization Study of the SBIR Program*. Washington, DC: U.S. Government Printing Office.

Smilor, R., G. Kozmetsky, and D. Gibson, eds. 1988. *Creating the Technopolis: Linking Technology Commercialization and Economic Development*. Cambridge, MA: Ballinger Publishing Company.

Southern Growth Policies Board. 2001. Conference to focus on building tech-based economies. *Friday Facts* 13 (28).

Stephan, P., A. Sumell, G. Black, and J. Adams. 2004. Firm placements of new PhDs: Implications for knowledge flows. In *The Role of Labor Mobility and Informal Networks for Knowledge Transfer*. Jena: Max-Planck Institut zur Fakulstat Kehrstuhl VWL.

Tibbetts, R. 1998. *An Analysis of the Distribution of SBIR Awards by States, 1983–1996*. Prepared for the Small Business Administration, Office of Advocacy. Washington, DC: U.S. Government Printing Office.

U.S. Public Law 97-219. 97th Cong., 2d sess., 22 July 1982. *Small Business Innovation Development Act of 1982*.

Zerbe, R. 1976. Research and development by smaller firms. *Journal of Contemporary Business* (Spring): 91–113.

# PART 2

# SHAPING SCIENCE

Science and technology policy does not simply promote science by providing a neutral cultural and institutional medium on which knowledge creation feeds. Instead, policy shapes science in concrete, if subtle, ways by enabling or disabling certain research agendas, collaborations, facilities, programs, and so forth. Science and technology policy also legitimates or delegitimates certain research through the procedures, requirements, and other structural elements of the policy process.

The chapters in this section trace the consequences of policies on the science that gets made or accepted as science. For these authors, the contours of science can be both a consequence of policies intended to shape it just so, as well as the accident of policies designed for other goals, closely or distantly related.

Pamela Franklin, whose service as both an environmental fellow and congressional fellow has given her detailed insight into the U.S. Environmental Protection Agency (EPA), analyzes aspects of the "sound science" debate through a case study on developing a health goal for chloroform—a demonstrated animal carcinogen and a suspected human carcinogen—in drinking water. For the EPA, sound science requires expertise that is legally defensible, often chemical-specific, and expedient. The agency relies on a variety of peer review processes and ad hoc expert panels, which may blur the distinction between expert and stakeholder. The EPA announced in March 1998 that it was considering revising the nonenforceable public health goal for chloroform from zero to 300 parts per billion. As the first nonzero drinking water health goal for a carcinogen, this proposal set a key precedent. The agency's final rule reverted to a zero-level goal, which the chlorine and

99

chemical industries challenged. The court ruled that the agency had failed to use the "best available, peer-reviewed science" in its decision. The chapter examines the use of private sector research in this regulatory process, focusing on the EPA's resolution of key scientific debates and its interpretation of appropriate and relevant expertise. The analysis evaluates the agency's assessment process according to criteria of scientific justification, balance, and efficiency, and shows how the requirements of the regulatory process shape what "sound science" is.

Philosopher of science Kevin Elliott tackles a dual challenge in his chapter by elaborating the controversy over chemical hormesis, which is both an intellectually vexing scientific anomaly and a politically vexing regulatory challenge. Hormesis is the phenomenon, itself controversial, in which substances that produce toxic effects at high doses produce beneficial effects at low doses. Elliott taps contemporary research and regulatory discussions concerning hormesis as a case study to offer, first, a brief description of the process of scientific change associated with hormesis as an anomaly and, second, an elucidation of the policy consequences of this description, including the pitfalls facing policy decisions that respond to novel, changing science. Elliott argues that in many cases of scientific change, including the case of hormesis, researchers create multiple characterizations of novel phenomena. Among other policy consequences, this multiplicity lends support to the interests of multiple stakeholders in ongoing political debates. The subsequent deliberation among multiple stakeholders, in turn, ensures ongoing consideration of multiple perspectives, facilitates more careful choices among those perspectives, and promotes public consent to their influences. Anomalies thus help shape a policy environment that, in turn, helps shape the process of scientific development.

Both Franklin and Elliott allude to the role of stakeholders' interests in shaping what gets counted as science. Economist Abigail Payne investigates what happens when those stakeholders have geographic preferences, and those preferences—rather than traditional peer review—are institutionalized in special programs to fund research. By the late 1970s, a majority of public research funding was concentrated in only a few universities located within a few states. Politicians sought ways to reduce this concentration without reducing the effectiveness of the research funded. Payne's chapter explores how the distribution of research funding across universities has changed—or not changed—over the last thirty years. It also explores the effects of two types of funding strategies to combat concentration—earmarking and set-asides—particularly their consequences for the quantity and quality of research they induce. Using thirty years of panel data on earmarks, federal funding, and academic publications for most research and doctoral universities, Payne assesses the way the geographic preference programs shape the distribution, quality, and quantity of research that universities perform. She finds a modest change in the distribution of research funding across research and doctoral universities, especially since 1990. She also finds that earmarks tend to increase the quantity but decrease the quality of academic publications, as measured by citations per publication, while set-aside programs like EPSCoR increase the quality of publications while decreasing their quantity.

Sheryl Winston Smith, a public policy scholar, raises the geographic question to a higher level of aggregation by asking about the relationship between international trade and the conduct of research and development (R&D). A central tenet of economic theory holds that openness creates opportunities for knowledge to flow across international borders. Yet the relationship between trade and new knowledge remains incompletely understood. Smith's chapter considers how flows of knowledge in an increasingly global economy shape innovation in the R&D intensive computer industry. Smith develops a historical narrative of innovation in this industry, and then, using changes in total factor productivity as the measure of the beneficial outcomes of innovation, she probes the hypothesis that trade and the ability to appropriate ideas from foreign sources (absorptive capacity) are important determinants of domestic innovation. The chapter concludes that R&D is neither undifferentiated nor isolated. International trade shapes both domestic and foreign knowledge creation in particular industrial contexts, and in positive and negative ways. Strong domestic R&D investment leads not only to enhanced innovation directly, but indirectly by contributing to absorptive capacity as well.

Together, these authors demonstrate a certain and—one suspects—increasing confidence that one can understand how various policy instruments shape science, directly and indirectly. Doing so, they up the ante on the stakes laid down by the previous section on shaping policy, and they predicate some additional optimism about understanding the shapings of technology and life in the subsequent sections.

# 5

# EPA's Drinking Water Standards and the Shaping of Sound Science

PAMELA M. FRANKLIN

## Introduction

Charged with incorporating "sound science" into its policy decisions, the U.S. Environmental Protection Agency (EPA) has been embroiled in intense scientific disputes about the regulation of chemicals to protect public health. These conflicts underscore the dependence of the agency's reputation on its scientific credibility based on the presumed authority of science (Smith 1992). In the regulatory arena, scientific expertise is often idealized: "Science seeks truth; science pursues objective knowledge; science is not influenced by political interests or short-term considerations or emotion. Science, in other words, is supposed to be everything that politics is not" (ibid. 74).

Operating under tensions and constraints imposed by its regulatory environment, EPA uses a type of science referred to as "regulatory science"—scientific knowledge created specifically for regulatory purposes, under time and political restrictions (Jasanoff 1990). Unlike so-called research science, which is open-ended with extended timelines, regulatory science is subject to statutory and court-ordered timelines.

While research science is based on peer-reviewed papers published in esteemed journals, regulatory science often relies on unpublished studies or "gray literature." Yet regulatory science is held accountable to Congress, courts, and the general public.

Within the American regulatory milieu, three key constraints directly impact EPA's ability to incorporate science into regulatory development: (1) regulatory procedures (2) requirements for openness and transparency, and (3) judicial intervention. As this chapter illustrates, these constraints hamper the agency's ability to assess science and make regulatory decisions.

Most EPA rulemaking follows the so-called informal or notice and comment procedures specified in §553 of the Administrative Procedures Act (APA). These requirements include timelines, public notification, and public comment periods. They prolong the rule-making process, which averages four years at EPA (Kerwin and Furlong 1992).

American requirements for openness in the regulatory process are rooted in the need for public accountability. William Ruckelshaus, EPA's first administrator, stated, "I am convinced that if a decision regarding the use of particular chemicals is to have credibility with the public and with the media . . . then the decision must be made in the full glare of the public limelight."[1] EPA is subject to the Federal Advisory Committee Act (FACA), which requires that advisory committee activities be open to the public.

Perhaps the single most significant aspect of the American regulatory context is the activist role of the courts, which have the ability and the willingness to intervene in agency rule making (Jasanoff 1995; Kagan 2001). The courts ensure that agencies interpret statutes according to legislative intent, properly follow procedural requirements, and do not exceed statutory authority (Levine 1992). Under the APA (§706), agency rule making is subject to judicial review if agency actions are (1) "unlawfully withheld or unreasonably delayed;" (2) "arbitrary, capricious, an abuse of discretion, or otherwise not in accordance with law;" (3) "without observance of procedure required by law;" or (4) "unwarranted by the facts."

Traditionally, the courts granted administrative agencies substantial deference in technical matters. For example, Judge Bazelon advocates judicial deference to agencies in technical matters: "Where administrative decisions on scientific issues are concerned, it makes no sense to rely upon the courts to evaluate the agency's scientific and technological determinations; and there is perhaps even less reason for the courts to substitute their own value preferences for those of the agency" (Bazelon 1977, 822). In contrast, the "hard look" doctrine supports greater judicial oversight over agency decisionmaking (McSpadden 1997). For example, Judge Levanthal believes that the courts have "a supervisory function" over agency decisionmaking, including review of the rationale for technical decisions (Leventhal 1974, 511).

Since EPA's founding, the courts have played an increasingly active role in regulatory actions (O'Leary 1993; Jasanoff 1995; Kagan 2001). Litigation is "an ordinary, rather than an extraordinary, part of the overall administrative process" (O'Leary 1993). A significant fraction of EPA's major regulations are challenged in courts (Coglianese 1996; Kerwin 1994). Two Supreme Court cases have established important, yet somewhat contradictory, precedents regarding regulatory decision-making. The landmark 1984 Supreme Court decision in *Chevron v. NRDC* holds that if a statute is ambiguous, the judiciary must defer to any reasonable interpretation offered by the agency.[2] Yet in *Daubert v. Merrill Dow Pharmaceuticals*, the Court ruled that judges, rather than juries or scientific institutions, should determine the appropriateness of expertise.[3]

This chapter describes one controversial decision-making process, EPA's development of a nonenforceable public health goal (also known as the maximum contaminant level goal, or MCLG) for chloroform levels in drinking water. EPA's development of the chloroform drinking water health goal illustrates how the agency, faced with the constraints of its regulatory environment, uses its operational definition of "sound science" to shape its scientific-regulatory decisions.

## Chloroform Drinking Water Standard

Chloroform, a prevalent by-product of drinking water chlorination, causes cancer in laboratory mice and rats. Studies in humans have linked water chlorination by-products with increased rates of bladder and other cancers. Table 5.1 presents key regulatory and research events in the chloroform rule development chronology.

EPA regulates drinking water contaminants through the Safe Drinking Water Act (SDWA). In March 1998, the agency announced that it was considering raising chloroform's MCLG from zero, the default

Table 5.1. Timeline: Chloroform in drinking water

| Regulatory action | Year | Research progress |
|---|---|---|
| Congress passes Safe Drinking Water Act (SDWA) | 1974 | Trihalomethanes found in chlorinated drinking water |
| | 1976 | National Cancer Institute finds chloroform carcinogenic in lab animals |
| EPA proposes "interim standard" for total trihalomethanes in drinking water | 1979 | |
| Congress amends SDWA | 1986 | EPA Carcinogen Risk Assessment Guidelines: linear dose-response model for carcinogens |
| EPA begins disinfection by-product rulemaking | 1992 | Morris et al. meta-analysis of epidemiology studies suggests link between bladder cancers and chlorinated water |
| EPA proposes disinfection byproduct standards: zero MCLG for chloroform | 1994 | |
| Congress amends SDWA, requires "best available, peer reviewed" science, sets Nov. 1998 deadline for disinfection by-product Final Rule | 1996 | EPA revises Proposed Cancer Guidelines: nonlinear models allowed |
| | 1997 | ILSI expert panel report: chloroform case study for proposed cancer guidelines |
| (Mar.): EPA proposes 300 ppb MCLG for chloroform | 1998 | |
| (Dec.): EPA publishes Final Rule: zero MCLG for chloroform | | |
| (Dec.): Chlorine Chemistry Council files lawsuit challenging chloroform MCLG | | |
| | 1999 | (Dec.) SAB completes report addressing chloroform mode of action |
| (Mar.): DC Circuit Court of Appeals overturns zero MCLG and remands to agency | 2000 | (Feb.): SAB draft review of chloroform risk assessment |
| | | (Apr.): SAB final report: chloroform health risk assessment |

policy for carcinogens, to 300 parts per billion (ppb). This proposal set an important precedent, because it was the first proposed *nonzero* drinking water health goal for a carcinogen. EPA based this proposal on new research—much of it funded by the chemical industry—about chloroform's cancer-causing mechanism.

## The Science Debates

Between 1992 and 2000, two debates dominated EPA's deliberations about setting a drinking water health goal for chloroform, reflecting broader conflicts in the public health and risk management communities. The first debate concerns whether the linear or the threshold model is the more appropriate dose-response model for carcinogenicity. The second concerns whether epidemiological or toxicological data is more relevant.

### DEBATE 1: DOSE-RESPONSE MODELS FOR CARCINOGENICITY

The mechanism through which chloroform causes cancer and the effects of carcinogens at very low doses have been debated extensively. Scientists use dose-response models to extrapolate from effects observed at high doses in laboratory animals.

Figure 5.1 illustrates a hypothetical dose-response curve for a linear model and a threshold model. Originally developed to describe the effects of radiation, *linear dose-response models* indicate that even very small doses of a chemical cause an adverse response. For carcinogenic chemicals, a linear dose-response model is based on the premise that carcinogens are genotoxic: cancer occurs through damage to cellular genetic material. In theory, even just one molecule can damage DNA or cause a mutation that initiates cancerous cell growth. There is no "safe" level of exposure, implying the need for zero-level health goals for carcinogens.

According to a *cytotoxic model* of carcinogenesis, cancer is caused by repeated injury to cells or tissues from exposure to a chemical or its metabolites. Repeated cellular regeneration to repair damaged tissue can lead to uncontrolled cellular growth from replication of existing mutations, or from the increased probability of new mutations (ILSI 1997). There can be a "safe" threshold for exposure to cancer-causing chemicals.

EPA's official position regarding use of linear or threshold models for carcinogens generally changed as scientific understanding evolved. Based on congressional intent in the 1974 SDWA, EPA established zero-level MCLGs for all carcinogens as a matter of policy (Cotruvo and Vogt 1985). In its 1986 Cancer Risk Assessment Guidelines, EPA prescribed linear dose-response models for carcinogens in the absence

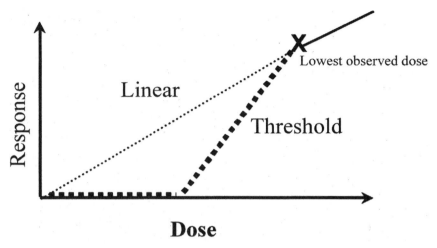

Figure 5.1. Hypothetical linear and threshold dose-response models for carcinogenic chemicals

of data demonstrating otherwise (US EPA 1986). In 1996, the agency proposed new cancer guidelines that allowed for the possibility of thresholds and nonlinear dose-response behavior at low doses.

Substantial scientific evidence suggests that chloroform is not mutagenic, a trait highly correlated with genotoxicity (Butterworth et al. 1998; Melnick et al. 1998). However, EPA's Science Advisory Board (SAB) notes that some evidence suggests a possible genotoxic contribution to chloroform's carcinogenicity (SAB 2000, 2). Animal laboratory evidence suggests that chloroform's metabolites cause cancer by an incompletely understood cytotoxic mechanism (Dunnick and Melnick 1993). Yet it is difficult to conclusively demonstrate cytotoxicity as a causal mechanism (SAB 2000), since there is no consensus regarding the appropriate standard of evidence. Furthermore, a cytotoxic mechanism does not necessarily imply that there is an observable threshold (e.g., Melnick et al. 1998).

DEBATE 2: EPIDEMIOLOGY VERSUS TOXICOLOGY

A second scientific debate centered on the two principal types of evidence that EPA used to assess chloroform's human health effects: epidemiological and toxicological. These disciplines differ in important ways (Evans 1976; Foster et al. 1999).

Laboratory-based toxicology studies usually involve exposing mice or rats to high doses of pure chloroform, through direct injection or via drinking water. The most frequently observed endpoints are kidney or liver tumors. Laboratory research uses a controlled, experimental environment to identify dose-response relationships, metabolic pathways, and carcinogenic mechanisms.

The controlled conditions that allow researchers to establish causation make it difficult to extrapolate to complex, real-world exposures. Most laboratory research focuses on the effects of pure chloroform, but in reality humans are exposed to drinking water containing complex mixtures of contaminants (US EPA 1998c; SAB 2000). Metabolic pathways in animals may be irrelevant for humans, and the impacts of extremely high (sometimes lethal) doses administered to animals in a short-term study are not clearly related to typical low-dose, long-term human exposures. It is uncertain how to extrapolate animal studies to sensitive human subpopulations such as pregnant women, the elderly, children, or those with compromised immune systems (US EPA 1996).

Epidemiology studies evaluate patterns of human disease in order to correlate exposure factors with disease outcomes (Evans 1976; US EPA 1994b; Foster et al. 1999). They have the advantage of directly observing health effects in human populations. Epidemiology studies show that exposure to chloroform or drinking water disinfection by-products is associated with bladder or colorectal cancers (Morris et al. 1992).

Epidemiology has a number of important drawbacks, including lack of experimental control, as well as the difficulty and expense of observing sufficiently large populations to obtain statistically meaningful results (US EPA 1998a). These results may be difficult to interpret due to confounding factors, including human behavior, genetics, and uncontrolled variables. Epidemiological studies of chloroform in drinking water frequently fail to quantify actual human exposures accurately and may not consider all relevant routes of exposure, such as inhalation during showering (ILSI 1998). Furthermore, epidemiology studies cannot determine the effect of chloroform alone, because so many disinfection by-products are present in chlorinated water (Dunnick and Melnick 1993; Schmidt 1999).

These distinct disciplinary perspectives have critical implications for the production of regulatory-relevant research on chloroform. Epidemiology studies are typically conducted by, sponsored by, and affiliated

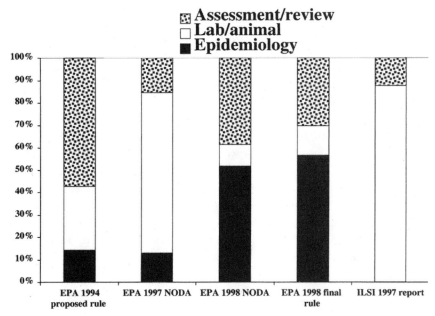

Figure 5.2. Types of studies cited in key EPA assessments

with public organizations at the national or state level because they may require access to public health records, their duration is incompatible with private sector timelines, and their results have direct implications for public health. Industry researchers engage almost exclusively in laboratory-based toxicology studies, which entail shorter time scales and more specific results with greater potential relevance to industry. Academic and government institutions also conduct a significant number of laboratory studies.

In its chloroform rule making from 1994 to 1998, EPA primarily cited laboratory studies as the basis for proposing the MCLG of 300 ppb (figure 5.2). In the 1994 proposed rule making, the majority of studies the agency cited were assessments or reviews (US EPA 1994a). More than 70 percent of literature studies cited in EPA's 1997 Notice of Data Availability (NODA) were laboratory studies, as were all of the citations by the 1997 ILSI panel (US EPA 1997; ILSI 1997). Members of the ILSI panel included experts in a number of disciplinary specialties but excluded epidemiologists; subsequently, the panel contended that EPA's charge to the panel explicitly excluded epidemiology (ILSI 1997). In its 1998 NODA and final rule, EPA cited several epidemiology studies but discounted their findings in justifying its final decision (US EPA 1998a; 1998b).

EPA's assessment of scientific evidence related to chloroform strongly favored toxicological data on grounds that epidemiological data were irrelevant. The agency's 1997 *Research Plan* found that "[a] number of epidemiology studies have been conducted but they have generally been inconclusive" (US EPA 1997b, 25). In its final rule, the agency found a suggestive link between epidemiological evidence and bladder cancer, but found it insufficient to establish a causal link (US EPA 1998a).

Although too late to affect EPA's 1998 chloroform rule making, the SAB endorsed the agency's preference for chemical-specific toxicology data. "The extensive epidemiologic evidence relating drinking water disinfection (specifically chlorination) with cancer has little bearing on the determination of whether chloroform is a carcinogen or not. . . . [t]he reason for the lack of epidemiology research on chloroform in drinking water and cancer is that humans are not exposed to chloroform alone in chlorinated drinking water so it cannot possibly be studied" (SAB 2000, 2).

However, the exclusion of epidemiological data in the agency's expert panel reviews, health risk analysis, and rule makings evoked strong criticism from the medical, public health, and environmental communities. These groups emphasize the extensive uncertainties associated with using toxicology data to predict the impacts on sensitive human subpopulations without considering relevant epidemiology data. A molecular and cell biologist from the National Institute for Environmental Health Science (NIEHS) argued that EPA's dismissal of human data weakened the ILSI panel's overall conclusions and the agency's proposal to set a nonzero MCLG (Melnick 1998). The National Institutes of Health (NIH) questioned EPA's selective use of toxicology studies, reproaching EPA for emphasizing animal studies over epidemiological data (NIH 1998). An academic scientist who worked closely with NRDC, an environmental advocacy organization, asserted, "There's absolutely no scientific reason for excluding epidemiology data. That decision was policy-driven."[4]

Thus, in its attempts to resolve these two scientific debates at the heart of the chloroform controversy, EPA was confronted with conflicting imperatives. It was required by statute to access the best available scientific evidence, in scientific fields in which a single study can take years. At the same time, the agency was required to meet congressionally imposed schedules for rule making as well as consult with its Science Advisory Board.

## Epilogue: The Role of the Courts

In response to reactions from the environmental and public health communities opposed to the proposed 300 ppb MCLG, EPA's final rule abruptly reverted to a zero-level MCLG (US EPA 1998a). The chlorine and chemical industries promptly challenged the final rule. The Circuit Court of Appeals for the District of Columbia ruled in industry's favor, finding that the "EPA openly overrode the 'best available' science suggesting that chloroform is a threshold carcinogen."[5] In vacating the final rule as arbitrary and capricious and in excess of statutory authority, the court was unsympathetic to EPA's reluctance to act in the face of scientific uncertainty: "We do not doubt that adopting a nonzero MCLG is a significant step, one which departs from previous practice. . . . The fact that EPA has arrived at a novel, even politically charged, outcome is of no significance either for its statuary obligation or for fulfillment of its adopted policy" (ibid., 14).

The court's emphatic rejection of EPA's final rule illustrates its veto power over the agency. It also indicates a shift in the court's role from merely an enforcer of statutory obligations to an active overseer of scientific decisionmaking. Far from deferring to the agency's expertise, the court substitutes its own interpretation of "best available" science.

# EPA's Operational Definition of Sound Science

In its assessment process, EPA critically examines available scientific evidence to ensure its compatibility with norms of sound science as well as its relevance in the regulatory process. Three criteria shape the agency's evaluation of scientific expertise: What is *good* science? What is *usable* science? And finally, *whose* science is used? (Franklin 2002). As shown below, the agency's definition of sound science significantly affected its resolution of scientific debates in the chloroform rule making.

## What Is Good Science?

Basing regulatory decisions on "good" science is the prevailing concern at the agency and in the scientific community at large. Smith (1992, 78) notes, "Every EPA Administrator, as well as other regulators, has embraced the concept of good science at least as a rhetorical goal." Typically, good science signifies objectivity (neutrality with respect to outcomes), independence (institutional separation of researchers from vested interests), and peer review (Ozawa 1996; Herrick and Jamieson 1995; Latour 1986; Jasanoff 1990).

Objectivity and neutrality provide science with its epistemological authority. "The authority of scientific expertise rests significantly on assumptions about scientific neutrality. . . . The interpretations and predictions of scientists are judged to be rational and immune from political manipulation because they are based on data that has been gathered through objective procedures" (Nelkin 1994, 111). Of course, absolute neutrality on the part of scientists is impossible since scientists are fallible human creatures: "No person is ever entirely free of potential conflicts, including one who is professionally active in research, consulting, or other noncommercial activities that are . . . compensated" (Smith 1992, 36). Sarewitz (1996, 78) notes, "Scientists themselves may have a political, intellectual, or economic stake in the outcome of a political controversy, and they may therefore interpret their scientific information in a way that favors their own predisposition." The trustworthiness of individual scientists depends upon the greater scientific community, rather than on individual scientists: "It is a community, unlike most others, where the shared ethos, developed over centuries, demands that every individual's claim is subject to open inspection, reexamination, and revision" (Guston and Keniston 1994, 59).

Independence is especially important for regulatory agencies such as EPA. The possibility of agency "capture" by regulated industry poses a grave threat to the sovereignty of scientific authority. Since EPA's earliest history, independent science advice became critical to the agency, arising from concerns that technical advisory groups within EPA were too closely tied to the industries and constituencies they served (Smith 1992).

Peer review, the hallmark of "good science," is a process in which researchers expose their work to the scrutiny of colleagues. Under the tacit "social contract" among scientists, scientists obey implicit rules in the production of knowledge, such as accurate and truthful reporting of observations and acknowledging the intellectual contributions of others (Guston and Keniston 1994). Peer review is intended to ensure adherence to norms of professional behavior, fidelity to scientific merits, and the openness of science as a public, verifiable enterprise (Merton 1942; Chubin and Hackett 1990).

Peer review is regarded as a highly effective mechanism for regulating quality and adherence to professional standards among scientists. Yet peer review remains enigmatic and unregulated: "The system that integrates the results of research—publication in scholarly journals, etc.—is loosely managed by the scientists themselves. It is very little studied, not well understood, and totally unregulated. Scientific publishing is presided over by a mixture of commercial and academic interests that are not accountable in the short run to any of the interested parties. . . . There is no official certification or licensing system for scientific research, with the possible exception of the doctoral degree. There are no rating agencies, no balance sheets, and few annual reports" (Woolf 1994, 83). Furthermore, as the scientific community's responses to the case *Daubert v. Merrell Dow Pharmaceuticals* illustrate, there is a critical lack of standardization in scientific review: "Scientists cannot even agree on standards for assessing the legitimacy of scientific information. . . . [or on] the fundamental question of what constitutes valid data and conclusions" (Sarewitz 1996, 79).

EPA considers peer review to be a critical component of its assessment process for validating scientific studies and determining what constitutes "good" science (US EPA 1997a; 1997b; SAB 1999a). EPA peer reviews range from relatively informal in-house reviews to reviews by agency-designated expert panels, the SAB, and the National Research Council (NRC). The National Academy of Science (1977) recommended that SAB's role should be that of a "neutral umpire" between EPA's tendency to act as an advocate and industry's use of information to bolster its own positions. "Much of the process by which EPA makes regulatory decisions is adversarial, and often scientific information is provided by one of the principals. Similarly the agency itself is sometimes placed in an advocate role. In either case, review can help to ensure a balanced treatment of scientific and technical information" (cited in Smith 1992, 88).

Because its reputation depends heavily on its scientific credentials, EPA strives to emphasize that good science at the agency is unrelated to policy decisions. The literature in science and technology studies suggests that science and policy overlap considerably; yet scientists and agency regulators repeatedly attempt to demarcate the worlds of science and policy into separable, and separate, spheres. Gieryn (1983;

1995; 1999) describes the arbitrary distinctions made between science and policy as "boundary work." Both political and scientific actors engage in boundary work either to take advantage of the epistemological authority of science, or to distance themselves from "policy decisions" (Guston 1999; Jasanoff 1990; 1995). Several studies have shown that boundary work is pervasive at EPA (e.g., Jasanoff 1990; 1992; Keller 2001; Powell 1999). EPA scientists and policymakers carefully distinguish the previous "default policy" (zero MCLGs for carcinogens) from policy based on "sound science." (For instance, a senior Office of Water official described the default linear model as "historically developed as a purely policy decision" that was "intended to be conservative, protective, not necessarily scientific.")[6]

To ensure the use of good science throughout the chloroform rulemaking, EPA sought a range of expert reviews of a controversial epidemiology meta-analysis (Morris et al. 1992; Poole 1997), of reproductive and carcinogenicity data by two ILSI panels (1993; 1997), and of chloroform and threshold carcinogenicity policy by SAB (SAB 1999; 2000). The extent and rigor of EPA's review processes were commensurate with the highly politicized nature of the chloroform debate.

## What Is Usable Science?

In the regulatory context, EPA considers good science necessary but not sufficient. EPA also requires *usable* science that meets the special demands of the regulatory arena: expediency, specificity, and legal defensibility (Jasanoff 1995). For agency scientists, what is useful is "not necessarily what [is] judged best by the academic peer review process" (Smith 1992, 84).

Expediency is critical because research timelines are often incompatible with regulatory deadlines, which typically do not allow time for a comprehensive, definitive scientific study. "In the case of highly complex and comprehensive social issues, scientific consensus over policy-relevant questions is rarely achieved on a time scale of even a decade or two . . . and indeed such consensus may never be achieved at all. Political action, in contrast, must often be taken more rapidly, both to forestall uncertain but conceivable consequences and to meet the responsibilities of representative democracy" (Sarewitz 1996, 77). In addition, usable science must be specific enough to meet statutory requirements for single-chemical standards. Finally, it must be characterized by a high degree of certainty to withstand legal challenges. The imperative for legally defensible standards arises from the legal-adversarial American political milieu, in which many regulations are subject to judicial review (Kagan 2001).

The agency's quest for usable science in developing the chloroform MCLG, in terms of expediency, chemical specificity, and legal defensibility, favored toxicology studies over epidemiology studies. Toxicology studies, while problematic for understanding human carcinogenic mechanisms, are relatively rapid and therefore useful for regulators. In contrast, epidemiology data, while providing relevant information for actual human exposures and responses, are time-consuming. In the chloroform regulatory development, EPA battled with statutory deadlines, complicated by the need to obtain scientific results in a compressed timeframe. EPA claimed that it reverted to a zero MCLG in the final rule because it had inadequate time to consult the SAB. Yet, in *Chlorine Chemistry Council*, the court insisted on adherence to strict timelines:

> But however desirable it may be for EPA to consult an SAB and even to revise its conclusion in the future, that is no reason for acting against its own science findings in the meantime. The statute requires the agency to take into account the "best *available*" evidence. 42 USC §300g-1 (b)(3)(A) [emphasis added]. EPA cannot reject the "best available" evidence simply because of the possibility of contradiction in the future by evidence unavailable at the time of action—a possibility that will *always* be present. . . . All scientific conclusions are subject to some doubt; future, hypothetical findings always have the potential to resolve the doubt (the new resolution itself being subject, of course, to falsification by later findings). What is significant is Congress' requirement that the action be taken on the basis of the best available evidence *at the time* of the rulemaking.

The court is emphatic that peer review and resolution of uncertainty must be subjugated to regulatory expediency: rulemaking deadlines trump scientific review. *Usable* science, therefore, becomes even more important than *good* science in the regulatory context.

EPA scientists favor chemical-specific data for use in health characterization, risk assessment, and standard development. In this respect too, epidemiology studies that evaluate human exposures to complex mixtures of drinking water contaminants are not as useful to the agency as rodent studies evaluating the effects of exposure to chloroform alone. For instance, in its March 1998 proposal for a 300 ppb MCLG, EPA emphasized the strength of the toxicology evidence supporting a threshold model of carcinogenicity for chloroform (US EPA 1998b). Based on the agency's issue definition as an assessment of chloroform alone rather than complex mixtures of water disinfection by-products, EPA emphasized the inability of epidemiology studies to generate chloroform-specific, regulatory-appropriate results.

In light of potential court challenges, the agency needs legally defensible scientific evidence with a high degree of certainty. While many details remain unknown about its carcinogenic mechanisms and other health effects, chloroform is a relatively well-studied chemical, with a body of mechanistic and animal data from toxicology studies. Thus, as the chloroform rule making clearly illustrates, the type of scientific evidence upon which EPA decisionmakers rely clearly depends on its usefulness in the regulatory context.

## Whose Science?

Finally, agency regulators face the question of *whose science* is appropriate for resolving regulatory decisions. Subject to stringent statutory and court-ordered deadlines, EPA often cannot generate sufficient in-house research to support regulatory development. Consequently, EPA relies upon external researchers, including regulated industry, to supply scientific evidence that shapes regulatory decisions. Private sector science provides critical advantages for the agency: it costs virtually nothing and provides otherwise inaccessible information.

However, private sector research has potentially serious drawbacks stemming from its independence from agency oversight, especially the possibility of bias. "Scientific experts, even while striving to be disinterested, are human beings with their own interests and frailties. . . . Experts may confuse public and private ends or find a close affinity between what is good for themselves and the larger public good" (Smith 1992, 21). When potential institutional biases are more blatant, there is grave danger for impacts on regulatory scientific research. Well-publicized incidents of industry malfeasance in sharing technical knowledge with regulatory agencies—for example, cases involving tobacco, asbestos, and tetra-ethyl lead—have undermined the public credibility of industry research.

The question of *whose* research the agency uses matters because it is the basis of science's authority and legitimacy, and that is its neutrality.

Regulators often invoke, at least implicitly, the institutional credibility of research organizations upon whose work they rely. Reputations clearly matter. In fact, with regard to the credibility of science and scientific experts, the perception of scientific truth matters just as much as, or even more than, the "actual" truth as determined by scientists (Sarewitz 1996, 75). In reports and rule making, EPA often refers to internationally renowned experts and to elite scientific institutions.

In developing the chloroform MCLG, EPA depended extensively upon industry-sponsored toxicology studies. Two chemical industry organizations, the Chemical Industry Institute for Toxicology (CIIT) and the International Life Sciences Institute (ILSI), played critical roles in both conducting and evaluating research. Describing EPA's reliance on industry studies, an SAB official says, "[The agency gives] more weight to the peer-reviewed literature. They are generally loath to rely on the gray literature, especially alone. This attitude is partly from a scientific standpoint and partly from a political standpoint. You get a lot of flak if you rely on it [industry-sponsored gray literature] too much."[7] The environmental organization NRDC expressed concern, however, with the objectivity of studies and reviews conducted by organizations associated with or funded by industry (Olson, Wallinga, et al. 1998).

To ensure a panel's objectivity, integrity, and credibility, EPA selected expert panels with a balance of perspectives. According to an SAB official, institutional affiliation plays a limited role in the selection of review panels: "I believe good scientists wear a different hat than just their organization's perspective when they sit at the table. . . . Obviously, if someone is a world famous scientist, at whatever organization, you'd be much less concerned about his/her potential bias of the organization than a lesser-known scientist."[8] A senior Office of Water official believes that the public nature of these panels acts as a check: "I think there is a tremendous regard for one's peers in this process. . . . It's a very public process; what you say is written down. If you did run amok, pretty soon you wouldn't be on any more panels."[9]

In the chloroform rule making, EPA's decisions about scientific expertise depended extensively on the opinion of panels that blurred the distinction between stakeholders and technical experts. For instance, the 1997 ILSI panel review of EPA's health risk assessment of chloroform formed the basis for the agency's decision to set a nonzero MCLG. Members of the 1997 ILSI panel were known for their standpoints favoring toxicology and cytotoxicity. A senior NIEHS scientist notes,

"Depending on who's on your panel, I can tell you the conclusion before they give you the data. This [ILSI] panel had certain members who were strong believers in cytotoxicity."[10] In addition, several members of the steering committee were affiliated with chemical industries and CIIT. NRDC criticized the ILSI expert panel's influence in the chloroform rulemaking, describing ILSI as "a largely industry-funded organization whose panels tend to be heavily biased in favor of scientists with views and backgrounds acceptable to industry" (Olson et al. 1998). Furthermore, several chemical industry groups, including the Chlorine Institute and the Chlorine Chemistry Council, were co-sponsors of the expert panel (ILSI 1997). NRDC expresses concern about implications of EPA's reliance upon unaccountable expert panels: "Expert panels such as the one appointed by ILSI generally are not balanced. They go about their business without public disclosure or scrutiny, and thus do not comply with the Federal Advisory Committee Act (FACA). Legally, EPA cannot rely upon them for advice. Yet this is clearly what the agency has done here, in relying heavily upon ILSI's assessment to come to its conclusions on chloroform" (Kyle et al. 1998).

While EPA recognizes that researchers' affiliations and funding affect credibility in the scientific community, the agency often goes out of its way to enlist participation from industry-sponsored groups in the production and assessment of research. Yet when stakeholders are included in such "expert" reviews, they are granted de facto status as experts, effectively shifting the scientific review process from an outcome-neutral process toward a negotiated process that accommodates interested parties. Ultimately, the principal actors in the chloroform rule making did affect EPA's disposition of scientific evidence and, perhaps more importantly in the long term, the public's faith in the agency's decision. Thus, EPA's ability to critically assess privately funded science is essential both to ensure its scientific validity and to preserve the agency's scientific credibility (Franklin 2002).

## Conclusion

As the chloroform MCLG development illustrates, EPA's regulatory context strongly affects how the agency actively shapes "sound science." First, the agency is captive to regulatory procedure: statutory mandates from Congress, combined with restrictive rule-making processes, dictate regulatory timelines that are completely unrelated to—and out of sync with—relevant scientific research timelines. Second, requirements to ensure openness and transparency impinge on agency resources by requiring timely responses to all public comments. They also subject the agency to public and media scrutiny, not just of regulatory decisions but also of scientific inputs to those decisions. Third, the active role of the judiciary underscores its effective veto power over EPA rule making and reinforces the agency's tendency to make regulatory decisions defensively, with future lawsuits in mind (Jasanoff 1995; Landy et al. 1994). As a consequence of the courts' prominent role, agency resources are siphoned away from scientific issues and directed to respond to legal challenges.

The chloroform case study illustrates three distinct facets of EPA's operational definition of sound science in the regulatory regime. To ensure *good* science, EPA relies on peer review for objective, neutral (or at least balanced), independent science to legitimize its scientifically based decisions. To reinforce their message of good science, agency policymakers and scientists repeatedly distinguish between science and policy decisions, asserting their reliance on unadulterated science.

In the regulatory context, EPA requires not just good science but also *usable* science. Due to multiple constraints imposed by Congress, political actors, and the courts, expediency is essential. As the *Chlorine Chemistry Council* case illustrates, legal defensibility is also a critical consideration. Thus, agency decisions favor quantitative, rather than qualitative, information that can be expressed with concrete estimates of uncertainty. In developing the chloroform MCLG, EPA found that certain types of scientific studies (toxicology) were more usable than others (epidemiology).

Finally, the agency considers the *source* of scientific expertise. Private sector researchers have made important contributions to scientific fields as the bases for regulatory decisions. Yet their role in the regulatory arena creates epistemological and political problems as the regulatory agency struggles to meet the demands of its political principals without sacrificing its scientific integrity.

NOTES

1. September 13, 1971, speech to American Chemical Society. As cited in Smith (1992, 80).

2. *Chevron USA, Inc. v. Natural Resources Defense Council* (1984). 467 US 837, 104 S.Ct. 2778.

3. *Daubert v. Merrill Dow Pharmaceuticals, Inc.* (1993). 113 S. Ct. 2794.

4. Personal interview, March 7, 2001, Berkeley, CA.

5. *Chlorine Chemistry Council and Chemical Manufacturers Association v. EPA.* (2000). US Ct. Appeals DC Cir. 98–1627.

6. Personal interview, 21 March 2001, Washington, DC.

7. Personal interview, 4 August 2000, Edgewater, MD.

8. Ibid.

9. Personal interview, 21 March 2001, Washington, DC.

10. Telephone interview, 15 March 2001.

REFERENCES

Bazelon, D. L. 1997. Coping with Technology Through the Legal Process. *Cornell Law Review* 62: 817.

Butterworth, B. E., G. L. Kedderis, and R. Conolly. 1998. The Chloroform Cancer Risk Assessment: A Mirror of Scientific Understanding. *CIIT Activities* 18 (4): 1–9. April.

Chubin, D. E., and E. J. Hackett. 1990. *Peerless Science: Peer Review and U.S. Science Policy.* Albany: State Univ. of New York Press.

Coglianese, C. 1996. Litigating within Relationships: Disputes and Disturbance in the Regulatory Process. *Law & Society Review* 30 (4): 735–65.

Cotruvo, J. A., and C. D. Vogt. 1985. Regulatory Aspects of Disinfection. *Water Chlorination: Environmental Impact and Health Effects.* R. L. Jolley et al., eds. Ann Arbor, MI: Lewis Publishers, 91–96.

Dunnick, J. K., and R. L. Melnick. 1993. Assessment of the Carcinogenic Potential of Chlorinated Water: Experimental Studies of Chlorine, Chloramine, and Trihalomethanes. *Journal of the National Cancer Institute* 85: 817–22.

Evans, A. S. 1976. Causation and Disease: The Henle-Koch postulates revisited. *Yale Journal of Biology & Medicine* 49: 175–95.

Foster, K. R., D. E. Bernstein, and P. W. Huber, ed. 1999. *Phantom Risk: Scientific Inference and the Law.* Cambridge, MA: MIT Press.

Franklin, P. M. 2002. Is All Research Created Equal? Institutional credibility and technical expertise in environmental policymaking. PhD diss., Univ. of California, Berkeley, Energy and Resources Group.

Gieryn, T. F. 1983. Boundary-Work and the Demarcation of Science from Non-science: Strains and Interests in Professional Interests of Scientists. *American Sociological Review* 48: 781–95.

———. 1995. Boundaries of Science. *Handbook of Science and Technology Studies,* ed. S. Jasanoff et al. Thousand Oaks, CA: Sage, 393–43.

————. 1999. *Cultural Boundaries of Science: Credibility on the Line.* Chicago: Univ. of Chicago Press.

Guston, D. H. 1999. Stabilizing the Boundary between U.S. Politics and Science: The Role of the Office of Technology Transfer as a Boundary Organization. *Social Studies of Science* 29 (1): 87–111.

Guston, D. H., and K. Keniston. 1994. Introduction: The Social Contract for Science. *The Fragile Contract: University Science and the Federal Government.* D. H. Guston and K. Keniston, eds. Cambridge, MA: MIT Press, 1–41.

Herrick, C., and D. Jamieson. 1995. The Social Construction of Acid Rain. *Global Environmental Change* 5 (2): 105–12.

ILSI. 1997. *An Evaluation of EPA's Proposed Guidelines for Carcinogen Risk Assessment Using Chloroform and Dichloroacetate as Case Studies: Report of an Expert Panel.* Washington, DC: International Life Sciences Institute, Health and Environmental Sciences Institute.

————. 1998. *The Toxicity and Risk Assessment of Complex Mixtures in Drinking Water.* Washington, DC: ILSI Risk Science Institute.

————. Risk Science Institute and Environmental Protection Agency. 1993. *Report of the Panel on Reproductive Effects of Disinfection Byproducts in Drinking Water.* Washington, DC: U.S. EPA Health Effects Research Laboratory, ILSI Risk Science Institute.

Jasanoff, S. 1990. *The Fifth Branch: Science Advisors as Policymakers.* Cambridge, MA: Harvard Univ. Press.

————. 1992. Science, Politics, and the Renegotiation of Expertise at EPA. *OSIRIS:* 195–217.

————. 1995. *Science at the Bar: Law, Science, and Technology in America.* Cambridge, MA: Harvard Univ. Press.

Kagan, R. A. 2001. *Adversarial Legalism: The American Way of Law.* Cambridge, MA: Harvard Univ. Press.

Keller, A. C. 2001. Good Science, Green Policy: The Role of Scientists in Environmental Policy in the United States. PhD diss. Univ. of California, Berkeley, Department of Political Science.

Kerwin, C. M. 1994. *Rulemaking: How Government Agencies Write Law and Make Policy.* Washington, DC: Congressional Quarterly Press.

Kerwin, C. M., and S. R. Furlong. 1992. Time and Rulemaking: An Empirical Test of Theory. *Journal of Public Administration Research and Theory* 2 (2): 113–38.

Kyle, A. D., D. Wallinga, and E. D. Olson. 1998. *Comments of the Natural Resources Defense Council on the U.S. Environmental Protection Agency Notice of Data Availability & Request for Comments: National Primary Drinking Water Regulations: Disinfectants and Disinfection Byproducts Notice of Data Availability.* Natural Resources Defense Council. OW-Docket #MC-4101. Washington, DC. June 9, 1998.

Landy, M. K., M. J. Roberts, and S. R. Thomas. 1994. *The Environmental Protection Agency: Asking the Wrong Questions from Nixon to Clinton.* New York: Oxford Univ. Press.

Latour, B. 1986. *Science in Action.* London: Open Univ. Press.

Leventhal, H. 1974. Environmental Decisionmaking and the Role of the Courts. *University of Pennsylvania Law Review* 122: 509–55.

McSpadden, L. 1997. Environmental Policy in the Courts. *Environmental Policy in the 1990s.* N. J. Vig and M. E. Kraft, eds. Washington, DC: CQ Press, 168–85.

Melnick, R., National Institute for Environmental Health Sciences. 1998. Letter to EPA Office of Water Docket Commenting on March 1998 NODA. 27 April.

Melnick, R., M. Kohn, J. K. Dunnick, and J. R. Leininger. 1998. Regenerative Hyperplasia Is Not Required for Liver Tumor Induction in Female B6C3F1 Mice Exposed to Trihalomethanes. *Toxicology and Applied Pharmacology* 148: 137–47.

Merton, R. K. 1942. Science and Technology in a Democratic Order. *Journal of Legal and Political Sociology* 1: 115–26.

Morris, R. D., A.-M. Audet, I. F. Angelillo, T. C. Chalmers, and F. Mosteller. 1992. Chlorination, Chlorination By-Products, and Cancer: A Meta-analysis. *American Journal of Public Health* 82 (7): 955–63.

National Institutes of Health. 1998. *Report on Carcinogens,* 11th Edition. Chloroform CAS No. 67–66–3. Available at ntp.niehs.nih.gov/ntp/roc/eleventh/profiles/s038chlo.pdf.

Nelkin, D. 1994. The Public Face of Science: What can we learn from disputes? *The Fragile Contract: University Science and the Federal Government.* D. H. Guston and K. Keniston, eds. Cambridge, MA: MIT Press, 101–17.

O'Leary, R. 1993. *Environmental Change: Federal Courts and the EPA.* Philadelphia: Temple Univ. Press.

Olson,, E. D., D. Wallinga, and G. Solomon. 1998. Comments of the Natural Resources Defense Council on the EPA "Notice of Data Availability" for the "National Primary Drinking Water Regulations: Disinfectants and Disinfection Byproducts." OW Docket MC-4101. 30 April 1998.

Ozawa, C.P. 1996. Science in Environmental Conflicts. *Sociological Perspectives* 29 (2): 219–30.

Ozonoff, D. 1998. The Uses and Misuses of Skepticism: Epidemiology and Its Critics. *Public Health Reports* 113: 321–23. July–August 1998.

Powell, M. R. 1999. *Science at EPA: Information in the Regulatory Process.* Washington, DC: Resources for the Future.

Sarewitz, D. 1996. *Frontiers of Illusion: Science, Technology, and the Politics of Progress.* Philadelphia: Temple Univ. Press.

Science Advisory Board. 1999. *An SAB Report: Review of the Peer Review Program of the Environmental Protection Agency.* Research Strategies Advisory Committee of the Science Advisory Board. EPA-SAB-RSAC-00–002. Washington, DC. November.

———. 1999. *Review of the Draft Chloroform Risk Assessment and Related Issues in the Proposed Cancer Risk Assessment Guidelines.* EPA-SAB-EC-LTR-00–001. Washington DC. 15 December.

———. 2000. *Review of the EPA's Draft Chloroform Risk Assessment.* EPA-SAB-EC-00–009. Washington, D.C. April.

Smith, B. L. R. 1992. *The Advisers: Scientists in the Policy Process.* Washington, DC: The Brookings Institution.

U.S. Environmental Protection Agency. 1986. Guidelines for Carcinogen Risk Assessment. *Federal Register* 51 (185): 33992–4003.

———. 1994. National Primary Drinking Water Regulations: Disinfectants and Disinfection Byproducts Proposed Rule. *Federal Register* 59(145): 38668–829. 29 July.

———. 1994. *Workshop Report and Recommendations for Conducting Epidemiologic Research on Cancer and Exposure to Chlorinated Drinking Water.* Washington, DC. 19–21 July 1994.

———. 1996. Proposed Guidelines for Carcinogen Risk Assessment. *Federal Register* 61 (79): 17960–8011.

———. 1997. National Primary Drinking Water Regulations: Disinfectants and Disinfection Byproducts Notice of Data Availability. *Federal Register* 62 (212): 59387–484. 3 November.

———. 1997. *Research Plan for Microbial Pathogens and Disinfection By-Products in Drinking Water.* US EPA, Office of Research and Development. EPA 600-R-97–122. Washington, DC. December.

———. 1998. National Primary Drinking Water Regulations: Disinfectants and Disinfection Byproducts Final Rule. *Federal Register* 63 (241): 69390–475. 16 December.

———. 1998. National Primary Drinking Water Regulations: Disinfectants and Disinfection Byproducts Notice of Data Availability. *Federal Register* 63 (61): 15674–92. 31 March.

———. 1998. *Health Risk Assessment / Characterization of the Drinking Water Disinfection Byproduct Chloroform.* Office of Science and Technology, Office of Water. PB99-111346. Washington, DC. 4 November.

Woolf, P. 1994. Integrity and Accountability in Research. *The Fragile Contract: University Science and the Federal Government.* D. H. Guston and K. Keniston, eds. Cambridge, MA: MIT Press, 59–81.

# 6

# The Case of Chemical Hormesis

## *How Scientific Anomaly Shapes Environmental Science and Policy*

KEVIN ELLIOTT

## Introduction

The low-dose biological effects of toxic and carcinogenic chemicals are a matter of heated debate. On the one hand, researchers such as Theo Colborn claim that extremely low doses of many chemicals may mimic hormones such as estrogen and be responsible for dramatic declines in animal populations. Some claim that these "endocrine-disrupting" chemicals may also be related to human reproductive cancers, immune disorders, and declining male sperm counts (Colborn et al. 1996; Krimsky 2000). Nicholas Ashford adds that approximately 5 percent of the U.S. population may suffer from extreme sensitivity to toxic chemicals. This phenomenon, which is frequently called "multiple chemical sensitivity" (MCS), may be linked to "Gulf War syndrome," "sick-building syndrome," and other environmental sensitivities (Ashford and Miller 1998).

On the other hand, the influential toxicologist Edward Calabrese suggests that low doses of many toxins may actually have beneficial effects.[1] He claims that these beneficial effects, which he calls "chemical hormesis," are widely generalizable across different species, biological endpoints, and toxins, and he notes that this phenomenon "is counter to the cancer risk assessment practices by U.S. regulatory agencies . . . which assume that cancer risk is linear in the low-dose area" (Calabrese and Baldwin 1998, VIII-1; see also Calabrese and Baldwin 1997; 2001; 2003).

These debates concerning low-dose chemical effects are significant not only because of their obvious ramifications for policy but also because they provide an opportunity to study how scientific anomalies shape science and policy. Broadly speaking, anomalies are conflicts between the claims of scientific theories, models, or paradigms and the empirical findings actually obtained by researchers (see, e.g., Kuhn 1970; Laudan 1977; Darden 1991; Elliott 2004b). Beneficial effects produced by low doses of toxic chemicals are an example of a scientific anomaly. In order to predict the effects of toxins and carcinogens, toxicologists currently use models that predict either harmful effects at all dose levels or thresholds below which chemicals have no biological effect (NRC 1994; Calabrese and Baldwin 2003). Although researchers and policymakers recognize that these models provide only rough approximations of actual chemical effects, the models are part of a toxicological framework in which toxic chemicals are expected to produce harmful effects (if they produce effects at all) at low doses. Thus, the occurrence of chemical hormesis is unexpected, or anomalous, relative to current toxicological science. Many of the phenomena associated with endocrine disruption and MCS are also anomalous (see Elliott 2004b).

Anomalies already play an important role in studies of science and policy. Thomas Kuhn (1970), among other philosophers, suggested that anomalies are crucial to the development of science because they stimulate the exploration of novel models, theories, and paradigms (see also Hanson 1961; Wimsatt 1987; Darden 1991; Elliott 2004a). Furthermore, the literature on science policy indicates that anomalies are likely to be particularly prevalent and problematic in areas of science that have policy ramifications. For example, when a body of scientific evidence supports a particular policy that runs counter to a political actor's own agenda, he or she is likely to search for and emphasize anomalous scientific evidence as a way of resisting the evidence for the policy (Herrick and Sarewitz 2000; Sarewitz 2000; Fagin 1999; Wargo 1996). Furthermore, whereas anomalies in highly theoretical areas of science might "simmer" relatively unnoticed for an extended period of time, enabling researchers to develop increased understanding of them (Kuhn 1970), the policy process is likely to bring anomalies to the political front burner immediately, when scientists still know very little about their characteristics (Collins and Evans 2002).

This chapter seeks to extend previous studies of anomalies in science and policy by focusing specifically on the potential for researchers and policymakers to develop multiple concepts for anomalous phenomena and to shape subsequent science and policy by selectively emphasizing some concepts over others. Recent research in the social studies of science has already emphasized that scientific concepts and metaphors contribute to the framing of problems at the interface between science and policy (Jasanoff and Wynne 1998; Kwa 1987). These studies may already address the shaping power of anomalies in at least a tangential sense, because anomalies play a central role in many of the science-policy problems that they have examined. Nevertheless, in focusing more explicitly on the policy-shaping importance of anomaly concepts themselves, I hope to provide a distinctive lens for viewing such problems. Moreover, I will argue that hormesis represents a particular sort of anomaly—namely, one that may conflict with fundamental assumptions or paradigms of a scientific field—so it provides an especially good illustration of the wide variety of significant conceptualizations that some anomalies may display.

The chapter makes three central claims: (1) that researchers develop multiple, underdetermined concepts for describing anomalous phenomena; (2) that individuals and institutions contribute to the shaping of subsequent science and policy when they emphasize some of those concepts rather than others; and (3) that cultivation of representative, relevant deliberation may be a helpful response to this state of affairs. The first section below examines recent research concerning chemical hormesis and elucidates at least six concepts that contemporary scientists use to describe it. The next section proposes some general categories for thinking about the ways that anomaly concepts shape ongoing research and policy discussions concerning hormesis. The third section argues that deliberative processes, in which a representative array of interested and affected parties develop, choose, and evaluate anomaly concepts, can helpfully influence the ways that these concepts shape subsequent science and policy.[2] Finally, I draw some general conclusions about the importance of anomalies for policymaking, briefly examining the anomalies of MCS and endocrine disruption to elaborate my claims.

Although this chapter focuses primarily on the "one-way" influence of anomaly concepts on later science and policy, it is important to recognize at the outset that those concepts are themselves shaped by numerous factors, including the bureaucratic, economic, and civil cultures in which they are developed (Jasanoff and Wynne 1998). Thus, even though this analysis focuses primarily on the choices of individual scientists or political actors to employ particular concepts, these broader social factors play a crucial role in determining not only which concepts these individuals employ but also which ones become "stabilized" within the regulatory context. One of the merits of the deliberative evaluation of anomaly concepts proposed in this paper is that it is likely to promote critical reflection concerning these factors that are not always adequately recognized as important influences on scientific practice.

## Multiple Concepts of Chemical Hormesis

Alcohol illustrates the sorts of effects characteristic of hormesis. At high doses, alcohol increases human mortality, but at lower doses it can actually decrease the mortality rate below the level of controls (Gordon and Doyle 1987). These sorts of hormetic responses result in U-shaped dose-response curves (see figure 6.1), which can involve endpoints such as fertility, cancer incidence, growth, body weight, or enzyme activity, in addition to mortality.

Although hormesis was widely reported in the early decades of the twentieth century (see Calabrese and Baldwin 2000), it subsequently passed out of the mainstream literature until it received new attention in papers by A. R. D. Stebbing (1982), Calabrese et al. (1987), and Davis and Svendsgaard (1990). In the 1990s, Calabrese performed two extensive literature searches designed to uncover evidence for chemical hormesis in previous toxicology studies (see Calabrese and Baldwin 1997; 2001). Although these searches have methodological weaknesses (see Jonas 2001; Crump 2001; Menzie 2001; Elliott 2000a, 2000b), they have provided sufficient evidence for at least some other scientists to conclude that "[t]here can be no doubt about the reality of hormesis" (Gerber, Williams, and Gray 1999, 278). Calabrese and Baldwin's recent discussion of the hormesis hypothesis in the journal *Nature* (2003) and a review of their work in *Science* (Kaiser 2003) appear likely to promote and legitimate future discussion of the phenomenon.

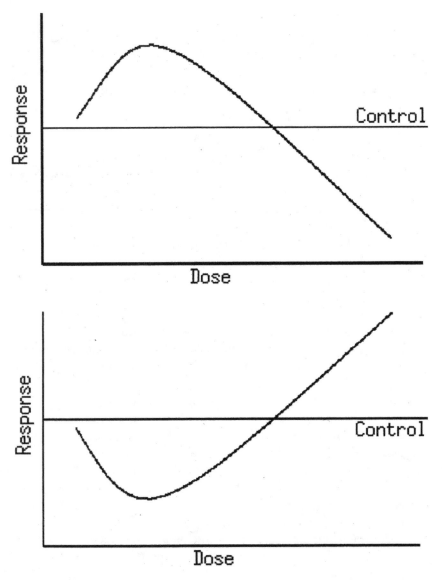

Figure 6.1. Examples of the general form of hormetic dose-response relationships. The bottom curve could represent the relationship between alcohol and human mortality, whereas the top curve could represent the hormetic effects of growth inhibitors on plant growth.

Although hormesis is characterized by U-shaped dose response curves in which opposite effects are observed for the same chemical and endpoint at low versus at high doses, researchers have struggled to specify a particular definition of the phenomenon. Part of the problem is that the production of beneficial effects by toxic chemicals runs sufficiently counter to current toxicological assumptions that researchers have to develop a significantly new conceptual framework in order to make sense of the phenomenon. Thus, although researchers are trying to describe roughly the same set of empirical data that lead to U-shaped dose-response curves, they have employed at least six distinct concepts of hormesis, none of which is fully supported by the available evidence: (1) U-shaped-dose-response-curve hormesis (2) low-dose stimulation/high-dose inhibition hormesis (3) beneficial hormesis (4) homeopathic hormesis (5) overcompensation hormesis, and (6) multiple-effects hormesis. The first three concepts are roughly operational, whereas the last three are broadly mechanistic (Elliott 2000b).

## U-Shaped-Dose-Response-Curve Hormesis

The first dominant concept, U-shaped-dose-response-curve hormesis, is defined as any nonspurious biological effect of a chemical that produces opposite effects at higher doses. The label reflects the effect's representation by a U-shaped dose-response curve, where the x-axis represents dose and the y-axis represents effect on a biological endpoint (see figure 6.1). The effects of lead on human brain activity illustrate this concept. Lead exposure normally increases the period of time between the occurrence of an auditory or visual stimulus and the appearance of electrophysiological responses in the nervous system, but children exposed to very low doses of lead exhibit decreased latency relative to controls (Davis and Svendsgaard 1990, 74). Studies of chemical hormesis in the past decade have revolved around the search for U-shaped dose-response curves of toxic chemicals, and the criteria used by Calabrese and Baldwin in their seminal study of hormesis were designed to show evidence for U-shaped dose-response curves (Calabrese and Baldwin 1997; 1998, III-4).

## Low-Dose-Stimulation/High-Dose-Inhibition Hormesis

The second concept, low-dose-stimulation/high-dose-inhibition hormesis, is limited to the stimulation of an endpoint at low doses and inhibition of the same endpoint at higher doses. For example, at very low doses the cytotoxic agent Adriamycin Doxorubicin inhibits cell growth, but at doses one or two orders of magnitude lower it stimulates cell growth (Vichi and Tritton 1989, 2679). This concept excludes U-shaped dose-response curves that involve low-dose inhibition and high-dose stimulation (e.g., low-dose inhibition of tumor formation). This distinction might not seem to be important, but this second concept has been particularly influential, as chemical hormesis was originally defined in 1943 as "the *stimulation* of biological processes by subinhibitory levels of toxicants" (Calabrese and Baldwin 1998b, 1, italics added; see Southam and Ehrlich 1943).

Beneficial Hormesis

A third operational concept, beneficial hormesis, can be defined as a beneficial low-dose effect caused by a chemical that produces harmful effects at higher doses. One example is the effect of alcohol mentioned above. A number of authors either explicitly or implicitly employ this concept of chemical hormesis, referring to hormetic effects as beneficial effects caused by low doses of toxins (e.g., Teeguarden et al. 1998; Paperiello 1998; Calabrese, Baldwin, and Holland 1999; Gerber, Williams, and Gray 1999).

　　All three of these concepts exclude some effects that many researchers regard as instances of chemical hormesis and include some effects that are often considered to be nonhormetic. On the one hand, Calabrese and Baldwin note that if the background level of a particular endpoint is particularly low or high, it may not be possible to observe inhibition or stimulation of the endpoint that occurs at low doses. So, for example, if tumor incidence occurs relatively rarely in a particular population, a hormetic effect that decreases tumor incidence might result in a curve with a threshold below which no effect is observed rather than in a U-shaped dose-response curve. Calabrese and Baldwin suggest that effects of this sort should be considered hormetic, but all three preceding concepts of chemical hormesis may exclude such phenomena. On the other hand, essential nutrients, especially metals, produce U-shaped dose-response curves that involve low-dose stimulation and high-dose inhibition. Such phenomena are included in the scope of these three concepts, but most researchers do not consider them to be instances of chemical hormesis, perhaps because they presume that hormetic chemicals produce their effects by some mechanism other than serving as essential nutrients in physiological processes (Davis and Svendsgaard 1990). Because of these difficulties involved in defining chemical hormesis operationally, broadly mechanistic concepts of chemical hormesis may be more helpful for future researchers.

## Homeopathic Hormesis

A fourth concept, homeopathic hormesis, is insignificant at present, but it was very influential during the investigation of hormesis at the beginning of the twentieth century. Current investigators frequently look back to the investigations of Hugo Schulz in the 1880s as the origin of the contemporary hormesis concept. He observed stimulatory effects on yeast fermentation when he exposed yeast to low doses of poisonous substances that inhibit fermentation at higher doses. He and the physician Rudolph Arndt became well known for the claim that toxins in general produce stimulation at low doses, which became known as the Arndt-Schulz law. Arndt was a homeopathic physician, however, and he used the Arndt-Schulz law to justify the use of homeopathic medicine, which seeks to treat illnesses by exposing patients to extremely small doses of substances that produce the symptoms of the illnesses. Because hormesis was widely regarded as a homeopathic phenomenon during the early decades of the twentieth century, and because homeopathy was not taken seriously in mainstream medicine, scientific interest in hormesis was itself inhibited (Calabrese and Baldwin 2000).

## Overcompensation Hormesis

A fifth concept, overcompensation hormesis, has been very influential in the past twenty years. This concept is defined as a biological response in which processes are stimulated to above-normal levels in an attempt to restore organismal homeostasis after alteration by a toxic chemical. For example, the growth retardant phosphon inhibits the growth of peppermint plants at all doses, but after two to five weeks of treatment, the plants exposed to low doses of phosphon appear to overcompensate to this stress, growing faster than controls (Calabrese 1999). As A. R. D. Stebbing (1982) has explained, the plausibility of this concept is based on the fact that multiple biological endpoints, including growth, are monitored and controlled by feedback processes. It seems evolutionarily

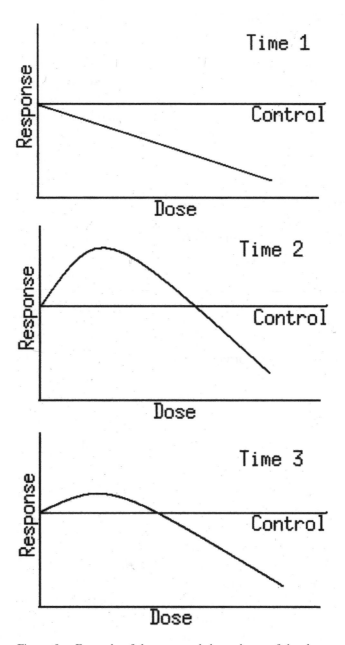

Figure 6.2. Example of the temporal dependence of the dose-response relationships characteristic of "overcompensation hormesis." Time 1 represents an initial period of inhibition at all dose levels. Time 2 illustrates the organism's overcompensation to the toxin at low doses. At time 3, the overcompensation effect has begun to subside.

advantageous for these feedback processes to respond to biological stressors by temporarily "overshooting" the return to homeostasis. In other words, this concept implies that, after a toxin is administered, a linear or threshold model is likely to capture the relationship between the dose of the toxin and its effect. After a period of time, the feedback processes might "kick in" and overcompensate for the stressor (at low doses), thus gradually changing the linear or threshold curve to a U-shaped curve for a period of time. Finally, the overcompensation re-sponse might wear off after an extended period of time, thus returning the U-shaped curve to a linear or threshold curve (see figure 6.2).

## Multiple-Effects Hormesis

The sixth concept is multiple-effects hormesis, defined as a low-dose ef-fect, opposite to that which occurs at higher doses, caused by a chem-ical's influencing the same endpoint in different ways at different dose levels because the chemical produces multiple biological phenomena. The U-shaped dose-response curve for alcohol consumption appears to be an instance of multiple-effects hormesis, as alcohol stimulates levels of HDL cholesterol at low doses, thus lowering risk of mortality by heart disease. This positive effect on mortality is offset at high doses because alcohol also produces other biological phenomena, including harm to the liver.

As with the three operational concepts, none of these mechanistic concepts is completely satisfactory. First, the mechanistic concepts are not precisely specified. For example, the causal processes underlying overcompensation hormesis are not completely understood. Thus, no criteria for distinguishing instances of overcompensation hormesis from multiple effects of a single chemical have been developed. A second, more important, problem is that any of the three mechanistic concepts is likely to exclude phenomena that some researchers would consider hormetic. Davis and Svendsgaard (1990, 77–78) report a variety of such phenomena. First, the interactive effects of some metals may inhibit carcinogenesis and produce U-shaped dose-response curves on cer-tain endpoints (Nordberg and Andersen 1981). Second, U-shaped dose-response curves could also result from the potential for organisms to adapt to chronic low-level exposures to a particular toxin (Smyth 1967). Third, organisms have numerous compensatory and protective mecha-nisms that defend the body against stressors, potentially producing above-normal effects on certain endpoints, at least for a time (Ishikawa et al. 1986). Such mechanisms do not appear to represent instances of the three preceding concepts of chemical hormesis. Thus, part of the reason for the diversity of concepts for the hormesis anomaly may be that none of the concepts currently in use is obviously satisfactory.

## Influences of Anomaly Concepts

The occurrence of multiple, underdetermined concepts for anomalous phenomena is significant because both scientific research concerning the phenomena and policy deliberation about them are shaped by the concepts that researchers and political actors choose to emphasize and the concepts that are "stabilized" by the surrounding social context. In the hormesis case, anomaly concepts have the potential to affect research and policy in at least four ways: (1) a research-interest influence (2) a research-guidance influence (3) a policy-emphasis influence, and (4) a public-attention influence. Some of the following examples of these influences have already occurred; others appear plausible but have not yet occurred.

The research-interest influence holds that, while many factors may affect the trajectory of scientific research (e.g., Jasanoff and Wynne 1998; Kitcher 2001), particular anomaly concepts may also powerfully increase or decrease research activities. One example of the research-interest influence is that the concept of homeopathic hormesis— because it linked hormesis with the maligned practice of homeopathy— encouraged scientists to be skeptical about hormetic research results throughout much of the twentieth century. Calabrese, in particular, laments the lack of attention that hormesis received from roughly the 1930s to the 1980s, and he suggests that a crucial factor was the conceptual linkage between hormesis and homeopathy (Calabrese and Baldwin 2000). The concept of overcompensation hormesis, in contrast, suggests that the phenomenon is the result of a fairly uniform biological process that would be a promising topic for future research. Emphasis of the concept of overcompensation hormesis, rather than homeopathic hormesis, might reasonably be expected to promote further research interest in the phenomenon (Elliott 2000b).

The research-guidance influence of anomaly concepts consists in helping scientists make choices about the direction of further research on an anomaly. In effect, one can regard concepts as very general hypotheses about the nature of an anomalous phenomenon. These hypotheses suggest a general direction of research that can eventually support or undermine the usefulness of the proposed concept. For example, Calabrese (1999) appeared to think (at least for a period of time) that overcompensation hormesis might be the most plausible way to conceptualize the hormesis phenomenon, encouraging him to study the temporal effects of hormetic chemicals. Calabrese found that at least some effects were time dependent, but others were not. Thus, overcompensation hormesis guided the selection of research problems, but eventually revealed itself to be conceptually inadequate for describing all instances of hormesis (see, e.g., Calabrese and Baldwin 2002).

One might criticize the epistemic significance of these first two influences of anomaly concepts by claiming that they can be localized to the "context of discovery" and that the "context of justification" eliminates any biases related to the discovery of scientific ideas. As Kathleen Okruhlik (1994) has argued, however, the context of justification consists primarily of choosing one hypothesis from among a small group of rivals. Thus, if particular anomaly concepts incline researchers to develop only particular sorts of hypotheses, then researchers are likely to choose a hypothesis of that sort as their favored explanation for the phenomenon, as long as it accords reasonably well with the available evidence.

A third influence of anomaly concepts would be the potential to increase or decrease the attention of policy makers toward particular policy-relevant features of an anomalous phenomenon. For example, policymakers would be unlikely to consider "multiple-effects hormesis" to be of immediate significance for risk assessment, because it suggests that hormetic effects are probably not uniformly beneficial from the perspective of the entire organism. The concept of beneficial hormesis, however, could encourage the notion that hormesis is indeed beneficial to the entire organism and should therefore challenge current risk-assessment models. Current research does not indicate whether "beneficial hormesis" or "multiple-effects hormesis" is a more appropriate concept for the phenomenon, but researchers and political actors may be able to influence preliminary policy discussions of the phenomenon by emphasizing one concept rather than another.

It seems all the more unwise to ignore the preliminary influences of anomaly concepts when one considers the public-attention influence, which consists in the power of concepts to shape public opinion, either by developing public interest in it or by creating the impression that the phenomenon is a significant source of risks or benefits. One should not assume that the preliminary effects of this influence can always be eliminated by future research because, as numerous authors have emphasized in their reflections on the Alar (daminozide) controversy, it can be quite difficult to temper public opinion once it has been mobilized (see, e.g., Jasanoff 1990; Whelan 1993). The concept of "beneficial hormesis" provides one potential example of the public-attention influence, as deployment of this concept may be more likely to rally public support for altering current risk-assessment practices. There is a fairly clear rhetorical advantage to claiming that toxic chemicals may produce beneficial effects at low doses, rather than the more complex claim that they may produce stimulatory effects that result from the production of multiple biological phenomena. While the public is likely to be fairly suspicious of any attempts to ease regulation of toxic chemicals (Renn 1998; Foran 1998), the concept of beneficial hormesis appears to be the choice most likely to bear fruit if interest groups do wish to influence public opinion in this direction.

## Responding to the Shaping Power of Anomaly Concepts

### Benefits of Deliberative Approaches

I have argued, first, that scientists and political actors may initially employ multiple, underdetermined concepts in response to anomalies and, second, that these concepts can shape further research and policy, depending on which ones are employed by these actors and stabilized by the broader social milieu. In light of these conditions, this third section argues that the development of representative, relevant deliberation is one advisable strategy for responding to the anomaly concepts' influences on science and policy. I define deliberation roughly the same way that the National Research Council (1996, 4) did, as "any formal or informal process for communication and collective consideration of issues . . . [in which participants] discuss, ponder, exchange observations and views, reflect upon information and judgments concerning matters of mutual interest, and attempt to persuade each other." Furthermore, deliberation that is "representative" and "relevant" incorporates, ideally, both the entire array of significant expert analyses of whatever phenomenon is under investigation and the full spectrum of per-spectives from stakeholders that are likely to be affected significantly by the phenomenon. Depending on the potential significance of a policy-relevant anomaly, deliberation may be warranted throughout at least three stages of scientific development, all of which have the potential to be affected by anomaly concepts: the proposal and design of research concerning the anomaly, the interpretation of the research, and the de-termination of the research's policy ramifications.

That researchers and political figures emphasize some concepts for anomalous phenomena rather than others is unavoidable. This choice is often largely underdetermined by available scientific evidence, but the choice can nevertheless shape the future course of research and pol-icy deliberations that are likely to have either an immediate or eventual effect on the public's well-being. The public therefore has a right to pro-vide some form of informed consent to the choice of anomaly concepts that affects their well-being (see, e.g., Shrader-Frechette 1991; Beau-champ and Childress 1994). Although deliberation may not be the only way to facilitate the public's consent, it appears to be one of the most promising approaches discussed in the current science-policy literature. Thus, there is a prima facie ethical case in favor of such deliberation.

Relevant deliberation concerning the choice of anomaly concepts also seems likely to promote substantively better policy decisions—in this instance through its thoughtful consideration of a particularly wide range of anomaly concepts. A number of philosophers of science have argued convincingly that the inclusion of a diverse variety of perspec-tives is crucial to scientific objectivity (see, e.g., Shrader-Frechette 1991; Solomon 2001; Longino 2002). They emphasize that individual scien-tists invariably approach their subject matter with a variety of biases and assumptions. Accordingly, the inclusion of individuals with many different perspectives is a valuable way to evaluate assumptions, elimi-nate biases, and thus improve scientific decision making. In the context of choosing anomaly concepts, for example, deliberation is likely to de-crease the extent to which promising anomaly concepts are ignored and to increase the extent to which subtle political and social influences of particular anomaly concepts are perceived (Shrader-Frechette 1991; Kitcher 2001; Longino 2002). Although consideration of a wide variety of anomaly concepts, together with their political influences, does not guarantee the development of successful science or policy, it does seem likely to promote both in the long run.

Finally, representative, relevant deliberation will probably promote policy decisions that are instrumentally valuable. Numerous authors have pointed out that stakeholders are likely to resist knowledge claims and decisions if they do not think that their perspectives received adequate representation in the deliberative process (e.g., Fiorino 1990; Shrader-Frechette 1991; NRC 1996; Botcheva 1998; Farrell et al. 2001). The public already harbors suspicion that special interests play an inordinately powerful role in many science and environmental policy issues (Wargo 1996; Fagin et al. 1999). Therefore, it seems advisable to promote deliberative processes as a means of persuading the public that policy-relevant anomalies are treated in a fair manner.

## Mechanisms for Deliberation

Given that representative, relevant deliberation is an advisable response to policy-relevant anomalies, this section briefly suggests some specific mechanisms and strategies for achieving it. The primary purpose of this chapter is not to develop novel strategies for deliberation but rather to reinforce the importance of thoughtfully employing existing deliberative processes for scrutinizing anomalies. Farrell, VanDeveer, and Jager (2001) argue that participation in the assessment process is one of several under-appreciated elements of environmental assessments. This chapter supports their claim that considerations of deliberation and participation are crucial to policymaking, and it emphasizes that these issues are especially important in response to anomalies. It may be helpful, however, to mention a few additional mechanisms that could complement current deliberative approaches for responding to anomalies.

An initial strategy would be to perform alternative assessments to indicate the scientific and policy consequences of choosing one conceptualization of an anomaly rather than another. Shrader-Frechette (1991) has suggested that when risk assessors are faced with methodological value judgments, it may be helpful for them to perform two or more risk assessments that employ different methodological choices (see also O'Brien 1999). For example, if one is not sure which model one should use for extrapolating the effects of toxic chemicals, one could perform several risk assessments, each using a different model. By comparing the results of the alternative assessments, one can probe the extent to which the methodological choice affects one's results.

Similarly, one might be able to estimate the significance of particular anomaly concepts by performing alternative assessments of the policy ramifications of the anomaly using different concepts. These assessments could involve formal methods of analysis in some cases, but they could also involve informal comparisons of the likely scientific and policy consequences of employing one concept rather than another. Performance of these alternative assessments could be incorporated into current institutional mechanisms for participation by encouraging members of scientific advisory boards or particular interest groups or NGO's to develop formal or informal assessments based on different significant conceptualizations of anomalies. This alternative-assessment strategy seems to be a reasonable way to begin deliberating about the consequences of conceptualizing an anomaly in one way rather than another. If, after employing the alternative-assessments strategy, policymakers determine that the choice between two or more plausible concepts could have significant consequences for the public's well-being, it might be worthwhile to initiate more extensive mechanisms for deliberation.

If policymakers determine that the choice to emphasize particular concepts for an anomaly is quite significant, a reasonable second step for promoting deliberation would be the development of "working groups" to analyze the anomaly. Policy analysts have already suggested a variety of specific forms that these groups might take, such as consensus conferences (Sclove 2000) or citizen advisory committees (Lynn and Kartez 1995). These groups could be designed to include representatives of multiple stakeholders (or at least individuals with a broad, representative array of perspectives concerning the anomaly; see also Brown this volume, chapter 1). They could also establish default rules for dealing with an anomaly, determining which concepts of the anomaly should be emphasized initially and establishing policies for iteratively adjusting the default rules as scientists gather further information.[3] This approach might provide an effective method for promoting deliberation, and the precise format could be adjusted in order to expend a degree of time and money that is commensurate with the significance of each anomaly.

Finally, in cases that involve especially high stakes and significant disagreements among stakeholders, it might be helpful to employ an adversary-proceedings approach to facilitate deliberation. Perhaps the most famous proposal for adversary proceedings in the United States came in 1976, when the Task Force of the Presidential Advisory Group

on Anticipated Advances in Science and Technology suggested establishing a "science court." The court would consist of a panel of impartial
scientific "judges" who would "rule" on debated scientific and technological questions.[4] The court would facilitate input from multiple stakeholders, because the "judges" would hear testimony from scientists who
espouse different positions on the disputed question. Shrader-Frechette
suggests a similar form of adversary proceeding, called a "technology
tribunal" (1985, 293ff; see also 1991, 216). The tribunal would be similar
to a science court, but it would address questions of science policy that
include both putatively factual and evaluative components. Furthermore, she suggests that the "judges" include informed nonscientists as
well as scientists. The evaluation of the strengths and weaknesses of different sorts of adversary proceedings is beyond the scope of this paper
(see Kantrowitz 1977; Michalos 1980; Shrader-Frechette 1985, 294ff;
Schneider 2000), but at least some forms might be able to promote effective deliberation in response to particularly recalcitrant debates about
anomaly concepts.

Other practical approaches may help facilitate deliberation, though
in less comprehensive ways. One approach is simply to maintain effective avenues for making stakeholders aware of anomalous scientific information that might concern them. Maintaining these avenues for
communication could be especially important in the case of anomalies,
because it might be important to gather stakeholders' input concerning
an anomaly long before it becomes a popular topic in the mainstream
media. Government agencies, including the U.S. Environmental Protection Agency, have recently been considering a variety of promising
mechanisms, including developing zip-code databases of interested individuals and organizations, posting online comments, developing online bulletin boards, and establishing online dockets (EPA Public Participation Policy Review Workgroup 2000). Providing financial support
for stakeholders who do not have extensive resources could also enhance deliberation. This idea may be necessary in order to implement
other approaches effectively; otherwise stakeholders with limited resources may find it difficult to contribute to deliberative proceedings
(Shrader-Frechette 1991; NRC 1996; Farrell et al. 2001).

These suggestions are admittedly sketchy. In order to put them into
practice, one would need to address a number of practical issues. One
important question is how significant an anomaly must be in order to
merit additional effort. Another issue is how to identify the particular

stakeholders to include. A third issue is the extent to which lay people, as opposed to scientists, should be involved. The answer to this question may vary depending both on the case at hand and on the particular goals that one is most interested in achieving. For example, one might be able to achieve substantively adequate deliberation by including only scientists in the process (as long as they have a sufficiently diverse array of perspectives and experiences related to the anomaly). It is less clear, however, that the participation of scientists alone could yield decisions that have the public's informed consent and that are instrumentally successful in practice. A final question is whether to structure deliberative mechanisms so that they are "binding" on the involved parties; otherwise they are likely to employ legal means to promote their interests even after participating in deliberative processes (Keating and Farrell 1999; Kleinman 2000).

## Conclusion

While this chapter has looked closely at only one example of an anomaly at the intersection of science and policy, an anomaly-based approach may have considerable explanatory power for other environmental controversies. A cursory inspection of the two other anomalies mentioned in the introduction (namely, MCS and endocrine disruption) suggests that the shaping power of anomaly concepts is readily discernible in other instances of anomaly. For example, Ashford and Miller (1998, 284) report that a group of policymakers and scientists at a 1996 conference in Berlin proposed a novel and very controversial concept of MCS, called "idiopathic environmental intolerance" (IEI). The concept of IEI is significant because it is somewhat more compatible with a characterization of the phenomenon as a self-caused or psychological disease, whereas the concept of MCS suggests that the problem is genuinely caused by toxic chemicals (insofar as the concept refers to a "chemical sensitivity"). Thus, the choice of concepts in the MCS case seems likely to have similar influences on research emphasis, research guidance, policy emphasis, and public attention to those observed in the hormesis case. In a similar fashion, the panel brought together to prepare the National Academy of Sciences report on endocrine disruption struggled to develop an appropriate concept and term for the phenomenon that they were instructed to investigate. They eventually replaced the term "endocrine disruptor" with that of a "hormonally active agent" (HAA), because "the term [endocrine disruptor] is fraught with emotional overtones and was tantamount to a prejudgment of potential outcomes" (NRC 1999, 21). In other words, the choice of concepts had strong public-attention and policy-emphasis influences.

Moreover, the likelihood that anomaly concepts can affect science and policy choices argues for the importance of broad deliberative processes in sorting out these concepts. Such processes can help expose the subtle ways that funding sources may bias scientific research (e.g., Fagin et al. 1999; Krimsky 1999), while also encouraging the integration of "lay expertise," whose deliberative value has been documented in an extensive literature (see Wynne 1989; Yearley 1999; Turner 2001; Collins and Evans 2002). While it may seem likely that most policy-relevant anomalies will receive deliberative consideration—given the growing emphasis on public participation in science policy—the effectiveness of these deliberative processes is likely to be affected by the extent to which the organizers and participants understand the reasons for deliberation and the crucial issues that need to be addressed (Farrell et al. 2001). The hormesis story helps to clarify the reasons that deliberation is important in response to anomalies and shows that the entire debate about an anomaly may be framed improperly if one does not begin by analyzing how and why various stakeholders choose and adhere to particular anomaly concepts.

Finally, there is the problem of how to distinguish legitimate anomalies, which merit deliberative analysis, from illegitimate anomalies defended by contrarians, charlatans, ideologues, or simple "nut-cases." There are no easy answers to the problem of distinguishing anomalies that will ultimately become significant from those that will later be dismissed as fluke results.[5] Nevertheless, a starting point is to evaluate the policy significance of anomalies that have some plausibility, by using the alternative-assessment strategy mentioned above, for example. Policymakers can then weigh the costs of initiating deliberative activity, on one hand, against the plausibility and potential significance of the anomaly, on the other hand. This weighing process needs to be informed by an awareness that anomaly concepts exert effective shaping influences on science and policy, and that these influences themselves can be shaped by deliberation among representative, relevant experts and stakeholders.

NOTES

I would like to thank Kristin Shrader-Frechette and a reader for the University of Wisconsin Press for helpful comments concerning this paper.

1. Calabrese, a researcher at the University of Massachusetts, Amherst, is the chairman of the Biological Effects of Low Level Exposures advisory board, a group of scientists organized to develop a better understanding of biological responses to low doses of chemical and physical agents. He is editor of the journal *Biological Effects of Low Level Exposures*, has organized several conferences related to the hypothesis of hormesis, and recently published a summary of the hormesis hypothesis in *Nature* (Calabrese and Baldwin 2003).

2. Numerous previous studies (e.g., NRC 1996; Kleinman 2000; Renn, Webler, and Wiedemann 1995; Funtowicz and Ravetz 1993) have recommended deliberation in response to policy-relevant science in general (or to specific areas of science, such as risk assessment). These have not, however, focused specifically on the importance of deliberative responses to multiple concepts of anomalies, particularly those anomalous phenomena that are difficult to conceptualize.

3. For an extensive discussion of default rules and their role in public policy concerning environmental hazards, see NRC (1994).

4. The original proposal of a science court (Task Force 1976) assumed that the court would address purely factual issues and leave evaluative questions for other organizations. Michalos (1980) and Shrader-Frechette (1985; 1991), among others, have questioned the possibility of distinguishing factual and evaluative issues in this way.

5. An instructive example is the "Benveniste" affair, in which the journal *Nature* struggled to determine whether to publish and how to evaluate anomalous findings that seemed to support the molecular "memory" of water (see Davenas et al. 1988; Benveniste 1988).

REFERENCES

Ashford, N. and C. Miller. 1998. *Chemical Exposures: Low Levels and High Stakes.* 2nd ed. New York: Van Nostrand Reinhold.

Beauchamp, T., and J. Childress. 1994. *Principles of Biomedical Ethics.* New York: Oxford Univ. Press.

Benveniste, J. 1988. Dr. Jacques Benveniste Replies. *Nature* 334: 291.

Botcheva, L. 1998. *Doing is Believing: Use and Participation in Economic Assessments in the Approximation of EU Environmental Legislation in Eastern Europe.* Global Environmental Assessment Project. Cambridge, MA: Kennedy School of Government.

Calabrese, E. 1999. Evidence that hormesis represents an 'overcompensation' response to a disruption in homeostasis. *Ecotoxicology and Environmental Safety* 42: 135–37.

———. 2001. Overcompensation stimulation: A mechanism for hormetic effects. *Critical Reviews in Toxicology* 31: 425–70.

Calabrese, E., and L. Baldwin. 1997. The dose determines the stimulation (and poison): development of a chemical hormesis database. *International Journal of Toxicology* 16: 545–59.

———. 1998. *Chemical Hormesis: Scientific Foundations*. College Station: Texas Institute for the Advancement of Chemical Technology.

———. 2000. Tales of two similar hypotheses: The rise and fall of chemical and radiation Hormesis. *Human and Experimental Toxicology* 19: 85–97.

———. 2001. The frequency of U-shaped dose responses in the toxicological literature. *Toxicological Sciences* 62: 330–38.

———. 2002. Defining hormesis. *Human and Experimental Toxicology* 21: 91–97.

———. 2003. Toxicology rethinks its central belief. *Nature* 42: 691–92.

Calabrese, E., L. Baldwin, and C. Holland. 1999. Hormesis: A highly generalizable and reproducible phenomenon with important implications for risk assessment. *Risk Analysis* 19: 261–81.

Calabrese, E., M. McCarthy, and E. Kenyon. 1987. The occurrence of chemically induced hormesis. *Health Physics* 52: 531–41.

Colborn, T., D. Dumanoski, and J. P. Myers. 1996. *Our Stolen Future*. New York: Dutton.

Collins, H., and R. Evans. 2002. The third wave of science studies: Studies of expertise and experience. *Social Studies of Science* 32: 235–96.

Crump, K. 2001. Evaluating the evidence for hormesis: A statistical perspective. *Critical Reviews in Toxicology* 31: 669–79.

Darden, L. 1991. *Theory Change in Science*. New York: Oxford Univ. Press.

Davenas, E., F. Beauvais, J. Amara, M. Oberbaum, P. Robinson, A. Miadonna, A. Tedeschi, et al. 1988. Human basophil degranulation triggered by very dilute antiserum against IgE. *Nature* 333: 816–18.

Davis, J. M., and W. Farland. 1998. Biological effects of low-level exposures: A perspective from U.S. EPA scientists. *Environmental Health Perspectives* 106: 380–81.

Davis, J. M., and D. Svendsgaard. 1990. U-shaped dose-response curves: Their occurrence and implications for risk assessment. *Journal of Toxicology and Environmental Health* 30: 71–83.

Elliott, K. 2000a. A case for caution: An evaluation of Calabrese and Baldwin's studies of chemical hormesis. *Risk: Health, Safety, and Environment* 11: 177–96.

———. 2000b. Conceptual clarification and policy-related science: The case of chemical hormesis. *Perspectives on Science* 8: 346–66.

Elliott, K. 2004a. Error as means to discovery. *Philosophy of Science* 71: 174–97.

———. 2004b. Scientific Anomaly and Biological Effects of Low-Dose Chemicals: Elucidating Normative Ethics and Scientific Discovery. PhD diss., Univ. of Notre Dame.

EPA Public Participation Policy Review Workgroup. 2000. *Engaging the American People*. Washington, DC: Environmental Protection Agency.

Fagin, D., M. Lavelle, and the Center for Public Integrity. 1999. *Toxic Deception*. 2nd ed. Monroe, ME: Common Courage Press.

Farrell, A., S. VanDeveer, and J. Jager. 2001. Environmental assessments: Four under-appreciated elements of design. *Global Environmental Change* 11: 311–33.

Fiorino, D. 1990. Citizen participation and environmental risk: A survey of institutional mechanisms. *Science, Technology, and Human Values* 15: 226–43.

Foran, J. 1998. Regulatory implications of hormesis. *Human and Experimental Toxicology* 17: 441–43.

Funtowicz, S. and J. Ravetz. 1993. Science in the post-normal age. *Futures* 25: 739–55.

Gerber, L., G. Williams, and S. Gray. 1999. The nutrient-toxin dosage continuum in human evolution and modern health. *Quarterly Review of Biology* 74: 273–89.

Gordon. T., and J. Doyle. 1987. Drinking and mortality: The Albany study. *American Journal of Epidemiology* 125: 263–70.

Hanson, N. 1961. Is there a logic of scientific discovery? In *Current Issues in the Philosophy of Science*, ed. H. Feigl and G. Maxwell. New York: Holt, Rinehart and Winston.

Herrick, C., and D. Sarewitz. 2000. Ex post evaluation: A more effective role for scientific assessments in environmental policy. *Science, Technology, and Human Values* 25: 309–31.

Ishikawa, T., T. Akerboom, and H. Sies. 1986. Role of key defense systems in target organ toxicity. In *Toxic Organ Toxicity*, vol. 1, ed. G. Cohen. Boca Raton, FL: CRC Press.

Jasanoff, S. 1990. *The Fifth Branch: Science Advisors as Policymakers*. Cambridge, MA: Harvard Univ. Press.

Jasanoff, S., and B. Wynne. 1998. Science and decisionmaking. In *Human Choice and Climate Change*, vol. 1, ed. S. Rayner and E. Malone. Columbus, OH: Battelle Press.

Jonas, W. 2001. A critique of "The scientific foundations of hormesis." *Critical Reviews in Toxicology* 31:625–29.

Juni, R., and J. McElveen, Jr. 2000. Environmental law applications of hormesis concepts: Risk assessment and cost-benefit implications. *Journal of Applied Toxicology* 20: 149–55.

Kaiser, J. 2003. Sipping from a poisoned chalice. *Science* 302 (17 October): 376–79.

Kantrowitz, A. 1977. The science court experiment: Criticisms and responses. *Bulletin of the Atomic Scientists* 133 (4): 44–47.

Keating, T., and A. Farrell. 1999. Transboundary environmental assessment: Lessons from the ozone transport assessment group. Knoxville, Tenn.: National Center for Environmental Decision-Making Research.

Kitcher, P. 1993. *The Advancement of Science: Science Without Legend, Objectivity Without Illusions.* Oxford: Oxford Univ. Press.

———. 2001. *Science, Truth, and Democracy.* Oxford: Oxford Univ. Press.

Kleinman, D. 2000. *Science, Technology, and Democracy.* Albany: State Univ. of New York Press.

Krimsky, S. 1999. The profit of scientific discovery and its normative implications. *Chicago-Kent Law Review* 75 (1): 15–39.

———. 2000. *Hormonal Chaos: The Scientific and Social Origins of the Environmental Endocrine Hypothesis.* Baltimore: Johns Hopkins Univ. Press.

Kuhn, T. 1970. *The Structure of Scientific Revolutions.* 2nd ed. Chicago: Univ. of Chicago Press.

Kwa, C. 1987. Representations of nature mediating between ecology and science policy: The case of the International Biological Program. *Social Studies of Science* 17: 413–42.

Laudan, L. 1977. *Progress and Its Problems.* Berkeley: Univ. of California Press.

Lave, L. 2000. Hormesis: Policy implications. *Journal of Applied Toxicology* 20: 141–45.

Longino, H. 2002. *The Fate of Knowledge.* Princeton: Princeton Univ. Press.

Lynn, F., and J. Kartez. 1995. The redemption of citizen advisory committees: A perspective from critical theory. In *Fairness and Competence in Citizen Participation,* ed. O. Renn, T. Webler, and P. Weidemann. Dordrecht: Kluwer.

Menzie, C. 2001. Hormesis in ecological risk assessment: A useful concept, a confusing term, and/or a distraction? *Human and Experimental Toxicology* 20: 521–23.

Michalos, A. 1980. A reconsideration of the idea of a science court. In *Research in Philosophy and Technology,* vol. 3, ed. P. Durbin. Greenwich, CT: JAI Press.

Nordberg, G., and O. Andersen. 1981. Metal interactions in carcinogenesis: Enhancement, inhibition. *Environmental Health Perspectives* 40: 65–81.

National Research Council (NRC). 1994. *Science and Judgment in Risk Assessment.* Washington, DC: National Academy Press.

———. 1996. *Understanding Risk: Informing Decisions in a Democratic Society.* Washington, DC: National Academy Press.

———. 1999. *Hormonally Active Agents in the Environment.* Washington, DC: National Academy Press.

O'Brien, M. 1999. Alternatives assessment: Part of operationalizing and institutionalizing the precautionary principle. In *Protecting Public Health and the Environment: Implementing the Precautionary Principle,* ed. C. Raffensperger and J. Tickner. Washington, DC: Island Press.

Okruhlik, K. 1994. Gender and the biological sciences. *Biology and Society, Canadian Journal of Philosophy*, supp. vol. 20: 21–42.

Paperiello, C. 1998. Risk assessment and risk management implications of hormesis. *Human and Experimental Toxicology* 17: 460–62.

Renn, O. 1998. Implications of the hormesis hypothesis for risk perception and communication. *Belle Newsletter* 7: 2–9.

Renn, O., T. Webler, and P. Wiedemann, eds. 1995. *Fairness and Competence in Citizen Participation*. Dordrecht: Kluwer.

Sarewitz, D. 2000. Science and environmental policy: An excess of objectivity. In *Earth Matters: The Earth Sciences, Philosophy, and the Claims of Community*, ed. R. Frodeman. Upper Saddle River, NJ: Prentice Hall.

Schneider, S. 2000. Is the "citizen-scientist" an oxymoron? In *Science, Technology, and Democracy*, ed. D. Kleinman. Albany: State Univ. of New York Press.

Sclove, R. 2000. Town meetings on technology: Consensus conferences as democratic participation. In *Science, Technology, and Democracy*, ed. D. Kleinman. Albany: State Univ. of New York Press.

Shrader-Frechette, K. 1985. *Science Policy, Ethics, and Economic Methodology*. Dordrecht: Reidel.

———. 1991. *Risk and Rationality: Philosophical Foundations for Populist Reforms*. Berkeley: Univ. of California Press.

Smyth, H. 1967. Sufficient challenge. *Food and Cosmetics Toxicology* 5: 51–58.

Solomon, M. 2001. *Social Empiricism*. Cambridge, MA: MIT Press.

Southam, C., and J. Ehrlich. 1943. Effects of extracts of western red-cedar heartwood on certain wood-decaying fungi in culture. *Phytopathology* 33: 517–24.

Stebbing, A. 1982. Hormesis: The stimulation of growth by low levels of inhibitors. *Science of the Total Environment* 22: 213.

Task Force of the Presidential Advisory Group on Anticipated Advances in Science and Technology. 1976. The science court experiment: An interim report. *Science* 193: 653–56.

Teeguarden, J., Y. Dragan, and H. Pitot. 1998. Implications of hormesis on the bioassay and hazard assessment of chemical carcinogens. *Human and Experimental Toxicology* 17: 454–59.

Turner, S. 2001. What is the problem with experts? *Social Studies of Science* 31: 123–49.

Vichi, P., and T. Tritton. 1989. Stimulation of growth in human and murine cells by adriamycin. *Cancer Research* 49: 2679–82.

Wargo, J. 1996. *Our Children's Toxic Legacy*. New Haven: Yale Univ. Press.

Whelan, E. 1993. *Toxic Terror: The Truth Behind the Cancer Scares*. 2nd ed. Buffalo, NY: Prometheus Books.

Wimsatt, W. 1987. False models as means to truer theories. In *Neutral Models in Biology*, ed. M. Nitecki and A. Hoffman. New York: Oxford Univ. Press.

Wynne, B. 1989. Sheep Farming After Chernobyl: A Case Study in Communi-
    cating Scientific Information. *Environment* 31: 10–39.
Yearley, S. 1999. Computer models and the public's understanding of science.
    *Social Studies of Science* 29: 845–66.

# 7

# Earmarks and EPSCoR

*Shaping the Distribution, Quality, and Quantity*
*of University Research*

A. ABIGAIL PAYNE

## Introduction

Since World War II, the U.S. federal government has played an important role in funding university research. Federal funding represents, on
average, more than 60 percent of total research funding used by universities. Congress and the federal agencies, through the research programs
and funding levels they support, play a critical role in shaping academic
research. To minimize the role played by politics in shaping the distribution of funding and to promote the funding of the best projects, most
agencies have adopted a peer-review process for distributing research
funds to universities (see Chubin and Hackett 1990). Peer review attempts to elicit information from researchers engaged in similar research
about the quality of the projects for which funding is sought.

Despite the interest in strengthening economic development
through R&D, between World War II and the early 1980s the federal
government rarely addressed issues surrounding which universities
would receive federal funds. As a result, by the late 1970s, most federal
R&D funding was distributed to a few universities located within a few

states. In 1978, for example, half of federal R&D funds allocated to universities was distributed to just six states and 80 percent to eighteen states.[1]

This high concentration of R&D funding attracted critics who accused the peer-review process of fostering an "old-boys" network in which the universities that historically received research funding continued to do so, regardless of the quality of proposals (see Lambright 2000). Another criticism of the high concentration of funding argues that without sufficient R&D funding, state or regional economic development is constrained. Given these concerns, politicians and universities have struggled with how, given limited resources, to reduce the concentration of funding while still promoting quality research.

This chapter explores the distribution of R&D funds across universities and how two types of funding schemes initiated in the early 1980s have shaped the distribution of federal funding and university research activities. The first type, not a formal program as such, is known as earmarking. The earmark, contained in federal budget documents that must be approved by Congress and the president, usually specifies a designated amount be allocated to a particular university. Universities use earmarks for a variety of research-related (and nonresearch) purposes. The second program targets universities in states that have historically received low levels of funding. Created by the National Science Foundation, the Experimental Program to Stimulate Competitive Research (EPSCoR) sets aside competitive grants designed to help build the infrastructure needed for such universities to compete effectively in the general grant programs. Because of EPSCoR's perceived success, most of the agencies responsible for distributing federal research funding have adopted similar programs.

The chapter addresses three questions. First, do more universities receive research funding? Given the historical development of research universities, it is very difficult for them to improve their comparative levels of research funding and research activities. Earmarks and set-aside programs are one way to affect the distribution of research funding across universities. Second, given that fully evaluating the quality of research proposals requires substantial technical knowledge, earmarks and set-aside programs may not fund high-quality research. Do these programs result in a lower quality of research activity at the universities funded? Third, given that set-aside programs attempt to create an environment in which proposed research is more fully evaluated and to give incentives to states to promote areas of research in which they show some strength, are set-aside programs more successful than earmarks in promoting quality research?

Using data from 1978 to 1998 (1980 to 1998 with respect to earmarks), this chapter examines how earmarked funding and set-aside programs shape the distribution of funding at universities and research activity there. During this period the federal government earmarked on average $3.6 million (all dollars reported as real dollars, using 1996 as the base year) in R&D related funds per university. Average federal research funding to universities located in a state qualified for EPSCoR was $16 million per university. Average federal research funding to universities not located in a state qualified for the set-aside program was $39 million.

The results suggest the distribution of research funding across research and doctoral universities has changed modestly, especially since 1990. In general, average funding per university increased at a slightly faster rate at universities not classified as Research I institutions. Institutions eligible to participate in the EPSCoR program also experienced substantial growth in federal research funding. The bulk of research funding, however, still remains with a small group of universities.

There are many ways to study the impact of politics on the distribution of research funding to universities. This chapter examines the effect of earmarks and EPSCoR on overall research funding and research activity as measured by academic publications, a traditional outlet for research activity. Thus, the questions raised here concern whether these two types of funding have shaped research activity within a university as identified by more traditional measures of research activity. The results further suggest that funding from earmarked appropriations increased the quantity of academic publications but decreased the quality of these publications as measured by citations per publication. Those universities that qualified under the set-aside programs, however, increased the quality of their publications while decreasing the quantity of publications produced.

Section one in this chapter reviews earmarks and the set-aside programs and explores how these programs have shaped the distribution of federal R&D funds. Section two analyzes changes in research funding using summary statistics and graphs. Section three presents the results from an econometric analysis of the effect of changes in funding on academic research activities under the two programs. The details of the analysis may be found in Payne (2002; 2003). Section four concludes, discussing the potential implications of the findings that earmarks and set-asides can shape the distribution, quality, and quantity of academic research.

## Earmarking, Set-Aside Programs, and the Distribution Federal Research Funding

### Congressional Earmarking

Apart from the usual budgetary process (see, e.g., Drew 1984; Geiger 1993; and Kleinman 1995) through which agencies receive funding to support their programs, members of Congress may specify that funds be appropriated for specific purposes. Such earmarks, found in the appropriations bills or their accompanying reports, designate funds to one or more universities. These funds become a part of a given agency's budget and the agency is expected to allocate these funds accordingly.[2]

Universities use earmarks for a variety of R&D activities, from large, capital-intensive projects to small, discrete research projects.[3] Some universities receive multiple years of earmarks for the same types of projects. Proponents claim earmarking allows universities that are not traditional recipients of federal R&D funds to build the infrastructure necessary to compete for peer-reviewed funding. Critics believe earmarks are pork-barrel allocations that are not used as productively as peer-reviewed funds.[4]

Universities, not specific researchers, seek earmarks. A university first identifies a set of potential activities for which it will seek earmarked funding. Such activities can include establishing a new academic program or school, building a research laboratory, renovating a dormitory, or funding a particular project. Given a set of activities, a university will then lobby members of Congress for earmarks to support them. The process of earmarking uses scarce university resources and raises the question of why a given project is funded by an earmark and not by some other type of revenue source. If universities view earmarks as a "last resort" venue for funding, this raises the question of whether universities use earmarks for less productive activities than the funding would otherwise be utilized if the government used another method for distributing the funding.

Earmarks have been measured several ways. Savage (1999) examines appropriations legislation and the accompanying reports to identify recipient institutions and the amounts allocated to them. The *Chronicle of Higher Education* identifies earmarks by asking the agencies responsible for distributing the earmarked funding. The agencies provide information on the amounts distributed, the recipient universities, and the reason for the earmark. These two methods differ insofar as the Savage data identify the earmarks that Congress intends to distribute and the *Chronicle* data identify the earmarks that are distributed. Although Congress expects agencies to distribute earmarked funding, in some instances the agency requires an institution to submit a proposal for the earmarked research, and it may reject the proposal under certain conditions. Similarly, the agency may "tax" part of the earmark amount to cover administrative costs.[5]

Figure 7.1 depicts the total level of real federal research funding allocated to the sample of research and doctoral universities, as well as the percentage of earmarked funding designated to this same group using the Savage dataset.[6] Until the mid-1980s, federal research funding remained fairly flat. Since 1985, total funding has gradually increased. Earmarks represent less than 10 percent of the total federal research funding at universities. In the early 1980s, the average level of earmarked funding received per university was quite low. This changed in the mid-1980s, when there was a sharp increase in earmarked funding. In the early 1990s earmarked funding dropped, but it has been on the rise again since 1996 (not shown in the figure).

Analyses conducted by Brainard and Southwick (2001) and the American Association for the Advancement of Science (AAAS 2001) found that the Departments of Agriculture and Commerce and NASA distribute the bulk of earmarks. There are no earmarks allocated through the NSF and very few, if any, through the National Institutes of Health (NIH). Although most earmarks cover activities concerned with R&D activities at the universities, they also cover such activities as distance learning projects, university transportation systems, dormitory renovations, and community outreach projects. Identifying these types of earmarks is more difficult using the Savage data set than the *Chronicle* data set.

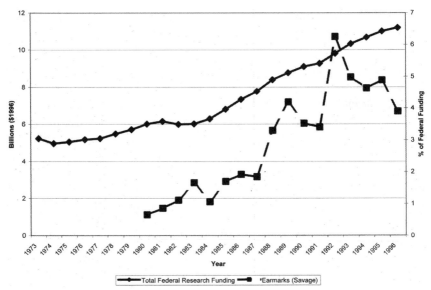

*Note:* The research funding dollars represent the total of all federal research funding allocated to institutions classified as a research or doctoral university as provided by the NSF WebCASPAR database. The earmarked funding represents the earmarks allocated in the congressional budget documents to research and doctoral universities. The percentage represents the earmarked funding allocated in one year divided by the total research funding for that year. This represents a rough estimate of the proportion of federal research funding being allocated for earmarked projects.

Figure 7.1. Total federal research funding and share of earmarks for research and doctoral universities

## Set-Aside Programs

NSF created EPSCoR in 1980. Politics, in part, drove the establishment of the program. In 1977, members of the House Committee on Science, Research, and Technology expressed concern that a high proportion of research funding was going to only a few states (see Lambright 2000). In response, NSF developed a program designed to stimulate more competitive research at universities in states with historically low levels of funding.

The essence of the program is as follows. Upon receiving EPSCoR status, a state identifies an "EPSCoR entrepreneur." This entrepreneur, from state government, industry, or a university, is responsible for working with partners in all sectors to develop a plan for improving the research infrastructure in the state. In addition, the entrepreneur must raise money from the state government and/or private industry to match EPSCoR funds.

In NSF's EPSCoR program, a state may submit only one proposal at a time. The program therefore encourages cooperation among universities within the state. Universities within the state can choose not to participate in the plan developed by the entrepreneur. Once NSF receives the proposal, it undergoes a competitive, peer-reviewed process in which proposals submitted by all EPSCoR states are evaluated based on their merits. In practice, the chance of success is dramatically greater for the proposals submitted to EPSCoR than under the non-EPSCoR programs. Although other agencies have adopted similar programs, there are differences in the treatment of states and universities that qualify for them.[7]

Initially, NSF designated Arkansas, Maine, Montana, South Carolina, and West Virginia for EPSCoR funding. In 1985, NSF added Alabama, Kentucky, Nevada, North Dakota, Oklahoma, Wyoming, and Vermont. In 1987, NSF added Idaho, Louisiana, Mississippi, and South Dakota, and in 1992 it added Kansas and Nebraska. Although not studied in this paper, NSF recently added Alaska, Hawaii, and New Mexico to the list of EPSCoR states, making a total of twenty-one states.

Figures 7.2 through 7.5 depict the average level of research funding and percentage change in funding by type of Carnegie (1994) classification for the following five periods: 1975–79, 1980–84, 1985–89, 1990–94, and 1995–98. Figure 7.2 depicts average federal research funding for the Research I and Research II institutions. Across all institutions, average federal research funding increased over the twenty-year period. For all Research I institutions, the average level of funding remains the highest. With respect to the Research I institutions in states that qualified during the period for EPSCoR status, federal research funding grew; however, the gap between the average level of funding for these institutions and the average level of funding for all Research I institutions widened during the sample period. There appears to be little difference in research funding between all institutions and the institutions in EPSCoR states with a Carnegie (1994) Research II classification.

Figure 7.3 depicts the percentage change in funding for each period using the 1975–79 period as the base. Given the base amount of funding to the institutions in the first period, the Research I institutions in states that qualified for EPSCoR status experienced the fastest growth in funding during the sample period. The growth in funding among the other types of institutions is similar.

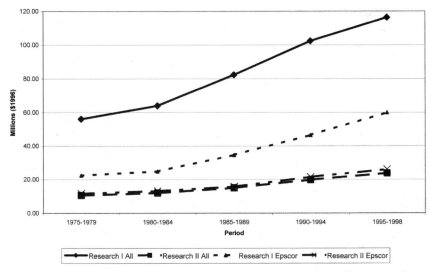

*Note:* This graph identifies the average level of funding per institution per type of classification (based on Carnegie classification and EPSCoR status) for each period. The EPSCoR status for this graph is based on whether the university is in a state that obtained EPSCoR status at some point during the entire sample period.

Figure 7.2. Average federal research funding per institution by Carnegie classification and EPSCoR status

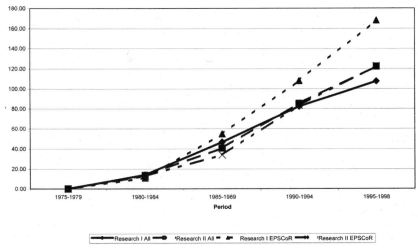

*Note:* The percentage change uses the average level of federal research funding per university between 1975 and 1979 as the base year. For each subsequent period, the average funding for that period is subtracted from the average for the 1975–79 period, divided by the average for the 1975–79 period and then multiplied by 100.

Figure 7.3. Percent change in research funding by Carnegie and EPSCoR status

Figure 7.4 depicts the research funding for all institutions and the institutions in EPSCoR states classified as a Doctoral I or Doctoral II institution under the Carnegie system. Despite the lower classification, across the sample period, the average level of funding for the Doctoral II institutions are higher than the average level of funding for the Doctoral I institutions. The institutions in states that qualified for EPSCoR status received slightly more federal funding than all institutions.

Figure 7.5 depicts the percent change in research funding from the first period, 1975–79. The greatest amount of growth was experienced by the Doctoral I institutions in states that qualified for EPSCoR status. Overall, however, the growth experienced by the Doctoral institutions during the sample period was substantial.

The figures suggest that while the bulk of funding still is being distributed to a few institutions in a few states, all types of institutions have experienced a growth in funding. Institutions in states with EPSCoR status have experienced substantial growth in funding.

Summary Statistics

Table 7.1 depicts the distribution of universities that received at least one earmark in any given year based on the Carnegie (1994) classification

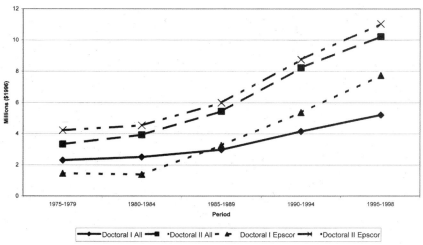

*Note:* See note to figure 7.2.

Figure 7.4. Average federal research funding per institution, by Carnegie status and EPSCoR status

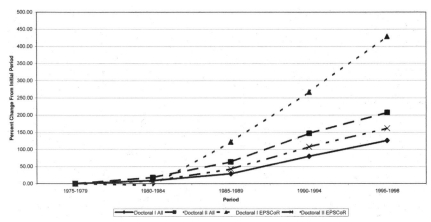

Figure 7.5. Percent change in research funding by Carnegie type and EPSCoR status

and EPSCoR status. With the exception of the Doctoral I universities, at least 75 percent of the universities within each classification received at least one year of earmarked funding. More than 50 percent of the Research I and Research II universities received more than five years of earmarked funding. Of those universities receiving five or more years of earmarks, 80 percent of the Research I, 90 percent of the Research II, 60 percent of the Doctoral I, and 83 percent of the Doctoral II universities are public universities. Thus, it appears that the bulk of earmarks are awarded to public universities.

Columns 5 through 7 report the distribution of universities in states that were eligible to participate in EPSCoR and other set-aside programs by 1998. Interestingly, twenty of the thirty-six universities are classified as a Research I or II university. All of the universities in states that became eligible for EPSCoR status received at least one year of earmarks, and most of them received five or more years of earmarks.

Table 7.2 depicts the average federal R&D expenditures, earmarks, articles published, and citations per publication across the four groups of universities. In Panel A, I report the summary statistics for all of the universities. Not surprisingly, Research I universities receive the bulk of federal R&D expenditures. On average, these universities received $83 million or $99,000 per faculty member over the sample period. These universities received an average of $4.4 million in research oriented earmarks, which represents only 5 or 6 percent of total federal research funding.[8]

Table 7.1. Distribution of earmarks by Carnegie classification and EPSCoR status

| Carnegie classification | # of universities | % public universities | % At least 1 earmark | % At least 5 earmarks | # in EPSCoR states | % At least 1 earmark | % At least 5 earmarks |
|---|---|---|---|---|---|---|---|
| | (1) | (2) | (3) | (4) | (5) | (6) | (7) |
| Research I | 86 | 67.4 | 84.9 | 57.0 | 7 | 100 | 85.7 |
| Research II | 36 | 75.0 | 77.8 | 55.6 | 13 | 100 | 92.3 |
| Doctoral I | 43 | 65.1 | 44.2 | 11.6 | 3 | 100 | 66.7 |
| Doctoral II | 52 | 65.4 | 75.0 | 23.1 | 13 | 100 | 46.2 |
| Total | 217 | 67.7 | 73.3 | 39.6 | 36 | 100 | 72.2 |

*Note:* The universities are classified under the Carnegie (1994) classification system. EPSCoR status is based on the state having been designated by the National Science Foundation or other federal agency as one with historically low levels of research funding. At least one earmark means at least one year of earmarked funding, which could include earmarks for more than one project. Similarly, at least five earmarks means at least five years of earmarked funding.

Research II universities received, on average, $16 million in federal research funding, or approximately $26,700 per faculty. Earmarks represent approximately 15 percent of federal expenditures at these universities. Earmarks seem to have the biggest impact for the Doctoral I universities, which received the lowest level of federal expenditures, $3.4 million or $8,000 per faculty. Earmarks represent approximately 86 percent of this funding. Doctoral II universities received an average of $6.2 million in federal expenditures, with earmarks representing approximately 51 percent.

Panel B of table 7.2 reports the summary statistics for the group of universities in states that are eligible to participate in the set-aside programs. Except for the Research I institutions, the average level of federal expenditures is similar to the average level for all universities (reported in panel A). Earmarks appear to represent a greater percentage of federal expenditures at these institutions than at the institutions that are not located in a state eligible to participate in the set-aside programs.

Tables 7.1 and 7.2 reveal that most institutions rely on earmarked funding but universities that are not Research I universities rely most strongly on earmarks. The tables also suggest that the universities that are the worst off in terms of receiving research funding are Research I universities located in EPSCoR-eligible states.

Table 7.2. Summary statistics: Research funding

| Carnegie classification | Federal research funding (millions) | Per faculty (1000s) | Research oriented earmarked funding (millions) |
|---|---|---|---|
| **Panel A: All universities** | | | |
| Research I | 82.91 | 98.94 | 4.41 |
| (standard deviation) | (59.68) | (103.50) | (7.58) |
| Research II | 15.95 | 26.69 | 2.47 |
| (standard deviation) | (8.83) | (17.50) | (3.52) |
| Doctoral I | 3.38 | 8.05 | 2.92 |
| (standard deviation) | (3.75) | (9.98) | (3.97) |
| Doctoral II | 6.15 | 19.49 | 3.11 |
| (standard deviation) | (8.18) | (40.16) | (3.42) |
| Total | 39.32 | 51.79 | 3.63 |
| (standard deviation) | (53.69) | (80.73) | (6.05) |
| **Panel B: Universities in EPSCoR states** | | | |
| Research I | 36.75 | 43.60 | 6.50 |
| (standard deviation) | (23.81) | (36.65) | (8.97) |
| Research II | 17.24 | 25.02 | 2.31 |
| (standard deviation) | (8.63) | (13.05) | (3.32) |
| Doctoral I | 3.68 | 6.25 | 3.27 |
| (standard deviation) | (3.58) | (4.81) | (4.67) |
| Doctoral II | 6.76 | 17.51 | 2.97 |
| (standard deviation) | (6.46) | (19.02) | (3.02) |
| Total | 16.42 | 24.58 | 3.34 |
| (standard deviation) | (16.89) | (23.98) | (5.11) |

*Note:* All dollars are constant with 1996 as the base year. Federal research funding represents the average expenditures per university per classification during the sample period (1980–98). Research earmarked funding represents earmarks identified to be associated with a research project, laboratory, or institute in the *Chronicle of Higher Education*'s data base.

# Impact on Research Activities

## Measures of Research Activities

To measure the effect of earmarks and set-asides on research activities, I focus on academic publications, a very traditional measure of research activities. The advantage of using publication measures is that we can identify both the number of publications produced and the number of times they are cited. Thus, we have both a quantity measure, and reasonable proxy (citation impact) for quality of research activity. One

Table 7.3. Summary statistics: Publications

| Carnegie classification | Articles published | Per faculty | Citations per article |
|---|---|---|---|
| **Panel A: All universities** | | | |
| Research I | 1838.73 | 2.08 | 18.93 |
| (standard deviation) | (1089.06) | (1.95) | (9.96) |
| Research II | 504.18 | 0.79 | 12.60 |
| (standard deviation) | (197.26) | (0.37) | (6.73) |
| Doctoral I | 244.79 | 0.51 | 9.95 |
| (standard deviation) | (344.14) | (1.05) | (6.56) |
| Doctoral II | 201.47 | 0.56 | 10.04 |
| (standard deviation) | (257.83) | (0.73) | (7.03) |
| Total | 915.46 | 1.20 | 14.01 |
| (standard deviation) | (1046.22) | (1.55) | (9.18) |
| **Panel B: Universities in EPSCoR states** | | | |
| Research I | 1046.48 | 1.05 | 12.51 |
| (standard deviation) | (414.85) | (0.29) | (5.00) |
| Research II | 530.33 | 0.73 | 10.45 |
| (standard deviation) | (201.80) | (0.26) | (5.25) |
| Doctoral I | 160.08 | 0.29 | 8.56 |
| (standard deviation) | (119.83) | (0.13) | (4.15) |
| Doctoral II | 169.34 | 0.42 | 9.19 |
| (standard deviation) | (123.48) | (0.32) | (5.23) |
| Total | 455.81 | 0.64 | 10.19 |
| (standard deviation) | (389.49) | (0.37) | (5.26) |

*Note:* Data on publications are from the Institute for Scientific Information and cover the period 1981–98. The number of publications is calculated based on the year of publication. The number of citations per article is a running total of citations from the year of publication to the year of data collection (2000).

disadvantage, however, is that universities are also involved in research that produces outputs other than academic publications, and measuring them is more difficult. This chapter focuses solely on academic publications and leaves for future research the effect of the program on other activities.

The data on publications were obtained from the Institute for Scientific Information (ISI). ISI provides data on the number of articles per year and the number of citations to articles published in a given year for most research and doctoral institutions as classified under the Carnegie system from 1981 to 1998. The citations are a cumulative sum of the citations starting from the year of publication and ending in 2000. Thus, articles published in early years will, on average, have more citations than articles published in later years.

Table 7.3 reports the summary statistics for the public measures based on Carnegie classification and EPSCoR type. Column 1 reports the mean number of articles published, column 2 reports the mean number of articles published per faculty, and column 3 reports the mean number of citations per article. Panel A reports the statistics for all universities and panel B reports the statistics for the universities in states that are eligible for the EPSCoR programs.

Publication activity is positively correlated with Carnegie classification. The means for the universities in the EPSCoR states, however, are consistently lower than the means for all universities. For example, across all Research I universities, on average, there are 2.1 articles published per faculty and there are 18.9 citations per article. Across the Research I EPSCoR universities, there are only 1.1 articles published and there are 12.5 citations per article. Thus, it appears that over the sample period, there are fewer publications and these publications have a lower impact, and arguably, therefore, a lower quality, for the EPSCoR institutions than for all universities.

## Congressional Earmarking

In deciding how to model the effect of earmarks on publications, it is important to consider the intended effect on research. Given that most earmarks are used for infrastructure, we should expect a long-term effect insofar as the funding improves research at the university by enabling the university to hire better researchers, have better facilities, and the like. I have, therefore, summed the earmarks from the first year of each data set to represent a stock of accumulated funding. The average stock of funding for the Savage data set is $21 million; for the *Chronicle* data set, $13 million.

To analyze the effect of the stock of earmarked funding on publication activity, I used an instrumental variables regression analysis (see Payne 2003 for more details). The regression uses either the number of publications or the number of citations per publication as the dependent variable and the accumulated earmarked funding, lagged one year, as the key regressor.[9] To control for nontime-varying factors that contribute to how earmarks affect publications (e.g., reputation, type of

Table 7.4. Effect of stock of earmarked funding on publications

| Dependent variable | # of publications | Citations per publication |
|---|---|---|
| **Two- stage least squares specification** | | |
| Earmarked funding | **22.32** | **-0.74** |
| (summed from 1st year of funding, lagged one year) | (3.30) | (0.11) |
| F-statistic on Instruments | 11.75 | 11.75 |
| p-value from 1st stage instruments | (0.000) | (0.000) |
| p-value from over-identification test | (0.620) | (0.094) |
| p-value from Hausman test | (0.000) | (0.000) |
| # of observations | 1683 | 1683 |

*Note:* A two-stage least squares estimation is used. Instruments for earmarked funding were based on a set of measures that reflected the average level of research activities by universities with the same Carnegie classification but located outside of the region in which the university under study is located. Also included in the regression are the following state-level political, economic, and demographic measures: state population, percentage of the state population under the age of eighteen, unemployment rate, dummy variable equal to one if the governor is affiliated with the Democratic Party, and measures of the level of political competition in the state upper and lower legislatures. Sample period covers 1990–98 and reports the results using data obtained from the *Chronicle of Higher Education*. The results are similar to those reported if the Savage data set is used in the analysis. See the text for an explanation of the overidentification and Hausman (1978) tests. Standard errors reported in parentheses unless otherwise stated. Coefficients in bold significant at p<0.05.

university, location), the regression includes a set of dummy variables to identify the university under study. The regression also uses a set of dummy variables to control for year effects. For example, if the level of earmarking is increasing across all universities, the year effects will control for this. Finally, the regression uses state-level political, economic, and demographic measures to control for changes in the socioeconomic and political environment in which the universities operate.

Table 7.4 reports the results from the instrumental variables regressions. The coefficient on the earmarked funding measure reflects the potential substitution and complement effects associated with earmarked funding and other types of funding on the publication measure. Column 1 reports the results when the number of publications is the dependent variable. An extra $1 million in earmarked funding results in an additional twenty-two publications. Given that average earmarked funding is $3.5 million, this suggests that a university will produce seventy-seven more articles per year, an increase of 7 percent, on average.

Column 2 reports the results when the number of citations per publication is the dependent variable. An extra $1 million in earmarked funding results in a reduction of citations per article by 0.74. Given the average level of earmarked funding, this suggests a decline of 2.59 citations per article, a decrease of 31 percent, on average.

The results in table 7.4 suggest earmarked funding increases the number of articles published but decreases the impact or quality of those articles, as measured by the number of citations per article. Thus, these results suggest that politically motivated funds, which lack the structure of peer-review or a thorough evaluation of the proposed research activities, do not support quality research as measured by a traditional research activity output. Alternatively, these results suggest that the motivations of the members of Congress that support earmarked funding may not be focused on promoting basic research as measured by these same traditional publication measures.

## Set-Aside Programs

To study the effect of the set-aside programs on research activities, we want to compare two things. First, we want to compare how, within a given university, being in a state that is eligible for set-aside funding affects research activities before and after the state becomes eligible for set-aside funding. Second, we want to identify two sets of universities where the first set are those that are in states eligible for the set-aside programs, and the second set are those that are not in states eligible for the set-aside programs. We construct these two sets with a regression analysis that contains multiple observations from universities that qualify ("treatment group") and do not qualify for the set-aside programs ("control group").

One issue that arises in constructing a set of universities that are not eligible for the set-aside program is whether these universities adequately reflect characteristics of the universities that become eligible for the set-aside programs. Payne (2003) discusses the intricacies of this issue in more detail. In essence, following Heckman et al. (1998), I have attempted to match the universities that are eligible for the set-aside programs with universities that are not eligible for the funding but possess similar qualities with respect to their research activities. Because qualification in the set-aside programs is a function of whether the state has historically received low levels of funding, there are many universities that have historically received low levels of federal funding but are not located in states that have qualified for the set-aside programs.[10]

One hundred universities are used in the regression analysis. Sixty-six universities are in the control group, of which forty-two are public and twenty-four are private. Thirty-four universities are in the treatment group, of which thirty-two are public and two are private.

Table 7.5 reports the results from a regression analysis that uses the publication measures as the dependent variable and two dummy variables to identify the effect of the set-aside programs. The value of the first dummy variable is equal to one once a state becomes eligible for the set-aside programs. This variable is lagged three years to account for the time it may take for the state to start receiving EPSCoR funding. The second dummy variable is equal to one if the university is in a state that is eligible for the set-aside programs subsequent to 1992. This second measure reflects the effect associated with the expansion of the set-aside programs in the early 1990s.

Also included in the regression are federal research funding (lagged three years), state level economic, political, and demographic measures, and school and year dummy variables. The regression analysis uses an instrumental variables technique whereby the measure of federal research funding is instrumented using as instruments a similar set of measures as reported above in the analysis of earmarked funding. Details on the specification can be found in Payne (2003).

Column 1 of table 7.5 reports the results when the number of publications is the dependent variable. EPSCoR universities produce fewer publications than non-EPSCoR universities. On average, qualifying for the set-aside programs results in a decline of thirty-one articles, or 5 percent. The effect subsequent to 1992 is imprecisely measured.

The results suggest that, after the expansion of the program, the quality of the publications improved. When the number of citations per article is used as the dependent variable, column 2, the overall effect of the set-aside programs is negligible. Subsequent to 1992, there is an increase in the citations per article by 1.1, representing, on average, an increase of more than 10 percent.

The results in table 7.5 suggest that the set-aside programs have a positive effect on university research activities. On average, the quantity of publications has decreased but their citation impact, and thus, arguably, their quality, has increased as a result of being in a state that participates in the set-aside programs. If more effort is required to produce a high impact/high quality publication than a low impact/low quality

Table 7-5. Effect of set-aside programs on publications

| Dependent variable: | Articles (1) | Citations/Article (2) | Engineering articles (3) | Life Sciences articles (4) | Engineering citations/article (5) | Life Sciences citations/article (6) |
|---|---|---|---|---|---|---|
| **Two-stage least squares specification** | | | | | | |
| Dummy for EPSCoR participation | -37.80 | -0.23 | -2.20 | -24.73 | **1.69** | 0.03 |
| (lagged 3 years) | (13.40) | (0.50) | (2.62) | (9.12) | (0.59) | (0.58) |
| Dummy for EPSCoR participation post-1992 | -9.05 | **1.08** | 2.34 | **-19.94** | **1.45** | **1.85** |
| | (14.07) | (0.33) | (2.81) | (8.97) | (0.64) | (0.57) |
| Federal research funding | **21.00** | **-0.15** | **8.93** | **15.35** | -0.24 | **-0.24** |
| (lagged 3 years) | (1.42) | (0.04) | (1.39) | (1.49) | (0.34) | (0.10) |
| F-Statistic from first-stage instruments | 36.07 | 30.23 | 13.57 | | 11.84 | |
| p-value from first-stage instruments | (0.00) | (0.00) | (0.00) | | (0.00) | |
| p-value from overidentification test | (0.61) | (0.64) | | | | |
| p-value from Hausman test | (0.00) | (0.03) | | | | |
| # of observations | 1497 | 1290 | 1302 | 1302 | 1266 | 1266 |

*Note:* A two-stage least squares estimation is used where federal research funding is treated as endogenous. Instruments for federal research funding were based on a set of measures that reflected the average level of research activities by universities with the same Carnegie classification but located outside of the region in which the university under study is located. Also included in the regression are the following state-level political, economic, and demographic measures: state population, percentage of the state population under the age of eighteen, unemployment rate, dummy variable equal to one if the governor is affiliated with the Democratic Party, and measures of the level of political competition in the state upper and lower legislatures. Also included in the regressions are the average state gross domestic product for states located in the same region as the state under study for the following industries: agriculture, health, chemicals, and electrical equipment. Sample period covers 1981–98. See the text for an explanation of the overidentification and Hausman (1978) tests. Standard errors reported in parentheses unless otherwise stated. Coefficients in bold significant at $p < 0.05$.

publication, the results suggest the EPSCoR program has helped to improve the effectiveness and value of research activity in states with historically low levels of research funding. Thus, these results suggest that set-aside programs provide a viable vehicle for helping universities with historically low levels of funding to improve the quality of their research. To the extent that universities with a higher quality of research will be able to attract better researchers and more funding, they may benefit from the set-aside programs in additional ways.

Are these results consistent across disciplines? Given that federal agencies are responsible for funding different types of research, we might expect the results to vary based on academic discipline. In columns 3 to 6 in table 7.5, I report the results from the examination of the effect of the EPSCoR program on the disciplines of engineering and life sciences.[11]

Across the two disciplines, the relationship between federal research funding and publications differs substantially. On average, the additional number of publications from an additional $1 million in federal funding is 8.9 in engineering and 15.4 in the life sciences. These numbers reflect, in part, differences in the expectations regarding the number of publications produced by a given researcher in any given year.

The set-aside programs affect publication activity at the EPSCoR universities in both disciplines. This result should not be surprising given that engineering is one of the major disciplines funded by the NSF (although other agencies also fund engineering projects). The life sciences have experienced the most dramatic rise in research funding to universities over the sample period. Given this rapid increase in funding, universities including those in states that qualify for set-aside funding, may benefit greatly.

For both disciplines, on average, the set-aside programs correlated with increased number of citations per publication. The effect on the number of publications, however, differs across the two disciplines. For engineering, the coefficients on both EPSCoR measures are imprecisely measured. The effect of the set-aside programs on the number of publications, thus, is unclear. The results for the life sciences suggest that they drive the overall results for the effect of set-asides on academic publications. On average, the effect of the state's becoming eligible to participate in the program is negative, suggesting on average a decline in twenty-five publications (8 percent). The effect after 1992 is worse, suggesting an average decline of forty-five publications (14 percent).

For both engineering and the life sciences, citations per article increase as a result of being in a state that is eligible to participate in the set-aside programs. There is a positive and significant effect for both of the EPSCoR coefficients with respect to engineering. The effect of the university's being in a state that is eligible for the funding increases citations per article on average by 1.7 (26 percent). The total effect after 1992 is an average increase of 3.2 citations per article (50 percent). Given the relationship between engineering funding and the NSF, these results suggest a dramatic improvement on publication quality by the NSF program. With respect to the life sciences, only the coefficient that represents the additional effect of the set-aside programs subsequent to 1992 is statistically significant. The results suggest that the citations per article in the life sciences subsequent to 1992 increased by 1.9, or 14 percent.

## Conclusion

Direct funding of research activities by Congress through the appropriations process has increased at research and doctoral universities over the last decade. Similarly, programs at the agency level designed to improve the research infrastructure at universities that historically have received low levels of research funding have also expanded. Previous research has concentrated on the role played by politics on the distribution of funding, as well as on how the universities have reacted to the set-aside programs. This chapter studied how these two different types of funding have shaped research activity at universities.

Using a panel data set that contains measures of federal research funding, earmarks, and academic publications, this chapter looked at the effect of these funding sources on the quantity and quality (as measured by citation impact) of publications. With respect to earmarked funding, the results suggest that while the quantity of publications increases, the quality decreases. In contrast, the results suggest that set-aside programs help to improve research quality at universities in states eligible for the set-aside programs, while reducing their quantity.

The results suggest that, despite the fact that politics initially drove both earmarking and set-aside programs, there are differences in the effect of these two sources of funding on research activity. One explanation for the difference is that with earmarked funding, universities lobby a politician for funds for a specific project. The lobbying and the decision making involved in the appropriations process may not allow the politicians to receive all of the information needed to evaluate fully the quality of the research funding. Additionally, the politician's underlying motivation in awarding the earmark may be different from the motivations of the bureaucrats responsible for evaluating and awarding funding under the set-aside and other research funding programs.

In contrast, the set-aside programs are designed to foster cooperation among research actors within the state eligible to compete for the funding under the set-aside. Thus, these universities must develop a high quality proposal that will be competitive with other universities seeking funding from the same source. This added competition and the more rigorous review of the research proposals may help to explain why the set-aside programs appear to be more successful in improving research quality than earmarks. Another factor that may play a role in the increase in the number of citations, but a decrease in the number of publications, is the requirement that institutions collaborate in their efforts to seek funding under the set-aside programs. The added requirement of collaboration may raise the profile of the researcher and her work but also take away from the time to conduct research.

Many additional questions remain unanswered. One such question concerns how these different funding mechanisms shape research activities other than academic publications. (A related issue is whether different funding mechanisms tend to support different *types* of research, e.g., less or more "applied," or less or more "high risk.") Another question concerns how the set-aside programs affect the economic growth in the states that have had historically low levels of research funding. A final question is how the funding of the set-aside program has enabled universities to become more competitive in seeking other types of federal research funding.

NOTES

1. The top six states were: California, New York, Massachusetts, Pennsylvania, Texas, and Illinois.

2. In rare instances, earmarks are used to resurrect research programs that were agency sponsored research projects in prior years but the agency discontinued. In these instances, the agency may award the funding under a peer-reviewed process. Often this occurs in the area of agriculture, where some grants are distributed using a formula.

3. Some earmarks are designated for nonresearch activities. For example, earmarks have been used to renovate dormitories and to develop an on-campus transportation system.

4. See Teich (2000) and Feller (2000) for a discussion of these issues.

5. For exploring the politics of earmarks, the Savage data set is superior because it identifies the appropriations subcommittee responsible for proposing the earmark. See Savage (1991; 1999) for a more complete description of the politics of earmarks.

6. Federal research funding was obtained from NSF's data base known as WebCASPAR and can be found at webcaspar.nsf.gov. The federal research funding measure reflects expenditures of federal funds by universities on an annual basis from 1972 to 1998. The data covers 218 research and doctoral universities, representing more than 90 percent of the research and doctoral universities in the United States. This paper does not analyze data for Alaska and Hawaii. With respect to data on federal research funding and earmarks, it is important to note that they are not measured exactly the same way. The former reflect annual expenditures and include university overhead and other benefits. In addition, federal R&D expenditures to universities include funding received by the universities for research-related earmarks. In contrast, earmarks could be expended over several years (rather than annually) and may or may not include any agency administrative costs or other amounts that might alter the size of the appropriation prior to the university's receiving it. Thus, while we can compare the level of earmarks and total federal expenditures at universities using simple statistics, it would be inappropriate to use these two measures together in more complex analyses.

7. For more detailed information on the various set-aside programs, see www.epscorfoundation.org.

8. Looking only at public universities, earmarks represent approximately 8 percent of funding at Research I universities, 26 percent at Research II universities, 41 percent at Doctoral I, and 72 percent at Doctoral II universities.

9. The measure of earmarked funding is instrumented to control for several potential measurement issues in the estimation. First, because earmarked funding is not the only type of funding that is available to a university, one should control for the fact that the earmarked funding is correlated with the other types of research funding. Payne (2003) discusses other reasons for using an instrumental variables specification. Under an instrumental variables specification, the instrumented measure, earmarked funding is predicted using a set of measures directly correlated with earmarked funding but indirectly corrected with the publication measure. In this case, I use four measures that reflect the research activities at universities located outside of the region in which the university under study is located and with the same type of Carnegie (1994) research or doctoral institution. These measures help to proxy the level of competition for

federal research funding and the level of productivity the university under study faces from other universities. The four measures reflect the following disciplines: social sciences, engineering, life sciences, and agricultural sciences.

10. To construct a control group of universities, I used the level of federal research funding in 1978 to identify universities. This year is prior to the initiation of EPSCoR and similar programs. I ranked the universities in three parts: first, I ranked the total federal research funding of those universities whose portion of federal research funding attributable to the life sciences was more than 60 percent; second, I ranked the total research funding of those universities whose portion of federal research funding attributable to engineering was more than 25 percent; third, I ranked the total federal research funding of the remaining universities. Thus, in constructing the control group, I control for differences in the research needs of different academic disciplines. Given the two disciplines with the largest amount of research funding are the life sciences and engineering, the method used to create a control group matches universities based on whether they have emphasized their research on life sciences, engineering, or something else.

11. I use a three stage least squares technique, allowing the error terms across the two disciplines to be correlated. The f-statistics reported for the instruments in the first stage regression reflect the joint test of the instruments in the two equations in the first stage regression.

### REFERENCES

American Association for the Advancement of Science. 2001. House and Senate Earmark R&D Funds in USDA, NASA, and DOE Budgets, mimeo, www.aaas.org.

Brainard, J., and R. Southwick. 2001. A record year at the federal trough: Colleges feast on $1.67-billion in earmarks: Budget surplus feeds Congress's pork-barrel spending, intensifying criticism. *Chronicle of Higher Education* 47 (48).

Carnegie Foundation. 1994. *Classification of the Higher Educational Institutions.* Carnegie Foundation.

Chubin, D., and E. Hackett. 1990. *Peerless Science: Peer Review and U.S. Science Policy.* Albany: State Univ. of New York Press.

Drew, D. E. 1985. *Strengthening Academic Science.* New York: Praeger.

Feller, I. 2000. Strategic options to enhance the research competitiveness of EPSCoR universities. In *Strategies for Competitiveness in Academic Research,* ed. J. S. Hauger and C. McEnaney. Washington, DC: American Association for the Advancement of Science.

Geiger, R. L. 1993. *Research and Relevant Knowledge: American Research Universities Since World War II.* Oxford: Oxford Univ. Press.

Hausman, J. A. 1978. Specification tests in econometrics. *Econometrica* 46: 1251–71.

Heckman, J. H., Ichimura, J. Smith, and P. Todd. 1998. Characterizing selection bias using experimental data. *Econometrica* 66: 1017–98.

Kleinman, D. L. 1995. *Politics on the Endless Frontier: Postwar Research Policy in the United States*. Durham, NC: Duke Univ. Press.

Lambright, W. H. 2000. Building state science: The EPSCoR experience. In *Strategies for Competitiveness in Academic Research*, ed. J. S. Hauger and C. McEnaney. Washington, DC: American Association for the Advancement of Science.

Payne, A. A. 2002. Do congressional earmarks increase research productivity at universities? *Science and Public Policy* 29 (5): 313–400.

———. 2002. The role of politically motivated subsidies on university research activities. *Educational Policy* 17 (1): 12–37.

Savage, J. D. 1999. *Funding Science in America: Congress, Universities, and the Politics of the Academic Pork Barrel*. Cambridge: Cambridge Univ. Press.

———. 1991. Saints and cardinals in appropriations committees and the fight against distributive politics. *Legislative Studies Quarterly* 16 (3): 329–47.

Teich, A., ed. 2000. *Competitiveness in Academic Research*. Washington, DC: American Association for the Advancement of Science.

# 8

# Innovation in the U.S. Computer Equipment Industry

*How Foreign R&D and International Trade Shape Domestic Innovation*

SHERYL WINSTON SMITH

## Introduction

As the sources of innovation become increasingly far-flung and diverse, understanding the relationship between globalization and innovation is critical. The literature finds a connection between foreign research and development (R&D) and domestic innovation, suggesting that R&D in one nation can contribute to innovation in another through spillover. The different fortunes of high-technology industries on the global stage suggest, however, that the story is more nuanced. Indeed, foreign R&D can help, but it also can hurt, and international trade is an important part of this story. This chapter pinpoints and elucidates how foreign R&D and international trade shape domestic innovation in the computer equipment industry in the United States.

The computer equipment industry in the United States is R&D-driven, reflecting strong ties with science and engineering from its inception. Computer equipment has been a strong domestic industry, with average export shares of 24.4 percent, compared to the average of 11.5

percent for all manufacturing over the period 1973–96.[1] The importance of continued investment in R&D as a driver of innovation and a contributor to endurance in the global arena is particularly apparent in this industry. International trade has served as a conduit for both competitive pressures and knowledge flows through foreign R&D.

Importantly, foreign R&D reflects an array of related attributes: an educated technical workforce, a sophisticated technical market, a potential source of knowledge, and also a source of sophisticated competition. A unique contribution of this research is its quantifying the effect of foreign R&D on domestic innovation in the computer equipment industry (SIC 357; see table 8.1, p. 181 for composition) and investigating the implications of these related attributes.[2]

The next section summarizes the conceptual framework relating innovation to knowledge transmission through foreign R&D and international trade, and describes the empirical model, which is then estimated. The empirical results, discussed in the third section, suggest that foreign R&D has a measurable impact on domestic innovation through several mechanisms. On one hand, exports to technically sophisticated markets abroad can enhance domestic innovation; on the other, import competition starts to discourage domestic innovation as the level of technical sophistication abroad increases. The fourth section discusses the broader implications and context of the empirical results through consideration of the nature of technological advance, international connections, and international competition.

## Conceptual Framework

### Innovation, Foreign R&D, and International Trade

Innovation depends on developing and finding knowledge and embodying it in new or better products or processes. In a narrow sense, innovation can be thought of as the transformation of knowledge into something that provides additional, often economic benefit.[3] The sources of a firm's knowledge involve not just its own research but also its connections to other sources of knowledge, including university research and knowledge gained through interactions with suppliers, contractors, customers, and even competitors (Branscomb and Florida 1998). As technological innovation becomes increasingly distributed internationally, the sources of a firm's knowledge may follow.

A firm's R&D is essential to both creating knowledge and absorbing it. In economic theory, knowledge is a public good: it is nonrival—it can be shared by more than one user without diminishing the ability of anyone else to use it; and it is at least partially nonexcludable—its use by others is difficult to prevent. These properties, particularly nonrivalry, lead to spillovers in which a firm that does not create knowledge can still make use of it.[4] Absorptive capacity is an important component of R&D spillover, which is more likely to occur if the "receiver" is advanced enough to find new knowledge, recognize its importance, and otherwise prepared to incorporate this knowledge effectively (Cohen and Levinthal 1989).

Foreign R&D can influence domestic innovation through several mechanisms. First, foreign R&D increases the global pool of knowledge, at least some of which will be accessible to the domestic industry. Empirical studies suggest that R&D from other nations contributes in this way to domestic innovation (Coe and Helpman 1995; Keller 2001; 2002; Connolly 1998; Bernstein and Mohnen 1998; Nadiri and Kim 1996). Second, foreign R&D increases the absorptive capacity where it is conducted. Third, foreign R&D contributes to creating markets for technically advanced products abroad, while also contributing to the creation of technically sophisticated competition.

The theoretical relationship between innovation and exports is equivocal. On one hand, theory provides support for a positive relationship: access to a larger market allows economies of scale and potential gains in productivity. Exporting might also facilitate knowledge spillovers through contact with foreign R&D. However, knowledge spillovers are likely to be bidirectional, that is, knowledge might flow out as well as in, thus proving detrimental in a technically sophisticated market.[5] Empirical evidence on the relationship between exporting and innovation is mixed (e.g., Amato and Amato 2001; Bernard and Jensen 1999).

Nor is the relationship between imports and innovation straightforward. Knowledge embodied in imported final goods and intermediate inputs might facilitate innovation in a number of ways: through reverse engineering, proof of concept, or leapfrogging. As well, knowledge transmission might occur through contact with scientists and engineers abroad, with suppliers, and with customers. In such cases, the relationship between imports and innovation should be positive. Additionally, import competition can be expected to affect domestic innovation by influencing market structure. However, the relationship between market power and innovation is ambiguous (e.g., Scherer 1984; 1992; Barzel 1968; Geroski 1990). Empirical evidence suggests that import competition can inhibit or spur domestic innovation (e.g., MacDonald 1994; Scherer and Huh 1992; Lawrence 2000; Lawrence and Weinstein 1999).

## Empirical Framework

This discussion suggests several testable hypotheses about the relationship between domestic innovation and R&D and international trade:

1.  *Domestic R&D enhances domestic innovation.*
2.  *Foreign R&D may enhance or discourage domestic innovation.*[6] Greater stocks of foreign R&D should increase the stock of knowledge for U.S. firms to tap into. If R&D elsewhere is accessible and is easily absorbed, then we expect to see a positive relationship to domestic innovation. On the other hand, foreign R&D might inhibit domestic innovation as the foreign industry becomes more technically advanced and as foreign absorptive capacity increases.
3.  *Imports may enhance or discourage domestic innovation.* A positive relationship between imports and domestic innovation might be expected through knowledge embodied in imported goods and through contact. Imports might also allow domestic firms to effectively exercise their comparative advantage in other areas, maintaining high productivity by relying on low-cost imports in noncritical areas of the product line. As well, import competition might spur domestic innovation. On the other hand, import competition might hurt a weak domestic industry if it cannot compete in price or quality.
4.  *Exports may enhance or discourage domestic innovation.* Exports can be positively related to domestic innovation through scale effects from increased market size, greater opportunity for learning by doing, and through contacts with foreign users and suppliers. On the other hand, exports can create more technically advanced competition through knowledge flowing to competitors. As well, it can be difficult to succeed in sophisticated markets abroad.

To test these hypotheses, I employ a model that relates domestic innovation to domestic and foreign R&D stocks and international trade. This model is based on a theoretical framework developed by Grossman and Helpman (1991a; 1991b; 1994).[7] Innovation is measured in terms of productivity, a widely used measure of technological change. An increase in total factor productivity (TFP), or output after all of the factors of production have been taken into account, means that the same output is produced with fewer inputs, and thus is associated with beneficial economic outcomes of successful innovation (e.g., see Griliches 1994; 1998; Jaffe 1996).[8] The specification to be estimated can be expressed:

$$\ln\mathit{tfp3} = \beta_0 + \beta_1 \ln\mathit{rdst} + \beta_2 \ln\mathit{frdst} + \beta_3\ \mathit{expsal\_1} + \beta_4\ \mathit{impsal\_1} + \beta_5 (\mathit{expsal\_1})(\ln\mathit{frdst}) + \beta_6 (\mathit{impsal\_1})(\ln\mathit{frdst}) + \beta_7\ \mathit{time} + \epsilon$$

This specification allows us to separate the effects of domestic R&D, foreign R&D, international trade, and, importantly, interactions among them.[9]

To quantify the effects of foreign R&D and international trade on domestic innovation in the computer equipment industry, I estimate this equation using data aggregated at the three-digit industry level covering the years 1973–96. Summary statistics and a description of the variables are given in appendix table 8.A1. Briefly, the dependent variable, *lntfp3*, is the log of total factor productivity (TFP). TFP data is calculated from the NBER Manufacturing Productivity database (Bartelsman and Gray 1996). The independent variables are *lnrdst*, the log of domestic R&D stock; *lnfrdst*, log of foreign R&D stock, *expsal_1*, U.S. exports normalized by net sales, lagged by one year; and *impsal_1*, U.S. imports normalized by net sales, lagged by one year. Domestic and foreign stocks of knowledge are measured by cumulative domestic R&D investment (NSF, various years) and cumulative R&D investment in the European Union and Japan (OECD 1996). The international trade variables are U.S. exports to and imports from the EU and Japan in the previous year (Feenstra 1996; 1997; 2002). The interactive variables allow the effect of exports and imports to depend upon the level of foreign R&D. The time variable is included to minimize the effects of common trends over time.

## Empirical Analysis

What was the effect of international trade and foreign R&D on innovation in the U.S. computer equipment industry over this period? Figure

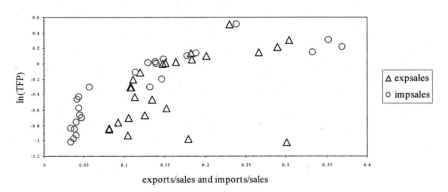

**ln(TFP) vs. exports/sales and imports/sales**

Figure 8.1. Total factor productivity and international trade, 1972–96

8.1 suggests that there is a positive relationship between TFP and both imports and exports over this period. Figure 8.2 suggests a positive relationship between TFP and both domestic and foreign R&D stock.

However, these simple relationships may be misleading. Separate consideration of trade and foreign R&D cannot tell the full story. How do trade and foreign R&D interact to influence innovation? I investigate this question by estimating the model described above using ordinary least squares analysis with robust standard errors. The regression analysis yields several insights about the role of foreign R&D and international trade in innovation in the computer equipment industry over this period.

The results of the ordinary least squares regressions are presented in the appendix table 8.A2.[10] The results support the hypothesis that domestic R&D enhances domestic innovation, as measured by TFP, in the computer equipment industry. The results in column 4 show that the elasticity of TFP with respect to domestic R&D is 1.07, indicating that for a 1 percent increase in domestic R&D stock there is an increase of 1.07 percent in domestic TFP. This effect is larger than those cited by Coe and Helpman (1995) for the U.S. economy and by Griliches and Mairesse (1998) for a subset of high-technology firms, but it seems reasonable given that this estimation is based solely on the computer equipment industry. The effect of foreign R&D stock on domestic TFP is negative; for a 1 percent increase in foreign R&D stock, there is a decrease of 1.66 percent in domestic TFP.[11] These results are significant at the 10 percent significance level.

**ln(TFP) vs. ln(domestic R&D stock) and ln(foreign R&D stock)**

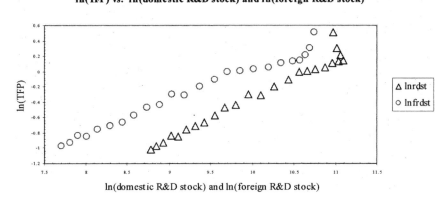

Figure 8.2. Total factor productivity and R&D stock, 1972–96

When we consider the interaction between international trade and foreign R&D stock in column 4, a more distinct picture emerges. When we include the interaction between exports and foreign R&D stock the coefficient on exports *(expsal_1)* is negative, while the coefficient on the interaction between exports and foreign R&D stock *(expsal_1)*(ln*frdst*), is positive. This relationship suggests that exporting can aid domestic innovation as foreign R&D stock increases, and thus the technical so-phistication of the market abroad increases. One explanation is that a technically sophisticated market will demand more sophisticated com-puter equipment, which encourages further domestic innovation, par-ticularly if this industry is especially adept at innovation. On the other hand, exporting to less sophisticated markets might involve more com-moditized computer equipment. In this situation, the importance of distribution channels and other local factors might be important. If we evaluate at the mean of foreign R&D stock, we can calculate the net re-lationship between exports and domestic TFP. In this calculation, we find that a 1 percent point increase in exports to sales is associated with a 2.27 percent increase in domestic TFP for this level of foreign R&D stock.

The relationship between imports and domestic TFP is more am-biguous when we account for the interaction between imports and for-eign R&D stock. In column 4, the coefficient on imports *(impsal_1)* re-mains positive, but the coefficient on the interaction between imports and foreign R&D stock *(impsal_1)*(ln*frdst*), is negative. If we evaluate at the mean of foreign R&D stock, we can calculate the net effect of imports: a 1 percent increase in imports to sales is associated with a 1.15 percent increase in domestic TFP. However, if we evaluate at the maximum of foreign R&D stock, the net effect of imports is negative: a 1 percent increase in imports to sales is associated with a 1.26 percent decrease in domestic TFP.

These results suggest that as imports are more technically sophisticated, there is a negative impact on domestic innovation, perhaps reflecting a greater competition for the high end of the market where goods are less commoditized and productivity gains are harder to achieve. Technically less sophisticated imports might affect domestic innovation differently. Import competition in the more commoditized segments of the industry might stimulate domestic manufacturers to seek more efficient methods of production. One possibility is that the more commoditized imports include many peripherals, which might enable domestic manufacturers to achieve higher TFP performance by using ready-made parts and peripherals while concentrating on other parts of the computer. Import competition on the commoditized end might also spur domestic manufacturers to introduce new and better products.

In all of the regressions, the coefficient on foreign R&D stock is negative and significant. When we account for the interaction between foreign R&D and international trade, this relationship remains. Evaluating at the mean of exports and imports, we can calculate that a 1 percent increase in foreign R&D stock is associated with a 1.75 percent decrease in domestic TFP.[12] The results together suggest that in the computer equipment industry, the net effect of foreign R&D on domestic TFP performance is negative. The results of these regressions are consistent with our expectations based on the greater context of the industry and strengthened by the details of the industry presented in the next section.[13]

## Characteristics of the Computer Equipment Industry

The results described above raise important questions about the role of foreign R&D and international trade in the U.S. computer equipment industry. We saw from the empirical analysis that foreign R&D seems to influence domestic innovation in this industry negatively, suggesting that the challenges of more technically sophisticated competition abroad might outweigh the expected positive effect of an increase in knowledge

Table 8.1 Composition of computer equipment industry (selected years)

| Year | Computers | Storage devices | Terminals | Peripherals | Calculating machines | Office mach. n.e.c. |
|------|-----------|-----------------|-----------|-------------|----------------------|---------------------|
| | 3571 | 3572 | 3575 | 3577 | 3578 | 3579[a] |
| 1972 | 0.451 | 0.086 | 0.041 | 0.187 | 0.078 | 0.158 |
| 1987 | 0.555 | 0.105 | 0.030 | 0.231 | 0.025 | 0.054 |
| 1994 | 0.610 | 0.138 | 0.018 | 0.172 | 0.017 | 0.045 |

[a] SIC 3579 office machinery "not elsewhere classified" includes typewriters; in the older SIC classification systems, typewriters were a separate SIC code. The relatively large change in this segment partially reflects the drop in output of typewriters in this time period.

*Source:* Computer equipment is defined by the 1987 SIC code 357. Calculated from NBER data (Bartelsman and Gray 1996). Shares calculated as fraction of value of shipments for SIC 357.

stock. The results in the previous section suggest that increasingly sophisticated foreign markets might aid domestic innovation through exports, but at the same time, the greater technical sophistication of imports might hurt domestic innovation. We will consider these findings in the context of the computer equipment industry's experience over this time period.

## Technological Advance and Industry Structure

From the beginnings of the computer industry, market structure evolved concomitantly with technological advance (Bresnahan 1999; Bresnahan and Malerba 1999), a dynamism characterized by periods of entry and exit. This volatility was largely driven by technological advance creating new market segments, for example, microcomputers giving way to PCs. The industry has followed a pattern of periods of high flux followed by consolidation into a few firms, a dynamic repeated multiple times in the industry history. In contrast to the more than seventy makers of personal computers, with no clearly dominant firm, in 1978, the five largest makers of computers controlled 54 percent of the U.S. market in 1998 (Crothers 1999).

Innovation in the computer equipment industry has been characterized by rapid, R&D-driven technological advance, derived from an extensive investment in R&D and a phenomenal rate of productivity (NSB 2000). The computer equipment industry exhibited a relatively high R&D intensity over the entire period, approximately four times greater than all manufacturing on average (NSB 2000).

Competition in the computer industry is rooted in technological advance, which may then erode the existing market for the predecessor. Ongoing technological displacement of older models and even classes of computers occurs with the introduction of more powerful, cheaper, and smaller ones. For instance, mainframe computers dominated computing equipment into the late 1980s but, by 1995, accounted for less than a third of the market as minicomputers and ultimately, personal computers, took a greater share (Warnke 1996).

Technological advance also expanded existing markets and created new ones sufficient to keep the price elasticity of demand greater than one. Thus, even as prices for a given level of computing performance fell, an even larger volume of computational capacity was expressed in market demand. A substantial base of scientists and engineers, small businesses, and perhaps to a lesser extent the vast potential market of home users played an important role in the rapid expansion of the computer industry in the United States (Bresnahan 1999), acting to both finance and motivate continued investment in innovation. Also driving demand was a vigorous and competitive software industry, creating a great range of applications and a degree of de facto standardization in operating system architecture that provided a common platform for these applications.

Standardization has shaped the evolution of technology and of market structure in the computer industry, beginning with widespread adoption of the IBM 360 and later models. IBM's decision to use open architecture instead of proprietary technology for the personal computer shaped the direction of the industry as IBM gained market share and set the industry standard (Bresnahan and Malerba 1999; NRC 1990).

## International Linkages and Competition

A combination of government, academic, and private sector resources resulted in a formidable advantage for the development of the U.S. computer industry, and the early years of its commercially viability are almost entirely a U.S. story. Initial high-risk purchases of high-speed computers by the U.S. government contributed heavily to the first computers and continued for several decades (Flamm 1988; MIT Commission 1989). In many respects, however, the computer industry has had international aspects from the start. From the origins in the tabulating machines of Hollerith (an antecedent of IBM) and of Powers (ultimately Remington Rand and its successor companies), international distribution was an essential component, establishing the broad global channels that remained important as the computer industry evolved. Even before the establishment of research facilities abroad, these international channels also formed the basis for global sources of incremental improvements to the products (Connolly 1967; Flamm 1988).

Important developments occurred in Europe, especially in Germany and England. Although the source of inventions was global, U.S. businesses were most successful in early commercialization. IBM drew on research from its English and German facilities for key aspects of the System 360 (Flamm 1988). This advantage in early commercialization by the U.S. industry continued into the 1980s and even later for some segments.

At a time of concerted effort in the United States to develop powerful computing machines, no comparable industry was developing in Europe or Japan. In Europe there were pockets of research, for instance in England, especially at Cambridge, but these were not well funded or extensive (Connolly 1967; Flamm 1988). Olivetti, the Italian typewriter company, was perhaps the most pioneering firm in Europe, establishing its Advanced Technology Center in California in 1972 and forming an alliance with AT&T in 1983 to produce PC-compatible computers (Olivetti Group 2002). Despite its early successes, Olivetti was unable to match the dynamic nature of innovation in the industry in the long term.

The Japanese government pursued a strategy of protecting its infant domestic industry from foreign competition but with extensive access to foreign technology through licensing agreements. Japanese trade restrictions severely limited the entry of foreign firms. IBM, which was in Japan in the 1930s and was "grandfathered" in, was pressured by the Japanese government to license state-of-the-art technology. The 1972 research alliance between Amdahl, Fujitsu, and Hitachi contributed to the reverse engineering of the IBM mainframe technology in Japan in the mid-1970s. Ultimately, the combination of protection and access to state-of-the-art technology proved successful in developing a globally competitive Japanese computer industry by the 1980s, with firms such as Toshiba and Sharp (Bresnahan 1999; Chandler 1997; Flamm 1988).

The U.S. computer industry began showing a trade deficit for the first time in 1991, the magnitude of which increased in subsequent years. However, the deficit to some extent reflects the shifting nature of the forefront of this industry. Several sectors remained domestic strongholds, while others saw a growth in imports. Peripherals stand out as a segment that became increasingly dominated by imports, and such growth reflects both shifts in production, particularly in the sourcing of components to lower-cost, off-shore suppliers, and in the sources of value added to software, as hardware increasingly became a commodity.

Competitive Response

What has been the role of foreign competition in innovation in the U.S. computer industry? From the 1970s onward, progressive editions of the *U.S. Industrial Outlook* suggest that the growing pressure of foreign competition, generally Japanese, was a force behind the industry's rapid technological change.

While the foreign market was an important source of demand for U.S. manufacturers and exports continued to grow considerably during the 1970s, imports—particularly peripherals—were growing faster (U.S. DOC 1976; 1983; 1995). In Europe and Japan, however, there were concerted efforts to erode this U.S. dominance (U.S. DOC 1973; 1976; 1977). The growth of the minicomputer market created greater opportunities for targeted European efforts to assume greater market share. Meanwhile, Japanese efforts were targeted at mainframes and peripherals (U.S. DOC 1977). Many European competitors teamed up with IBM competitors using proprietary systems, while the Japanese efforts focused on making machines equivalent in power to and compatible with IBM equipment.

During the early 1970s, the nature of foreign competition was largely a matter of seeking to erode U.S. market share abroad, especially in computers and particularly in the European market. American computer manufacturers had near complete control of the U.S. market for computers; U.S. companies had the greatest global presence in all the segments of the computer market: in mainframes, in minicomputers, and in the embryonic microcomputer sector. Thus the concern on the part of the U.S. industry was primarily over the loss of markets abroad as its foreign competitors became more technologically advanced.

By the late 1970s there was increasing competition for the U.S. domestic market as well, particularly from Japanese firms (U.S. DOC 1977). Peripherals were controlled more by the lower cost of production than by the rate of technological advance, enabling Japanese companies to produce large scale, IBM compatible peripherals from the 1970s onward. In printers, the Japanese market share rose from zero in 1979 to 70 percent just three years later; U.S. manufacturers responded in part by exiting the low-end of the market (U.S. DOC 1983). Even as the personal computer market was beginning, the prospect of significant foreign competition was feared (U.S. DOC 1980; 1982). Japanese competition in the personal computer market initially aimed at the lower-priced segment, but the entrance of Japanese companies into the market challenged U.S. supremacy in high-end supercomputers in the early 1980s (U.S. DOC 1982; 1984). By the start of the 1990s, there was concern about the perceived erosion of competitive advantage previously enjoyed by the U.S. computer industry and the implications for future competitiveness (Dertouzos et al. 1989; McKinsey Global Institute 1993; NRC 1990).

A decade later, however, many of these concerns were seen as contributing to the relative strengths of the U.S. industry rather than leading to its feared decline (Mowery 1999), regardless of whether technological competition arose from new segments within the industry or from foreign firms. Moreover, in the case of foreign competition, the competitive edge that the U.S. firms developed through their technological competence meant that foreign firms generally were trying to catch up. Nonetheless, the competition was sophisticated, which might have forced U.S. firms to continue to innovate (Dertouzos et al. 1989). This ability to respond with new products in new segments has been aided by aspects of the U.S. national innovation system that encourage fluidity, including the relative ease of entry and exit of new firms, the success of entrepreneurs in leaving established firms and starting new segments, and the supply of trained scientists and engineers (Bresnahan 1999).

## Conclusion

Taken together, the econometric results and the consideration of the industry experience suggest that sophisticated foreign competition has had both stimulating effects and deleterious impacts on domestic innovation in the U.S. computer equipment industry. Integrating the quantitative results and the lessons from the industry experience suggests the following inferences:

- The computer equipment industry has been characterized by rapid technological advance, which in some cases erodes the existing market and in other cases stimulates new segments. Investment in R&D in this industry has been an important driver of innovation. This influence is highlighted by the magnitude of the impact of domestic R&D stock on TFP performance that we see in the empirical results.

- The computer equipment industry has been international in scope. Important advances occurred globally, but the development of a commercial computer market developed earliest in the United States, enabling the domestic industry to be at the forefront. As the industry developed more completely in Europe and Japan, the U.S. industry faced greater challenges. The negative effect of foreign R&D on domestic TFP in the empirical results implies that as the computer equipment industry abroad became increasingly sophisticated, foreign competition presented a greater challenge to the U.S. industry.

- The U.S. industry has been relatively successful at the high end of the computer equipment market. U.S. exports benefit from a technically sophisticated foreign market that demands more sophisticated equipment. Increasing foreign R&D reflects a greater stock of knowledge that the U.S. industry might access through exporting. On the other hand, U.S. exports in more commoditized segments of the computer equipment industry have not been as strong. These relationships are underscored by the empirical results, in which we saw a positive effect on domestic TFP as the level of foreign R&D increases.

- The U.S. computer equipment industry traditionally dominated the most technically advanced segments of the industry. Greater competition for the high-end of the computer equipment market, such as in the supercomputer field, has been a challenge for the domestic industry. In part, this might be because it is harder to achieve productivity gains in the less commoditized segments of the computer equipment market. There might also be an element of complacency, as U.S. manufacturers may not expect competition in the most advanced areas and thus fail to respond with greater innovation. The empirical results, in which we see domestic TFP decreasing as imports reflect higher levels of foreign R&D, reflect this.

The dynamic nature of the computer equipment industry has meant that market success depends largely on continued innovation. With significant international trade and extensive commitment to innovation in the computer equipment industry in the United States, Europe, and Japan, domestic innovation cannot be fully understood without taking into account its connections to foreign innovation, especially via international trade. In the long term, the health of an industry depends not only on the flow of ideas but also the ability to utilize and appropriate knowledge from a variety of sources. The positive and negative consequences of international trade can play a critical role in domestic innovation. In the final analysis, an effective innovation policy will continue to depend on investment in the requisite domestic science and technology infrastructure, the sine qua non for meeting the challenge of competition.

One interpretation of innovation on the global stage is a paradigm of foreign R&D increasing both the pool of knowledge and the sophistication of competition. There are elements of collaboration, in the sense that if both domestic and foreign R&D increase, there is the possibility of mutual learning—at least to the extent that both are at or near the technological frontier. But there are also elements of competition. Domestic innovation in a global setting will continue to involve a balance between these two forces.

APPENDIX

Table 8.A1. Summary statistics

| Variable | Description | Observations | Mean | Std. Dev. | Min | Max |
|---|---|---|---|---|---|---|
| *lntfp3* | ln(TFP) | 24 | -0.2752 | 0.4351 | -0.9745 | 0.5135 |
| *lnrdst* | ln(domestic R&D stock) | 24 | 10.1094 | 0.7900 | 8.8452 | 11.0937 |
| *lnfrdst* | ln(foreign R&D stock) | 24 | 9.2906 | 1.0286 | 7.7016 | 10.7483 |
| expsal_1 | (exports/sales)$_{t-1}$ | 24 | 0.1588 | 0.0686 | 0.0811 | 0.3042 |
| impsal_1 | (imports/sales)$_{t-1}$ | 24 | 0.1201 | 0.1033 | 0.0344 | 0.3687 |
| *(expsal_1)(lnfrdst)* | (exports/sales)$_{t-1}$* ln(foreign R&D stock) | 24 | 1.5077 | 0.7515 | 0.6505 | 3.2700 |
| *(impsal_1)(lnfrdst)* | (imports/sales)$_{t-1}$* ln(foreign R&D stock) | 24 | 1.2025 | 1.1364 | 0.2652 | 3.9427 |

*Note:* Author's calculations are based on data from a variety of sources. The variable *lntfp3* is calculated from the NBER Manufacturing Productivity database (Bartelsman and Gray 1996). The author aggregated the NBER data to the three-digit level. The variable *lnrdst* is calculated from NSF R&D in Industry series. It is calculated using the perpetual inventory method as described in Griliches (1998). The variable *lnfrdst* is calculated from OECD. Variables *expsal_1* and *impsal_1* are created from 4 digit import (Feenstra 1996) and export (Feenstra 1997) data set. The author translated this data from 1972 SIC codes to 1987 SIC codes at the four-digit level and then aggregated to the three-digit level.

Table 8.A2. Regression results

| Dependent variable: $lntfp3$ | Domestic R&D stock | Domestic and foreign R&D stock | R&D stock and trade terms | R&D stock, trade terms, and interactive terms |
|---|---|---|---|---|
| | (1) | (2) | (3) | (4) |
| lnrdst | 0.0957 | 0.4745 | 1.031 | 1.0691 |
| | (0.69) | (2.48)* | (4.59)* | (1.99)** |
| lnfrdst | | -0.7532 | -1.5486 | -1.6580 |
| | | (-2.60)* | (-4.12)* | (-2.85)* |
| expsal_1 | | | 0.7136 | -7.4761 |
| | | | (2.96)* | (-2.61)* |
| impsal_1 | | | 0.0319 | 16.0341 |
| | | | (0.10) | (2.04)** |
| (expsal_1)(lnfrdst) | | | | 1.0495 |
| | | | | (2.84)* |
| (impsal_1)(lnfrdst) | | | | -1.6249 |
| | | | | (-2.06)** |
| time | 0.0505 | 0.1181 | 0.1684 | 0.1667 |
| | (3.08)* | (3.86)* | (4.315)* | (4.87)* |
| no. obs. | 24 | 24 | 24 | 24 |
| R2 | 0.9838 | 0.9876 | 0.9910 | 0.9934 |

*Note:* The dependent variable is the log of total factor productivity. Variables are defined in table A1. t-statistics in parentheses. Equations are estimated using ordinary least squares (OLS) with robust standard errors.

* significant at 5 percent level

** significant at 10 percent level. The equation estimated is:

$$lntfp3 = \beta_0 + \beta_1 lnrdst + \beta_2 lnfrdst + \beta_3\ expsal\_1 + \beta_4\ impsal\_1 + \beta_5 (expsal\_1)(lnfrdst) + \beta_6 (impsal\_1)(\ lnfrdst) + \beta_7\ time + \epsilon$$

## NOTES

1. This paper is part of a larger study of four high-technology industries selected to encompass a broad spectrum of industry characteristics and outcomes with regard to domestic innovation in the global context. In addition to the computer equipment industry (SIC 357) discussed in this paper, the other three industries are household audio and video equipment (SIC 366), communications equipment (SIC 365), and scientific instruments (SIC 381+382). See Smith (2004).

2. SIC 357 comprises: computers (3571), storage devices (3572), terminals (3573), peripherals (3574), calculating machines (3578), and office machinery

not elsewhere classified (3579). This classification does not include software, which is included in business services under SIC 737, computer programming, and data processing.

3. "Knowledge" encompasses many aspects including managerial and organizational skills in addition to scientific and technical information.

4. See the extensive discussion of knowledge spillovers in Griliches (1979; 1992) and Jaffe (1996).

5. This will be more likely if the exporting is via a foreign subsidiary—especially if the subsidiary does R&D—than if the export goes through foreign distributors and agents without a local institutional presence.

6. In many cases, both the positive and negative effects on domestic innovation might occur. Thus, the net result might reflect a balancing out of these opposite mechanisms. This consideration also applies to hypotheses 3 and 4.

7. In the Grossman and Helpman model, innovation occurs through investment in R&D, which results in an increase in either the number or the quality of available intermediate goods for the production of final products. International trade allows access to the cumulative stock of knowledge beyond domestic borders; thus, the cumulative stock of knowledge increases with the cumulative volume of international trade. See Grossman and Helpman (1991a; 1991b; 1994). For a complete presentation of this model and the econometric analysis see Smith (2004).

8. The conceptual framework is rooted in a well-developed theoretical model relating R&D to productivity through a production function, $Q = AK^{\alpha}L^{\beta}$, where $Q$ = output, $K$ = capital, $L$ = labor, and $\alpha$ and $\beta$ are the shares of capital and labor, respectively. Total factor productivity is then $A = Q / K^{\alpha}L^{\beta}$. While TFP is a good measure of particular types of innovation, such as cost-reducing innovation, it is less good at measuring other types of innovation, such as quality improvements and innovations leading to novel new goods. See Griliches (1979) for a detailed discussion of inherent problems in measuring output in R&D intensive industries and related issues. See Griliches (1998).

9. See Smith (2004) for details.

10. The results discussed here summarize the more detailed discussion and analysis presented in Smith (2004).

11. Engelbrecht (1997) found a smaller, negative elasticity of -0.10 of TFP with respect to foreign R&D stock in the United States at the level of the whole economy. Other typical values for elasticity of domestic TFP with respect to foreign R&D stock range from 0.04 in the United States at the level of the whole economy (Coe and Helpman, 1995) to 0.5 for an average of pooled industries in eight European countries and Canada, but not the United States (Keller 2002).

12. The net effect of foreign R&D stock remains negative at both the maximum and minimum values of exports and imports to sales.

13. Using ordinary least squares to estimate the regression equation poses several econometric problems that might lead to biased and inconsistent estimates due to correlation between the independent variables and the error term. These issues are discussed at length in Smith (2004). Nevertheless, ordinary least squares estimation of this equation yields important insights into the relationship between innovation, foreign R&D, and international trade in the computer equipment industry.

## REFERENCES

Amato, L. H., and C. H. Amato. 2001. The effects of global competition on total factor productivity in U.S. manufacturing. *Review of Industrial Organization* 19: 407–23.

Bartelsman, E. J., and W. Gray. 1996. *The NBER Manufacturing Productivity Database*. NBER Technical Working Paper T0205. Cambridge, MA: National Bureau of Economic Research.

Barzel, Y. 1968. Optimal timing of innovations. *Review of Economics and Statistics* 50 (3): 348–55.

Bernard, A. B., and J. B. Jensen. 1999. Exceptional exporter performance: Cause, effect or both? *Journal of International Economics* 47: 1–25.

Bernstein, J. I., and P. Mohnen. 1998. International R&D spillovers between U.S. and Japanese R&D intensive sectors. *Journal of International Economics* 44 (2): 315–38.

Branscomb, L. M., and R. Florida. 1998. Challenges to technology policy in a changing world economy. In *Investing in Innovation: Creating a Research and Innovation Policy That Works*, ed. L. M. Branscomb and J. H. Keller. Cambridge, MA: MIT Press.

Bresnahan, T. 1999. Computing. In *U.S. Industry in 2000: Studies in Competitive Performance*, ed. D. C. Mowery. Washington, DC: National Academy Press.

Bresnahan, T., and F. Malerba, eds. 1999. *Industrial Dynamics and the Evolution of Firms' and Nations' Competitive Capabilities in the World Computer Industry*. Cambridge: Cambridge Univ. Press.

Chandler, A. D., Jr. 1997. The computer industry: The first half-century. In *Competing in the Age of Digital Convergence*, ed. D. B. Yoffie. Boston: Harvard Business School Press.

Coe, D. T., and E. Helpman. 1995. International R&D spillovers. *European Economic Review* 39: 859–87.

Cohen, W. M., and D. A. Levinthal. 1989. Innovation and learning: The two faces of R&D. *Economic Journal* 99 (397): 569–96.

Connolly, J. 1967. History of computing in Europe. IBM internal document.

Crothers , B. 1999. Study: PCs in half of U.S. homes. *CNET News.com* [online]. 9 February 1999 [cited 17 September 2004]. Available at http://news.com.com/2100-1040-221450.html.

Dertouzos, M. L., R. K. Lester, and R. M. Solow. 1989. *Made in America: Regaining the Productive Edge.* Cambridge, MA: MIT Press.

Feenstra , R. C. 1996. *U.S. Imports, 1972–1994: Data and Concordances.* NBER Working Paper W5515. Cambridge, MA: National Bureau of Economic Research.

———. 1997. *U.S. Exports, 1972–1994, with State Exports and Other U.S. Data.* NBER Working Paper W5990. Cambridge, MA: National Bureau of Economic Research.

Feenstra , R. C., J. Romalis, and P. K. Schott. 2002. *U.S. Imports, Exports and Tariff Data, 1989–2001.* NBER Working Paper W9387. Cambridge, MA: National Bureau of Economic Research.

Flamm, K. 1988. *Creating the Computer: Government, Industry, and High Technology.* Washington, DC: The Brookings Institution.

Geroski, P. A. 1990. Innovation, technological opportunity, and market structure. *Oxford Economic Papers* 42: 586–602.

Griliches, Z. 1979. Issues in assessing the contribution of research and development to productivity growth. *Bell Journal of Economics* 10 (1): 92–116.

———. 1992. The search for R&D spillovers. *Scandinavian Journal of Economics* 94 (supp.): 29–47.

———. 1994. Productivity, R&D, and the data constraint. *American Economic Review* 84 (1): 1–23.

———. 1998. *R&D and Productivity: The Econometric Evidence.* Chicago: Univ. of Chicago Press.

Grossman, G. M., and E. Helpman. 1991a. *Innovation and Growth in the Global Economy.* Cambridge, MA: MIT Press.

———. 1991b. Trade, knowledge spillovers, and growth. *European Economic Review* 35 (2/3): 517–26.

———. 1994. Endogenous innovation in the theory of growth. *Journal of Economic Perspectives* 8.(1): 23–44.

Jaffe, A. 1996. *Economic Analysis of Research Spillovers: Implications for the Advanced Technology Program.* Washington, DC: National Institute of Standards and Technology.

Jaffe, A. B., and M. Trajtenberg. 1998. *International Knowledge Flows: Evidence from Patent Citations.* NBER Working Paper W6507. Cambridge, MA: National Bureau of Economic Research.

Keller, W. 2001. *The Geography and Channels of Diffusion at the World's Technology Frontier.* NBER Working Paper 8150. Cambridge, MA: National Bureau of Economic Research.

———. 2002. Geographic localization of international technology diffusion. *American Economic Review* 92.(1): 120–42.

Lawrence, R., and D. Weinstein. 1999. *The Role of Trade in East Asian Productivity Growth: The Case of Japan.* NBER Working Paper 7264. Cambridge, MA: National Bureau of Economic Research.

Lawrence , R. Z. 2000. Does a kick in the pants get you going or does it just hurt? The impact of international competition on technological change in U.S. manufacturing. In *The Impact of International Trade on Wages*, ed. R. Feenstra. Chicago: Univ. of Chicago Press.

M.I.T. Commission on Industrial Productivity. 1999. *The U.S. Semiconductor, Computer, and Copier Industries*. Cambridge, MA.

MacDonald, J. M. 1994. Does import competition force efficient production? *Review of Economics and Statistics* 76 (4): 721–27.

McKinsey Global Institute. 1993. *Manufacturing Productivity*. Washington, DC.

Mowery, D. C., ed. 1999. *U.S. Industry in 2000: Studies in Competitive Performance*. Washington, DC: National Academy Press.

Nadiri, M. I., and S. Kim. 1996. *International R&D Spillovers, Trade and Productivity in Major OECD Countries*. NBER Working Paper 5801. Cambridge, MA: National Bureau of Economic Research.

National Research Council, Computer Science and Technology Board, Commission on Physical Sciences, Mathematics, and Resources. 1990. *Keeping the U. S. Computer Industry Competitive: Defining the Agenda*. Washington, DC: National Academy Press.

National Science Board. 2000. *Science and Engineering Indicators 2000*. Vol. NSB-00–1. Arlington, VA: National Science Foundation.

National Science Foundation, Division of Science Resources Studies. Various years. *Research and Development in Industry*. Arlington, VA: National Science Foundation.

Organisation for Economic Co-operation and Development (OECD). 1996. *Research and Development Expenditure in Industry (ANBERD)*. Paris: OECD.

Olivetti Group. 2002. Olivetti Tecnost—News [online]. 2002 [cited 17 September 2004]. Available at http://www.olivetti.com/group/.

Scherer, F. M. 1984. *Innovation and Growth: Schumpeterian Perspectives*. Cambridge, MA: MIT Press.

———. 1992. Schumpeter and plausible capitalism. *Journal of Economic Literature*, September:1416–33.

Scherer, F. M., and K. Huh. 1992. R&D reactions to high-technology import competition. *Review of Economics and Statistics* 74 (2): 202–12.

Smith, S. W. 2004. Innovation and Globalization in Four High-Tech Industries in the U.S.: One Size Does Not Fit All. PhD diss., Harvard University.

U.S. Department of Commerce. 1973. *U.S. Industrial Outlook 1973*. Washington, DC: U.S. Government Printing Office.

———. 1976. *U.S. Industrial Outlook 1976; with Projections to 1985*. Washington, DC: U.S. Government Printing Office.

———. 1977. *U.S. Industrial Outlook*. Washington, DC: U.S. Government Printing Office.

———. 1980. *U.S. Industrial Outlook*. Washington, DC: U.S. Government Printing Office.

———. 1982. *U.S. Industrial Outlook.* Washington, DC: U.S. Government Printing Office.

———. 1983. *U.S. Industrial Outlook.* Washington, DC: U.S. Government Printing Office.

———. 1984. *U.S. Industrial Outlook.* Washington, DC: U.S. Government Printing Office.

———. 1995. *U.S. Global Trade Outlook, 1995–2000.* Washington, DC: U.S. Government Printing Office.

Warnke, J. 1996. Computer manufacturing: Change and competition. *Monthly Labor Review* 119 (8): 18–29.

# PART 3

# SHAPING TECHNOLOGY

Science and technology policy encompasses a variety of different policy instruments and domains of action. The social shaping of technology occurs not just directly through the commitment of funds and other material and human resources, but also through policies whose primary or even secondary functions are not specifically related to science and technology. It occurs not only in Congress and agencies in Washington, DC, but also in decentralized and informal settings, and by the accretion of decisions and nondecisions as well as by masterstrokes of design and implementation.

The chapters in this section exemplify this variety of instruments and action, even if not in technology. While three of the authors deal with new telecommunications technologies, and the fourth with older transportation technologies, it is worth noting that two of the telecommunications chapters build their arguments through historical analyses, which suggest that the shaping of even new technology is not a discrete event or choice. Rather, it is an ongoing and durable negotiation among stakeholders who have different visions for both the technologies and for the society in which they are embedded.

Patrick Feng, a science and technology studies scholar, examines technical standards, those often-obscure protocols, rules, and codes that specify how a given group of technologies should operate (or interoperate) and thus play a key role in shaping technology. From manufacturing to agriculture to computing, standards permeate almost every facet of modern life. Their importance to trade and commerce mean they are implicated in numerous political-economic processes, including globalization. In short, standards constitute a fundamental, though often invisible, framework for the contemporary, technologically mediated world. Feng's chapter looks specifically at the regrettably thin way one proposed World Wide Web standard for privacy incorporates the preferences of users. Although Feng bleakly concludes that, at least in the standards-setting arena, the goal of participatory design is unattainable because of barriers of knowledge, cost, geography, and interests, he does find that there are "user representations" applied by traditional participants in standard-setting deliberations. Because standards are such critical aspects of our technological future, and because of the democratic ethic that insists that persons who are affected by a decision have some input into its making, Feng concludes that a variety of participatory reforms are necessary to aid in the appropriate shaping of technical standards.

One challenge of decentralized and private decisionmaking about technology, as science and technology studies scholar Jason Patton puts it, is achieving the socially optimal rather than the technically possible and profitable. Few policy domains confront this challenge as starkly as transportation. Getting people out of automobiles is the holy grail of solving traffic congestion, air pollution, and social inequities in cities in the United States. While buses are often the most cost-effective form of public transit, they are typically the mode of last resort for people with transportation choices. Yet this stigma has less to do with inherent limitations of bus service than with a lack of the application of technological innovation and appropriate infrastructural design. The targeted development of new technologies and the incremental redesign of city streets could reduce the gap in practicality and status between private automobiles and buses. Patton describes how creating explicit incentives for bus travel but only implicit disincentives for driving—a scenario similar to the change by which automobiles became the most practical form of transportation for individuals—may render such an approach politically palatable. Steering design and innovation for marginal transportation modes reshapes technologies and infrastructures to facilitate forms of life congruent with widely held environmental values and the diverse needs and of heterogeneous urban populations.

Standards and decentralized design of infrastructure are both thematic for communication scholar Christian Sandvig, whose chapter reveals a debate in the netherworld between computer science and public policy that is shifting the state's role in shaping the benefits of the Internet for innovation. Some argue that the Internet's gift is found in a design feature called the "end-to-end principle." This principle promotes "stupid" networks in which the center lacks intelligence and performs only a few functions, while nodes at the edge of the network—the ends—build complex applications. The Internet, with a smart PC and dumb router, exemplifies end-to-end. Because of the intelligence at the ends, experiments like the World Wide Web can be deployed there without permission. However, commercial interests are currently deploying intelligence inside the network's core, speeding some traffic over others, blocking traffic, eavesdropping, and disguising some nodes as others. Such activities are so worrying for the end-to-end design that its proponents have asked the U.S. government to intervene to preserve the Internet's "natural" form. But Sandvig shows that the end-to-end ideal is just that, and his chapter outlines an alternative, based on historical observations of organized communication systems, which from their earliest manifestations, suggest that networks always develop complexity at their core as users ask more of them. The key to shaping innovation on the Internet, Sandvig concludes, is not the degree of logic within intermediary nodes, but which nodes are trusted.

The interrelation of networks and innovation is also central to the chapter by Carolyn Gideon, a public policy scholar interested in how regulatory policies can shape the development of technology in networked communications industries. She begins with a historical inquiry, noting that recent accounts of the development of telephony have challenged the notion that it was a natural monopoly rather than the result of policy decisions to sanction and protect it. Telephony is but one example of how policy may determine industry structure, and consequently the way technology is developed and deployed. Yet the technological consequences of such regulatory policies are not often considered. Using the telecommunications and Internet industries as examples, Gideon's chapter explores how regulatory policies—especially those promoting a particular industry structure—shape the development of technology. She develops "the network pricing game," a stylized model of competition designed for the special characteristics of a communications network industry. As an analytic tool for determining what market characteristics influence a network firm's exit decision, this model helps show what incentives influence a firm's technology investments. Based on historical analysis and the network pricing game, Gideon examines how the rules for leasing unbundled network elements are likely to shape the development of technology.

Together, these authors probe some of the most notoriously difficult aspects of modern technology, that is, their infrastructural and networked character. They offer some concrete hope that social values like privacy, participation, equity, environment, and even competition and innovation—although enmeshed in these networks—do not need to be entirely subordinate to them.

9

---

# Shaping Technical Standards

## *Where Are the Users?*

PATRICK FENG

---

## Introduction

On March 21, 2001, the *Wall Street Journal* reported that Microsoft was throwing its weight behind a new standard intended to address consumer concerns about online privacy (Simpson 2001). Dubbed the "Platform for Privacy Preferences Project" (P3P), this proposed standard would create a common syntax with which websites could encode their privacy policies. These machine-readable policies could then be automatically "fetched" by a web browser and passed on to users, allowing them to find out what personal information a site collects without having to read through a long, legalistic privacy policy. Supporters of P3P say that this technology will help "empower" Internet users by making it easier to find out what a website's privacy policy is and decide for themselves which sites they feel comfortable visiting. Critics say that P3P is just an industry initiative to make companies look good while delaying the passage of "real" privacy legislation in the United States. Regardless of which side of the argument one is on, few dispute the fact that P3P has the potential to dramatically change how websites and web users deal with privacy issues.

Despite its potential importance, most computer users have never heard of P3P, are unaware of who is designing this standard, have no idea how to participate in its design process, and would have neither the time nor the inclination to participate even if they did know how to get involved. In other words, P3P is pretty much the antithesis of user involvement in technological design: little public awareness, few opportunities for users to participate, and little desire to do so, even if when given the chance. Even though this proposed standard could have a major impact on web users' privacy, few will ever know about its existence until after it has been completed. This situation raises the question: Do users matter at all in shaping technical standards?

This chapter examines the role of users in the shaping of technical standards. I begin with a brief review of technical standards and the standards-setting world. I then look at the some of the challenges standards pose for user involvement, pointing to technical standards-setting as but one arena where direct user involvement is difficult—if not impossible—to achieve. Rather than focus on participation, I suggest that what we really need to look at is user representations. Although users rarely participate directly in standards-setting, they are nevertheless present at the design table in the form of opinion polls, usability studies, personal anecdotes, designers' biases, and so on. I conclude with a number of policy suggestions on how the idea of user representations might be mobilized to make standards-setting processes more transparent and accountable to the public.

## Standards and Standard-Setting

### Defining Standards

Technical standards are those protocols, rules, and codes that specify how a given set of technologies should operate and interoperate. For example, Hypertext Markup Language (HTML) is a standardized language that specifies how web pages should be coded and read, regardless of what computer you are using. Standards are common not only in computing but also in almost every facet of modern life. Standards serve to ensure product quality, create uniformity, assure compatibility between technologies, produce "objectivity" in measurement, normalize operating procedures, and more (Egyedi 1996; Porter 1995).

Traditionally, two forms of standard-setting have been predominant: *de jure* and *de facto*. De jure standards are those specified by law (e.g., food safety standards set by national governments); de facto standards arise from market monopolies (e.g., Microsoft's ubiquitous Windows operating system, which controls the bulk of the PC market). In the early part of the twentieth century, formal standards organizations began to take shape, creating a third way of standardization: cooperative, international standards negotiated among government representatives of various nation-states. An example of such an organization is the International Organization for Standardization (ISO), which produces standards in areas ranging from units of measurement to film speeds to environmental management practices (Loya and Boli 1999). More recently, regional associations and industry consortia have proliferated, creating a fourth mode of standardization that fits neither the de jure, de facto, nor quasi-judicial models mentioned above (Heywood et al. 1997; Updegrove 1995). In this chapter, I focus on the quasi-judicial methods and the fourth mode, which Egyedi (1996) refers to as "grey standardization."

Like other technologies, standards arise through a complex design process in which social and technical considerations meld together. Standards have technical requirements but are also influenced by social factors. While international standards organizations often describe their work as being "purely technical" in nature, participants in the design process quickly discover that, where technical battles are concerned, political and economic considerations are never far behind (Egyedi 1996; Schmidt and Werle 1998).

## How Standards Are Made

The number of international standards organizations now in existence is surprisingly large. At the end of the 1980s there were at least eighty international and regional standards organizations in the telecommunications sector alone (Macpherson 1990). The number has since increased. Even for a fairly new technology such as the World Wide Web, numerous standards organizations are involved in laying the groundwork for web technologies. Among the more important of these are: the Internet

Corporation for Assigned Names and Numbers (ICANN), responsible for maintaining the domain naming system that ensures each website has a unique name and IP address; the Internet Engineering Task Force (IETF), responsible for setting basic data transport protocols such as TCP/IP and HTTP; and the World Wide Web Consortium (W3C), responsible for document structure and style protocols such as HTML and CSS. This proverbial "alphabet soup" of organizations promulgates standards that are meant to ensure that the Web "works" the same way all around the world, regardless of local politics, economics, laws, or culture. In the words of Loya and Boli (1999, 176), today's global network of standards development organizations is "almost incomprehensibly complex."

Although these organizations are distinct, there are many similarities among the procedures of most Standards Development Organizations (SDOs). In general, standards are made by small groups of technical experts. While international organizations such as ISO have thousands of people working for them (either directly or indirectly), the number of people working on any given standards initiative is usually small (Loya and Boli 1999). For example, the working group responsible for designing the P3P standard has about thirty members, of which about half are regularly active in group teleconferences and meetings.[1]

The basic model for promulgating a standard is as follows: (1) a member of an organization (a country, a company/organization, or an individual, depending on the SDO) suggests an item to be standardized; (2) if there is sufficient interest, a working group is convened and charged with drafting a standard; (3) periodically, drafts of this group's work are made available to the rest of the SDO's membership for comments and feedback; (4) if and when the working group is satisfied with its work, it releases a final draft to the SDO's membership; (5) the draft standard is then ratified by the entire membership, sent back for further work, or rejected outright.[2] The bulk of the work in drafting a standard is done by a relatively small group of people, regardless of the size of the SDO.[3]

AN EXAMPLE OF STANDARDIZATION: THE PLATFORM
FOR PRIVACY PREFERENCES PROJECT

The Platform for Privacy Preferences Project, introduced at the beginning of this chapter, provides an illustrative example of many of the challenges facing user participation in standards development.

Table 9.1. Participants in P3P working groups, broken down by affiliation

| Number of members from... | P3P specification working group | | P3P policy & outreach working group | |
| --- | --- | --- | --- | --- |
| Academia | 4 | (10%) | 3 | (7%) |
| Government/government agency | 2 | (5%) | 4 | (10%) |
| Industry— | 28 | (68%) | 27 | (64%) |
|     Advertising companies | 9 | | 5 | |
|     Financial services | 3 | | 3 | |
|     Hardware companies | 2 | | 5 | |
|     Software companies | 6 | | 4 | |
|     Telecom companies | 2 | | 2 | |
|     Other | 6 | | 8 | |
| Nongovernmental/nonprofit | 3 | (7%) | 6 | (14%) |
| W3C staff | 4 | (10%) | 2 | (5%) |
| Total | 41 | (100%) | 42 | (100%) |

*Source:* W3C website and participant observation. The numbers above are based on the composition of P3P working groups as they were in January 2002. The category other includes organizations such as industry lobby groups and niche companies (e.g., start-ups focusing solely on online privacy services).

The story of P3P could be told in any number of ways. One version of the story might be as follows:

In 1995, a group of technology companies got together with a number of privacy advocates with the idea of creating a standardized mechanism for data exchange. Their plan was to create a standardized method for describing and controlling the flow of data over the Web; they believed that this would give users more control over their personal information as they surfed the Web, which in turn would build trust and help protect consumer privacy. Soon a new "'privacy activity'" began at the World Wide Web Consortium (W3C), an industry-led nonprofit organization responsible for developing web standards. After many twists and turns, a specification began to take shape; by November 1999, a "last call" draft had been issued. The draft standard, which was now known as P3P, began gaining support from key technology companies throughout 2000 and 2001. Finally, on April 16, 2002—nearly five years after the start of the project—P3P was approved by the W3C membership and became an official W3C recommendation.

Where are the users in this story? Largely invisible. While users' needs (e.g., concern over personal privacy) are presumably the driving force behind initiatives such as P3P, users themselves are conspicuously absent.

Instead, other actors—technology companies, privacy advocates, and industry consortia—are presumed to speak on behalf of users (see table 9.1). How and why this happens is the subject of the next two sections.

## Challenges to User Involvement

There are several characteristics of technical standards setting that make it difficult for users to directly participate in their design. These include the highly technical nature of discussions; the high cost of attending working group meetings; the large number of users potentially affected by standards decisions; and the fact that standards are less like products than infrastructure.

### Technical Expertise

Not surprisingly, most users lack the level of technical expertise that regular participants in standards committees have. Nor should they be expected to be experts. However, most SDOs have a de facto policy of requiring participants to have a high degree of technical expertise. For example, in the case of the IETF, that particular lingo centers around networks, IP addresses, routers, packets, and other technical details related to the Internet. And while IETF prides itself on its openness—in theory, anyone who is interested in a particular working group can just join in—this prerequisite for technical expertise presents a very real barrier to participation by lay members of the public. As one staff member explained, the IETF is "an open organization . . . open to anyone to play, but you play by joining a working group . . . [and] in order do that you have to speak the lingo" (personal observation, November 7, 1999).

### High Cost of Attending Meetings

Not only are participants expected to have a high degree of technical knowledge, they are also expected to bear the costs of attending committee meetings. The travel costs alone can easily run into thousands of dollars per year, depending on which working groups or committees a person is on. As an example, consider the P3P standard mentioned at the beginning of this chapter. In 2000, the P3P working group held several face-to-face meetings in Toronto, New York, and San Francisco. In addition, many members of the P3P working group attended the general

W3C meetings held in Amsterdam and Boston. The travel costs for attending just these five meetings would be beyond the budget of most laypersons. Moreover, there is the actual time taken up by such meetings. All in all, it is an expensive proposition—one that is too costly for most people.[4]

Who, then, attends these meetings? Mostly technical professionals from industry and government whose job it is to stay on top of the latest developments in their field. They are able to attend because their employers pay for their expenses (as well as their salaries). Technically, these people "volunteer" their time to participate in international standards work; in practical terms, of course, they are only volunteers insofar as their employers allow them to work on standards-setting, rather than other activities.[5]

In addition to meeting costs, participants must also stay abreast of what is going on between meetings. For example, if you had been a member of W3C's P3P working group, you would have spent at least two hours per week on weekly teleconference calls and several hours each week reading and replying to e-mails on the P3P working group mailing list. There were 1,545 messages sent to P3P specification working group mailing list in 2000—an average of 6.2 messages per work day, a nontrivial time commitment for working group members.[6] As one can see, the costs of participating in a technical working group are high, both in terms of time and money.

## Many, Heterogeneous, and Globally Dispersed Users

When the participatory design movement started, the focus was on specific projects with clearly definable sets of designers and users. For example, the UTOPIA (Usable Technology for Older People: Inclusive and Appropriate) project involved designers building a computer system for a specific group of people.[7] In this case, it was relatively easy to identify who the users would be, and it made sense to involve them in the design of this computer system. (After all, the intended users—mostly secretaries—had a deeper, more intimate knowledge of the work processes to be modeled than did the computer systems designers.) Unlike the UTOPIA project, however, standards usually do not have easily delimited groups of users. While decisions about technical standards do ultimately affect users, it is often difficult to point to a specific group of people as *the* key stakeholders with respect to a given standard. The term "users," in this case, represents a large and heterogeneous group of people. For example, potential users of P3P might include an American consumer shopping online at a major U.S. retailer, a group of children in Ireland browsing an educational website, a person in India trying to access his or her personal health information, and so on. Few commonalities exist among these different users aside from the fact that they are all using the Web as a medium for accessing information.

The challenge of size and diversity is especially true when it comes to global standards. Increasingly, standards are being set at the supranational and international levels. Consider, for example, the Internet, where standards need to be implemented uniformly all over the world if applications such as e-mail and web browsing are to work the same everywhere. Or consider food safety standards, which are increasingly important (and contentious) because of the global nature of today's food production and distribution systems. Globalization has not only raised the profile of international standards organizations, it has also expanded the definition of "users" to encompass a potentially huge group of people, far beyond what direct participation can handle.

## Standards as Infrastructure

Unlike stand-alone technologies, standards are rarely marketed as final products in themselves. Rather, they are meant to serve as common building blocks for the future products of many companies. Thus, the users of these standards may be hard to identify, since no actual products may exist yet. Instead of thinking of standards as products, one should think of them as a kind of infrastructure: they form a foundation upon which further technological innovation can be built. Unlike regular products, infrastructure can serve multiple purposes, have multiple users, and be appropriated for multiple functions and goals. Infrastructure does not serve a single group of users but rather a heterogeneous community of users who have different needs, values, interests, and so on. Given this, identifying relevant user groups becomes even more of a challenge.

## Participatory Design and User Representations

The nature of technical standards work poses formidable challenges to involving users in design processes. In this section I first review the rationale behind calls for "participatory design" and then argue that rather than focusing on direct participation, we should shift our attention to user representations.

This is not to say that direct participation is the only way in which users can influence the design of technology. The public can influence technologies through indirect means such as buying or not buying products, participating in marketing surveys and focus groups, appealing to legislatures or regulatory agencies, and seeking redress in courts. These acts attempt to influence designers of technology through market mechanisms and through existing institutions of social control. It is fair to say, however, that for members of the lay public, direct participation in technological design is the exception, not the norm. Nevertheless, in democratic societies it is taken as an axiomatic truth that *those who will be affected by a decision have an intrinsic right to have some nontrivial degree of involvement in the making of that decision.* If this does not take the form of direct participation, then it should—at the very least—take a form that is seen as politically legitimate (e.g., representation by elected officials). As we shall see, even this second, weaker condition is seldom met in standards-setting circles.[8]

## Participatory Design

Participatory design (PD)—the notion that those who will ultimately use a technology should be directly involved in its design—sounds fine in principle, yet is seldom seen in real-life technological projects. Why not? Before we tackle this question, let us briefly review some salient points about PD.

As a social movement, PD traces its roots to Scandinavia, where it grew out of a desire to make computer systems more responsive to user needs. As Schuler and Namioka (1993, xi) explain, PD represents "a new approach towards computer systems design in which the people destined to *use* the systems play a critical role in *designing* it" (emphasis in the original). PD differs from traditional design in several fundamental ways: it views users as "experts" and their needs as being paramount; it assumes that user perceptions are at least as important to success as any "technical" considerations; and it considers the design of technical systems within the broader social context. In short, PD views the development of technical systems as a process in which users are as important as designers, and the process just as important as the final product.

PD is also aimed at making the design of technology more democratic. Supporters argue that PD results in better, more responsive
systems and, at a moral level, acknowledges users' rights to participate
in decisions that will eventually affect their lives. This idea is appealing,
especially in light of democratic theories and cultures that tell us that
people who will be affected by a decision have a right to participate in
that decision. PD is thus grounded in democratic beliefs—it challenges
the user to be an active part of the design process rather than a passive
consumer, to shape technology and not just be shaped by it. While the
PD movement is still in its infancy, it holds the potential for a radically
different way of doing design, one in which social and ethical considerations are raised alongside technical ones (Greenbaum and Kyng 1991).

PD has, however, been slow to catch on in countries outside of Scandinavia, and in areas like standards-setting, it is often hard to envision
how PD could work. Rather than give up on PD, however, I would suggest that the case of technical standards challenges us to reconsider what
it means to "participate" in a decision. If direct participation must give
way to some other forms of user involvement, what might these be?

## User Representations (Or "If not participation, then what?")

Rather than focus on participation, we should instead focus on how
users are represented during the design process. For although users are
rarely directly involved in the design of technical standards, they are
nevertheless present, in the form of opinion polls, usability studies, personal anecdotes, and designers' biases and stereotypes. In other words,
"users" are invoked by designers as constraints to be obeyed (or ignored). Understanding how these user representations are constructed
may provide an opening for injecting "real" user needs into the design
of technology.[9]

### APPEALING TO STATISTICS: "STUDIES SHOW THAT . . ."

Online privacy is a hot topic: opinion polls consistently show widespread public concern as to the apparent erosion of privacy, as more
and more people take to the Internet (see, e.g., Cranor et al. 1999). But
what do these studies mean? That was the question members of the P3P
working group had on their minds during one of their face-to-face
meetings. The chair of the working group, trying to galvanize support
for a proposed privacy feature in the standard, made reference to a

number of public opinion polls that showed high levels of concern about privacy. "Studies show that privacy is a real concern." But others in the group disputed this claim: "People may say that [in surveys], but in real life they gladly trade away their privacy for a few extra air miles."

Both sides were constructing an image of users and attempting, in Latour's (1987) terms, to "enroll" users in their cause. On one side, users were being represented through surveys and public opinion polls: whatever users said in these surveys must be true. On the other side, users were being represented through their supposed actions, rather than their words: despite what people might say about preferences for privacy, their actions are contrary and so in reality they care much less than they say they do. Depending on which side ultimately won, "users" would end up being either privacy-conscious individuals, or individuals who really did not care about the issue. Such representations, without the users present, would, of course, influence the design of P3P.

APPEALING TO THE PERSONAL: "MY MOTHER WILL NEVER USE THIS FEATURE . . ."

On another occasion (during a P3P teleconference), a list of possible features was being discussed. One of the conference call participants suddenly spoke up: "You know, my mother will never use this feature. . . . Why are we discussing this?" His remark was soon followed by other comments about family members wanting or not wanting such a feature in their computer.

In an attempt to capture what the "typical" user wants, appeals are sometimes made to the personal—through anecdotes and stories, for example.[10] This rhetoric stands in some contrast to the "studies show" approach mentioned earlier. What is surprising, though, is how successful these anecdotal appeals are. Even for a group of technologists, the personal is often powerful: statements about one's mother or grandmother were often treated as valid arguments.

APPEALING TO TECHNOLOGY: "THIS WILL BREAK THE WEB!"

At yet another face-to-face meeting, a contentious proposal to change a section of the P3P proposed standard met resistance. But this time the resistance did not come from a user but rather an inanimate object. "This [proposed change] will break the Web!" exclaimed one of the meeting attendees. By this claim she meant that the proposed change, if allowed to proceed, would ruin some fundamental function of web technology, making it impossible to browse certain websites properly and threatening the very stability of the World Wide Web. The unspoken corollary to "this will break the Web" was, of course, "users will not stand for this." Thus, with a single statement, the whole user community (as well as the whole web community) had been mobilized to resist a proposed change in the P3P specification![11]

When people, represented through either statistical data or personal anecdotes, fail to persuade, one can always try mobilizing artifacts instead. The appeal to technology—this standard is no good because it will "break the Web"—depends on an unspoken premise that technological artifacts are rigid, inflexible, and predestined to follow a path of their own. In other words, such arguments rely on a belief in technological determinism, the idea that technology evolves according to its own internal dynamic, largely beyond the control of humans. If technology is so determined, then designers must follow the dictates of technology, rather than setting out to shape technology according to social, political, and ethical values (Feng 2000). Of course, such determinism obscures the fact that design is a value-laden activity—one that has profound implications for both users and nonusers alike.

## Invoking Users

These vignettes suggest that users are not so much represented in the design arena as they are invoked. That is to say, users are seen as something to be mobilized to support one position or another. Like "nature" in the world of science, "users" in the design arena do not speak for themselves but rather are spoken for by others. Lessons learned from science studies are therefore applicable here: like "nature," "users" are not predefined entities but things to be defined and mobilized in support of one side or another (Latour 1987). Thus, in the case of P3P, "users" becomes a malleable term that companies mobilize in their attempts to delay privacy legislation; they do so by promising that P3P will "put control back in users hand" and "empower users" to make choices about privacy for themselves, without the need for government legislation. Similarly, proponents of legislation construct an image of users, though in their case the image is one of people who support privacy legislation and have little need for technologies like P3P.

Obviously, these user representations are far from perfect—they fail to convey the needs and desires of "real" users. In particular, they appear to suffer from at least two major flaws. First, user representations are built up from the experiences of those designers active in the standards group. Designers, of course, are only human and thus suffer from biases no less than the rest of us. Therefore, their image of the "typical" user is often skewed (Norman 1988). Second, and perhaps more troubling, user representations seem to suffer from a systemic bias: since those who participate in standards-setting are mostly from industry, they are biased toward seeing the world in a certain way.[12] This systemic bias could mean, for instance, that almost all the standards that come out of today's standards organizations are biased toward business interests, possibly at the expense of "the public interest."

Given these deficiencies, one may be tempted to ask whether any alternatives exist to user representations. Might it be possible to lessen or even eliminate our use of them? The answer, likely, is *no*. If we compare standards-setting with, say, law-making in representative democracies, we see user representations occur there, too. While legislators are elected to represent the citizens of their district, they cannot avoid the use of representations. Indeed, every claim of "citizens want this" or "citizens do not want that" belies a political position—citizens are invoked by legislators, lobby groups, public advocacy groups, and others in order to support a specific cause. What, then, is the difference between the representations used in the legislative arena and those invoked in the design of technology? My answer is, "not much." The real question to consider is: How might user representations be made more appropriate and robust, so that the resulting technologies we build are truly reflective of the public interest?

## Conclusion: Making Users Visible Again

I have argued that standard setting is a politically charged activity that raises several normative issues. The first is that standards are usually set by small groups of experts, yet their impact on the public can be enormous. According to democratic theory, those who are affected by a decision have a right to participate in making that decision. Technical standards, which are often made invisible to the public, seldom achieve this democratic aspiration. Should the public participate in standards-setting, and if so, how?

The second dilemma also relates to expertise. Once upon a time, experts were believed to make objective decisions based solely on technical criteria. Students of technology, however, have argued that the

boundary between "technical" and "social" is illusory: technical decisions are always, in part, social decisions. Moreover, experts who participate in the standards-setting arena are usually drawn from industry,
and industry-led consortia (e.g., W3C) are increasingly involved in standards setting. Given such a population of experts, what then is their role
in the standards-setting arena and their responsibility to protect the
"public interest"? If experts could be counted on to make objective decisions based solely on technical criteria, perhaps the question of representation would become moot—we would expect the facts to "speak for
themselves," as it were, and expertise would merely be the means by
which objective facts were translated into practical action. But as scholars in science and technology studies have argued, nature does not
speak for itself. Nature must be interpreted, and experts, no less than
laypersons, are influenced by social considerations. If this claim holds,
shouldn't the public have as much a right to participate in technological
decisionmaking as any so-called expert?[13]

Third, issues arise concerning the institutions that set technical
standards. More and more, standards are being set by industry-led consortia. While organizations such as the W3C have done an admirable
job so far of engaging the public, the fact that control over the standards
arena is shifting from quasi-judicial bodies to industry-led groups may
be worrisome to defenders of the "public interest." Without the formal
public structure and accountability of traditional standards bodies such
as ISO, do these new consortia threaten to promulgate standards of
interest only to business? What responsibilities, if any, do those involved
in standards development have with respect to protecting the public
interest?

Tackling these issues will require some serious thought. One possible
starting point would be to increase participation by users—but not in
the classic PD sense of involving all users who might potentially be affected by a standard. Rather, the solution might be to include more public advocates at two critical phases: (1) during the design process itself, by
setting aside money for public interest groups to attend and participate
in standards meetings; (2) at some "mandatory review stage," during
which a proposed standard would be made open to review by members
of the general public, who then would have a chance to make binding
recommendations on the fate of that proposed standard. Other possibilities might include the use of consensus conferences and other forms of
"deliberative democracy" within the standards development process.[14]

These ideas may not be particularly welcomed by standards organizations, but these sorts of "institutional" reforms are necessary if
standards-setting processes are to be made more democratic and publicly accountable. Indeed, there are signs that at least some standards
organizations are becoming more open to the idea of increased public
involvement. At the W3C, for example, there has been talk of making
the work of technical working groups open to the public by default.[15]
Such considerations reflect an increased awareness on the part of the
organization that transparency—even for a supposedly "technical"
organization—is important.[16] It may well be, then, that increased user
involvement in standards-setting processes will depend as much on
changing the organizational culture of standards organizations as it will
on achieving changes from traditional policy actors such as national
governments. If so, democratic theorists will need to expand their gaze
beyond the traditional sphere of politics (i.e., electoral politics) to these
new institutions, such as international standards organizations, that are
playing an increasingly important role in the shaping of the material
world in which we live.

## NOTES

1. From my experience observing the standards process at W3C, this is not
atypical.

2. Of course, this description glosses over many of the details of standards
work, but it does convey the basic, idealized flow of events in designing a standard. A problem with "standard" accounts of the standards-making process is
that they divide the standards-making process into discrete steps and assume
rather naïvely that each step (e.g., "recognize need for a standard") is unproblematic. Needs are taken as given rather than constructed, and design is seen as
straightforward, rather than value-laden. These assumptions would be questioned by anyone who has actually spent time participating in standards working groups.

3. In this respect, standards-setting has some similarities with lawmaking.
Compare, for example, the process for drafting a legislative bill in the U.S.
Congress: small committees do the bulk of the work in hammering out a draft,
but this must then be passed by the entire legislature.

4. Some organizations, such as ICANN and IETF, broadcast parts of their
meetings via the Internet in order to facilitate participation by those unable to
attend the meetings in person. This is of some help but is a poor substitute for
actually attending the meeting and talking with people face-to-face.

5. Despite their global reach, international standards organizations generally have few paid staff. For example, ISO has about 155 full-time staff, W3C has about 65, and IETF has less than 10 (Loya and Boli 1999, 177; personal observation 1999). The bulk of standardization work is done by "volunteers," with costs being borne by member companies/organizations.

6. Calculated based on 250 work days per year.

7. UTOPIA was one of the first participatory design projects to be written about. The designers of this system argued that user involvement in the design process would benefit both designers and users (see Bødker et al. 1987).

8. Barbar (1984) criticizes the representative model of democracy as being weak in terms of actually representing/involving citizens. He argues for a shift to "strong democracy" in which citizens are much more actively involved in all facets of political decisionmaking.

9. The following vignettes are drawn from my participant observation of the P3P working group, from October 1999 to March 2001.

10. As Norman (1988) notes, designers are not typical users; their relatives apparently are, though!

11. In discussions with working group members after the meeting, it became clear that what the person actually had meant when she said "this will break the Web" was "this proposed change threatens our business model, and we aren't willing to change our model just to accommodate this change in the specification." But since economic arguments are frowned upon in technical bodies such as standards organizations, such economic concerns must be raised obliquely, under the guise of a technical concern that the proposed change would "break the Web."

12. From my own observations of an IETF meeting in Washington, DC, two years ago, participants in this standards organization are not very diverse: probably 80 percent+ of the attendees were white, 90 percent+ were male, and 95 percent+ were engineers or computer scientists.

13. For more on the question of expertise and technological decisionmaking, see Sclove (1995).

14. For a review of the deliberative democracy literature from a science and technology studies perspective, see Hamlett (2003).

15. In the past, most working groups adopted a "closed door" policy by default: while the output of the working group (e.g., technical specifications and prototype demonstrations) were made available to the public, "internal" documents such as minutes of working group meetings were not. The current talk at W3C is of reversing this position, so that working groups will be "open" by default and would have to justify decisions such as keeping the minutes of meetings secret.

16. This parallels a shift in the culture of the Internet from one where crypto-anarchist values ruled in the days of early Internet to one where mainstream values such as accountability, fiscal responsibility, and transparency have been demanded by businesses and government.

REFERENCES

Bødker, S, P. Ehn, J. Kammersgaard, M. Kyng, and Y. Sundblad. 1987. A UTOPIAN experience: On design of powerful computer-based tools for skilled graphical workers. In *Computers and Democracy: A Scandinavian Challenge,* ed. G. Bjerknes et al. Aldershot, UK: Avebury.

Cranor, L., J. Reagle, and M. Ackerman. 1999. Beyond concern: Understanding net users' attitudes about online privacy. At http://www.research.att.com/resources/trs/TRs/99/99.4/99.4.3/report.htm.

Egyedi, T. 1996. Shaping standardization: A study of standards processes and standards policies in the field of telematic services. PhD diss., Delft University.

Feng, P. 2000. Rethinking technology, revitalizing ethics: Overcoming barriers to ethical design. *Science and Engineering Ethics* 6 (2): 207–20.

Greenbaum, J., and M. Kyng. 1991. *Design at Work: Cooperative Design of Computer Systems.* Hillsdale, NJ: Lawrence Erlbaum Associates.

Hamlett, P. W. 2003. Technology theory and deliberative democracy. *Science, Technology & Human Values* 28 (1): 112–40.

Heywood, P., M. Jander, E. Roberts, and S. Saunders. 1997. Standards: The inside story. *Data Communications* (March): 13–17.

Latour, B. 1987. *Science in Action: How to Follow Scientists and Engineers through Society.* Milton Keynes, UK: Open Univ. Press.

Libicki, M. C. 1995. Standards: The rough road to the common byte. In *Standards Policy for Information Infrastructure,* ed. B. Kahin and J. Abbate. Cambridge, MA: MIT Press.

Loya, T. A. and Boli, J. 1999. Standardization in the world polity: Technical rationality over power. In *Constructing World Culture. Interorganizational Nongovernmental Organizations since 1875,* ed., J. Boli and G. M. Thomas. Stanford: Stanford Univ. Press.

Macpherson, A. 1990. *International Telecommunications Standards Organizations.* Norwood, MA: Artech House.

Norman, D. 1988. *The Design of Everyday Things.* New York: Doubleday.

Personal observation. 1999. Participant observation of Internet Engineering Task Force meeting held in Washington, DC, 7 November.

Porter, T. 1995. *Trust in Numbers: The Pursuit of Objectivity in Science and Public Life.* Princeton, NJ: Princeton Univ. Press.

Schmidt, S. K., and R. Werle. 1998. *Coordinating Technology: Studies in the International Standardization of Telecommunications*. Cambridge, MA: MIT Press.

Schuler, D., and A. Namioka, eds. 1993. *Participatory Design: Principles and Practices*. Hillsdale, NJ: Lawrence Erlbaum Associates.

Sclove, R. 1995. *Democracy and Technology*. New York: Guilford.

Simpson, G. 2001. As Congress mulls new privacy laws, Microsoft pushes system tied to its browser. *Wall Street Journal*, 21 March.

Updegrove, A. 1995. Consortia and the role of government in standard setting. In *Standards Policy for Information Infrastructure*, ed. B. Kahin and J. Abbate. Cambridge, MA: MIT Press.

# Technical Change for Social Ends

## Shaping Transportation Infrastructures in U.S. Cities

JASON W. PATTON

## Introduction

Faith in technical innovation is often coupled with hopes for social progress (Marx 1987; Pfaffenberger 1992; Smith and Marx 1994; Sarewitz 1996). New technologies are commonly expected to improve social circumstances while, in some instances, technical development is the very measure of progress (Adas 1989). Even though such hopes are often overblown, new technologies do enable particular social possibilities while simultaneously foreclosing others. This relationship between technical and social change offers a challenge and arguably an obligation to consider how technical innovation and implementation might be steered more intelligently. In this chapter, I argue that directed technical change can facilitate ameliorative social change by making particular everyday practices more appealing and convenient. I use the case of urban public transportation to show how design and policy can promote such change through the shaping of the built environment.

With his concept of the technological fix, Alvin Weinberg (1966) provides one such formulation of the relationship between the technical and the social. He argues that technology can provide solutions that are relatively easy to identify and implement compared to the complexity of social ills. Weinberg (1966, 37) writes: "The [social] problems are, in a way, harder to identify just because their solutions are never clear-cut: how do we know when our cities are renewed, or our air clean enough, or our transportation convenient enough?" A technological fix for a problem like traffic congestion might reduce people's commute times with some clever innovation. By avoiding a change in people's commuting practices, the fix is an expedient intervention because, as Weinberg explains, "[t]he resolution of social problems by the traditional methods— by motivating or forcing people to behave more rationally—is a frustrating business" (37). In Weinberg's formulation, the technological fix overcomes the urgency of social problems with breakthrough technologies that extend the viability of the social status quo. Such fixes are expedient because they mitigate a problem without disrupting the social patterns that create it. The technological fix does not solve social problems per se. Rather, it overcomes a limitation on social practice through technological innovation.

From regional, state, and federal perspectives, public transportation is a means for addressing traffic congestion, vehicle emissions, high levels of energy consumption, and social inequities in U.S. cities (Calthorpe 1993; GAO 2001; Kenworthy and Laube 1999a). While buses are often the most cost-effective form of public transit, they are generally the transportation mode of last resort (Transportation and Land Use Coalition 2002; Bullard and Johnson 1997; Wypijewski 2000). There is also widespread sentiment that modest numbers of people are willing to ride buses by choice. Few riders means fewer resources for improving service, and congested streets make it difficult to provide good service. With the notable exception of those influenced by ideology, who would choose a second-class technology for the arduous task of traversing the metropolis? This stigma shapes planning efforts and funding patterns, and it shades the possibility of people making bus-riding a part of their lives.

In this chapter, I offer a variation on Weinberg's technological fix that uses the steering of technical change as a means for shaping ameliorative social change. While my argument also considers breakthrough technologies, I emphasize how those innovations facilitate social change by redesigning what is possible and practical in people's daily lives. In contrast to Weinberg, I develop technical change as a means for addressing social problems through people's behaviors. Rather than forcing people to behave more rationally, I argue that the built environment can be designed such that the convenient choice is also—socially and environmentally—the right choice. This argument is embedded in the details of bus-riding on Telegraph Avenue in Oakland, California. I consider the promise of intelligent transportation systems and the emerging concept of bus rapid transit to explore how technical change creates social opportunities. Through the concepts of infrastructure and community of practice, I suggest that the technical levers for shaping social change are embedded in the details of how design and policy shape daily life.[1]

## Bus Riders as a Community of Practice

In most U.S. cities, the experience of bus-riding is little different today than it was fifty years ago. Catching the bus on Telegraph Avenue in Oakland is no exception. People wait at stops marked with signs listing numbered routes. Without maps or schedules, these numbers read like a code known only to the community of bus riders. One board member of the Alameda-Contra Costa Transit District characterized it as "the AC Transit lottery—choose a number and see where it takes you." The bus stop is typically a forlorn piece of sidewalk, occasionally graced with a bench, its low-budget advertising, and a garbage can. As the bus arrives, people follow the door, first with their eyes and then with their feet, anticipating where the bus might stop. Exiting passengers use both the front and rear doors while those waiting to board crowd around the front door for their turn. When the door is clear, the boarding passengers climb up onto the bus and present their cash, transfers, and passes. Each person stops at the fare box to be checked by the driver who also serves as banker, public relations representative, informational assistant, and disciplinary presence for the transit agency. After the last passenger has paid, the driver pulls the bus away from the curb. Known as "dwell time" in the bus world, the time spent at bus stops typically comprises up to 10 percent of a route's total running time (AC Transit 2002, 4:15). Two blocks away, the sequence repeats itself—and so on across the expanse of the city.

Despite such obstacles to quick and convenient travel, AC Transit served an average of 235,000 weekday riders in 2000. These riders are immigrants who speak seventy-five different languages, African

Americans, people with disabilities, school children, the urban poor, people who cannot drive, and those who have chosen not to drive. For many, their transportation marginality is a part of their social marginality. In urban areas across the United States, African Americans are eight times as likely as whites to take the bus. Similarly, people with household incomes of less than $20,000 are eight times as likely to take the bus as those with household incomes of $100,000 or more (Pucher and Renne 2003, 58, 67). As bus riders, they in fact live the tedious, mundane, and daily problems of public transit agencies that are politically marginal in funding decisions and planning processes. From the perspective of the transit agencies, their constituents disproportionately come from those groups with the least political clout.

Yet across the differences, bus riders share a great deal as a community based on their common activities and experiences in moving about the city. These common routines bind bus riders together as a community of practice (Wenger 1998; Lave and Wenger 1991). Bus riders are a community linked by similar routines of walking, waiting, and riding, as well as by their knowledge of stops, fares, routes, headways, and transfers. Bus riders are also linked by their participation in the shared social spaces of the bus and the bus stop that have their own norms and etiquette. To outsiders and newcomers, the seemingly simple task of riding the bus is a recurring source of confusion and frustration. By emphasizing such shared activities, the community-of-practice concept cuts across demographic divisions by race, class, gender, age, and physical ability.

The second-class status of buses and bus riders manifests in the design of transportation infrastructure, including city streets and the fixtures along those streets like traffic signals and bus shelters that shape how people move about the city. In general terms, I define infrastructure as a context, created by design and implemented in durable form, which provides a basis for human activities (Star 1999; Star and Ruhleder 1996). As a context designed to support particular human activities, an infrastructure will differentially support multiple communities of practice. For example, the bus on Telegraph Avenue must pull out of the traffic flow at each stop and then work its way back in. While bus stops are usually marked as no parking zones, bus drivers in congested areas negotiate double-parked cars and delivery trucks. On streets without curbside parking, a bus pull-out is often provided for the purpose of getting the bus out of the way. These bus pull-outs are street improvements that help ensure the free flow of private motor vehicles. The bus driver and riders bear the burden of this improvement. At each stop, the bus driver negotiates getting out of—and then back into—the travel lane, passengers lurch side-to-side as well as front-to-back, and the travel time gets longer with every stop. Because of traffic congestion and dwell times, travel speeds for AC Transit buses have decreased at the average rate of 1 percent per year for the last twenty years (AC Transit 2002, 4:21).

This largely negative sketch of everyday bus-riding shows how social stratification is built into public space through the design of transportation infrastructure. It also suggests the role of design in promoting the viability of bus-riding. While bus riders are a heterogeneous community, the commonalities provide a means for conceptualizing how buses might be moved toward the mainstream. The relative practicality of a transportation mode depends in part upon an infrastructure tailored to its needs. For example, how effectively a street supports driving versus bus-riding facilitates people's transportation choices. To grow the community of bus riders, that infrastructure could be redesigned to provide a material basis that better supports the social patterns of bus-riding. The inclusiveness of the community of practice concept suggests the potential for a broad coalition of people who would benefit from improved buses.

Such investments should not be thought of as subsidies for people of color or low-income. Such an approach risks perpetuating bus riders' social marginality by reinforcing their transportation marginality. It falls into a mind-set that the bus is a minimal social safety net that need not—and possibly cannot—provide high quality service. In contrast, better buses could provide improved services for marginalized groups while additionally bringing the practice of bus-riding closer to the mainstream. For the politician, engineer, planner, or academic who rides the bus, participation in this community is an entrée to a world of problems that are typically out of sight, experience, and mind for decisionmakers. Thus, the concept of bus-riding as a community of practice is a means for theorizing the permeability of social stratification by race, class, and gender through common activities in daily life. It also provides a means of strategizing social change with coalitions that bridge these demographic divisions.

## Just Add Silicon?

How might the second-class status of buses be overcome? Bus riders and potential riders want to get from point A to point B, they want to do so quickly, and they want a relatively positive experience along the way. Meeting these criteria requires a route network that effectively connects many destinations, travel times that can compete with the private automobile, and buses as well as stops that are reasonably comfortable and appealing for those with transportation choices. For attracting new riders, implicit in these criteria is overcoming the bus's reputation as second-class transportation. Attracting new riders requires technical changes and service improvements that are dramatic enough to break with a decades-old pattern of neglect for buses in the United States. Such change combines technical, social, and cultural elements. Change the infrastructure of buses to change people's bus-riding experiences and thereby change what buses mean to people. A new image for buses is part of the work of increasing this mode's share of users. The field of intelligent transportation systems (ITS) and the emerging concept of bus rapid transit (BRT) offer technical possibilities to explore these avenues for social and cultural change.

Intelligent transportation systems add information technologies (IT) to new and existing transportation infrastructures to create "smart" streets and vehicles that communicate with each other for improved efficiencies.[2] Through sensors, feedback loops, and web pages, proponents in research institutes, the IT industry, and the U.S. Department of Transportation (USDOT) hold out the promise of maximizing the efficiency of existing roadways to reduce environmental impact, increase safety, and ensure the competitiveness of American industry. Federal funding for ITS began in 1991 with the Intermodal Surface Transportation Efficiency Act (ISTEA) and, in 1998, the Transportation Equity Act for the Twenty-first Century (TEA-21) augmented it. In 1997, the USDOT estimated a twenty-year development period for metropolitan ITS in the country's seventy-five largest urban areas at a cost of $24 billion to the federal government. Even though most of the projects are implemented at the local or regional levels, federal support shapes system architectures to ensure software and hardware compatibility for data sharing and to reduce the cost of components through economies of scale.

Intelligent transportation systems bring the ebullience of the "IT revolution" to transportation's recalcitrant problems. The USDOT effuses that ITS "represent the next step in the evolution of the nation's entire transportation system. It involves the latest in computers, electronics, communications, and safety systems. These advancements can be applied to our vast transportation infrastructure of highways, streets, and bridges, as well as to a growing number of vehicles, including cars, buses, trucks, and trains. The future has arrived!"[3] While much of the future still needs to be funded, built, and lived, ITS does offer the unusual opportunity of adding a new dimension to transportation and doing so largely from scratch. While new silicon will not replace old concrete, the implementation of ITS is a chance to modify how existing infrastructures perform. Limited resources, rising land values, and the lack of space in urban areas are precluding the construction of new roads or the expansion of old ones. The shift in strategy is thus from increasing the amount of roadway to increasing the efficiency of existing roadways (USDOT 1996; 1998).

The development of intelligent transportation systems marks a significant change in strategy by the federal government for providing transportation infrastructure. Yet despite the environmental promise of increased efficiency through information technologies, the parallel between building new roads and many intelligent transportation systems is that both increase roadway capacity. For example, the purpose of many applications like traffic signal systems, automated guidance systems, and even automatic toll systems is to enlarge the capacity of roadways by increasing the speed and/or density of motor vehicle traffic. The most visible goal of ITS is to reduce congestion by coordinating the movement of private automobiles at the level of the transportation system. These applications are technological fixes in Weinberg's sense: they extend the viability of the private motor vehicle without disrupting the social practices of drivers. In one promotional brochure from the USDOT (1998), illustrations depict open freeways passing through verdant countryside. The cities are bounded by beltways and, curiously, the expansive suburbs of America's metropolitan areas are nowhere to be found (Lewis 1997; Bel Geddes 1940). While the artist struggles to make information technologies visible, the vision is clear for how driving ought to be: individualized mobility unfettered by congestion, sprawl, pollution, and stratification. Yet these applications of ITS offer more efficient means for pursuing well-established ends. They extend the nation's commitment to the private automobile through the investment in an electronic infrastructure that facilitates driving.


## Inducing Bus Riders with Improved Infrastructure

Urban transportation in the United States is dominated by the private automobile, which accounted for 85.9 percent of all trips in 2001. In comparison, transit accounted for 1.7 percent of all trips, foot travel accounted for 9.5 percent, and bicycle 0.9 percent (Pucher and Renne 2003). The private automobile is also the most energy-intensive form of wheeled transportation. In the San Francisco Bay Area, energy use per passenger mile is 70 percent higher for cars than for buses (Kenworthy and Laube 1999b). Comparing car travel in U.S. cities to bus travel in Canadian cities, energy use per passenger mile is 218 percent higher.[4] The difference in relative energy use for buses is best explained by higher levels of patronage for public transportation in Canadian cities, where 10 percent of miles traveled are made via public transportation versus 3 percent in U.S. cities. Kenworthy and Laube (1999b, 33–34) explain, "Such data show that the gains to be made in energy efficiency from better bus [engine] technology are dwarfed by the possibilities offered through patronage improvements." While buses are a more energy-efficient means of moving people, full buses are dramatically more efficient than private automobiles.

How might new technologies produce these larger energy savings by catalyzing a shift in people's practices to more energy efficient transportation modes? What if, for example, only 50 percent of U.S. urban trips were made by car, with 25 percent by transit and the remaining 25 percent by foot and bicycle? Despite the automobile orientation of many intelligent transportation systems, information technologies are an integral component for the emerging concept of bus rapid transit (BRT).[5] Proponents of BRT seek to combine the advantages traditionally associated with rail transportation and the cost-effectiveness of buses by introducing information technologies and redesigning city streets. For example, AC Transit estimated the capital cost of a proposed bus rapid transit line at 60 percent less than the cost of light rail for an eighteen-mile route from Berkeley, through Oakland, and to San Leandro.[6] The goal is a dramatically different yet cost-effective form of transportation that can compete with the convenience of the private automobile and overcome the cultural inertia of buses as second-class transportation.

This transformation could mark a coming renaissance in the bus world like the renaissance of the 1970s and '80s that transformed street-cars into light rail (Barry 1991). Beginning in the 1920s, streetcars were increasingly regarded as an antiquated technology that would be re-placed by the private automobile. By 1950, that transition was complete in the majority of U.S. cities (Jackson 1985; Foster 1981; McShane 1994). Light rail has since emerged—along with a growing middle-class interest in urban living—as a fashionable approach for reinvesting in cities. Since 1980, new light-rail lines have opened in more than fifteen U.S. metropolitan areas, with planning and funding initiatives under-way in additional cities (GAO 2001, 7). This renaissance combined tech-nical innovation with a new public perspective to re-imagine the con-temporary relevance of an otherwise dated technology.

Given their cost effectiveness, how might a comparable transforma-tion for buses attract new riders and bring bus-riding into the main-stream? "Induced traffic," an increasingly common concept in trans-portation debates since the early 1990s, offers some direction (Hills 1996). This idea is typically used with the following logic in evaluation of—or opposition to—the construction of new roads. As traffic congestion in a particular area gets worse, drivers will increasingly look for alternatives, either taking different routes, switching to different modes, or avoiding the travel altogether. Traffic congestion thus reflects a kind of market balance between the supply of transportation infrastructure available for drivers in a particular area and the relative demand for that infra-structure compared to its alternatives. Consequently, attempting to re-lieve congestion by increasing the capacity of a road will have the effect of inducing more drivers to use that road. Drivers who had found other routes or modes may return and people who had used other modes may begin driving. In the short term, adding capacity addresses the local problem of congestion. It also has the political advantage of addressing a visible problem that is close at hand. Yet it contributes to problems that are more easily ignored at the regional, national, and global levels (Dougherty 1998).

Might it be possible for intelligent transportation systems and the re-design of city streets to induce bus-riding by providing improved infra-structure? In 2001, AC Transit completed a major investment study that chose bus rapid transit as the preferred technology for the East Bay's major bus corridor (AC Transit 2001b). The proposal integrates intelligent transportation systems and street redesigns that would provide a different kind of infrastructure. Consider how AC Transit's vision would change the experience of bus-riding on Telegraph Avenue in ten years' time. People wait at BRT stations that include shelters, bus arrival information, and ticketing machines. Schedules are no longer needed because a bus arrives every five to seven-and-a-half minutes. A digital display provides real-time information on the number of minutes until the arrival of the next bus. The global positioning system that provides this information also ensures that buses are evenly spaced along the route. Bunching—when multiple buses arrive at the same stop at the same time and thus create gaps in service—rarely occurs. In the minutes before the next bus, people who do not have passes buy tickets from the machines at the station. When the bus arrives, its doors line up with marked locations on the platform because an optical guidance system guarantees that the bus always stops at the same location. Four sets of doors open and allow passengers to enter and exit the bus quickly. Because the bus floor is level with the platform and separated by only a few inches, seniors and people with disabilities ride the bus comfortably and confidently. With a proof-of-payment system using passes and time-validated tickets, the driver concentrates on driving and the dwell time at stops is reduced because people no longer line up at the fare box.

Travel from one station to the next is comfortable and quick. BRT stations are built at "bus bulb-outs" such that the platforms reach out to the edge of the travel lane. While providing space and amenities for waiting riders, the bus bulb-out also improves service and eases the driver's job. Because the bus must no longer merge with traffic, acceleration away from the station is smooth and rapid. The travel lane along the station platform is dedicated for the exclusive use of buses, and its pavement receives careful attention to ensure a smooth ride. Using information technologies, the bus communicates with traffic signals and receives priority treatment at intersections. With dedicated lanes and signal priority, the bus's travel is largely unconstrained by the surrounding automobile congestion. In the Telegraph Avenue corridor, fourteen stations connect downtown Oakland to downtown Berkeley. Compared to the current sixty-two stops spaced at two-block intervals, the bus moves across the city rapidly with a BRT station every third of a mile (AC Transit 2001a, 35). The infrastructure provides a well-designed base to support the community of bus riders. By providing a system that is more convenient, comfortable, and understandable, such improvements are projected to induce a 35 percent increase in ridership from 2001 to 2020 (AC Transit 2001b, 26–27). This increase is in addition to a projected 11 percent ridership gain given existing bus service, population growth, and in-fill development.

## A Cautionary Note on How Good Opportunities Go Bad

Technical change presents an opportunity—but no assurance—of ameliorative social change. Out of respect for the obstacles, I temper this sketch of a radiant bus-riding future with the following social fact. The innovation, design, and implementation of new technologies tend to reflect the contemporary social order (Cockburn and Ormrod 1993; Forty 1986). Bijker and Law (1992, 3) write, "It is sometimes said that we get the politicians we deserve. But if this is true, then we also get the technologies we deserve. Our technologies mirror our societies. They reproduce and embody the complex interplay of professional, technical, economic, and political factors." Technical change tends to follow a process of innovation directed by what is technically possible, profitable, and relevant to the short term. However, the deliberate steering of technology could exemplify a kind of "design by society" where the careful consideration given to the design of any particular product is applied to the overall process of shaping the material world (Woodhouse and Patton 2004; Whiteley 1993). Indicative of these obstacles are the trajectory of intelligent transportation systems in the United States and the implementation of bus improvements in Oakland.

While not all intelligent transportation systems are automobile-oriented, the marginality of ITS for buses reflects the marginality of bus riders and public transit agencies. Showing this bias, intelligent transportation systems were initially called intelligent vehicle highway systems (IVHS). The futuristic vision of intelligent vehicles and highways received its first major showing at the 1939 New York World's Fair in the General Motor's Futurama exhibit. Regarded as the fair's most popular pavilion with more than five million visitors, the exhibit design by Norman Bel Geddes depicted a United States in 1960 composed of rationally planned cities connected by high-speed motorways (Bel Geddes 1940; Smith 1993). That futuristic vision included automated driving through vehicles and highways that would communicate with each other using what are now known as information technologies. To this day, this kind of automated driving offers tremendous potential for increasing the capacity of existing roadways by having freeways full of cars moving at very high speeds with virtually no distance between them. When initial categories of research and development (R&D) were defined for ITS, they included advanced vehicle-control systems to develop automated high-speed driving. Tellingly, public transit was not included as an initial category of R&D (Whelan 1995).

Like the trajectories of technical innovation, the redesign of city streets encounters analogous obstacles to steering technical change for social ends. Preparing for bus rapid transit on Telegraph Avenue, AC Transit experimented with elements of BRT to gain experience in their deployment and to implement near-term bus improvements. The real-time bus arrival display was one technology on the cusp of implementation. The City of Oakland entered an agreement with Clear Channel, the international media corporation, for the installation of bus shelters in which revenue from advertising on the shelters would pay for the cost of the program. It stipulated that approximately 250 bus shelters would be installed throughout the city at selected stops and that each shelter would include a real-time bus arrival display. While AC Transit identified the 500 busiest bus stops, Clear Channel decided which 250 of those 500 stops would receive shelters. A programmatic goal of the company's contract suggested an obstacle: "Select an appropriate distribution of advertising locations that will generate the maximum amount of revenue from program inception in order to create a financially viable program" (City of Oakland 2002, 3). Because of the structure of a public/private partnership, the shelters were located to provide advertising first and bus improvements second. This familiar second-class status of the needs of bus riders had multiple sources: a transit agency that lacked jurisdiction over city rights-of-way; a municipality that saved money at the expense of bus riders; and a media corporation motivated by profit over the public good.[7]

In one respect, this story is an example of the persisting divide between haves and have-nots. One AC Transit board member explained, "If the map of shelter placement was overlaid on maps showing ethnicity and/or income, I fear the picture would be distressing." The densest areas of bus shelters correspond with the wealthiest areas of the city that have bus service. Not surprisingly, the majority of the busiest bus lines are located in the city's poorest neighborhoods. Yet, in another respect, this story shows how the transformative potential of a new technology is dulled by the pressures of the social status quo. Many of the city's busiest bus stops will not have shelters or displays. Pointing to the problems of this placement strategy, one bus advocate asked, "Does AC Transit know that the City is planning to put up shelters at stops that are exclusively used to drop off passengers on commute-hour only lines?" Some streets in affluent neighborhoods have shelters located every few blocks, whereas shelters on some of the busiest bus lines are few and far between. These busiest lines are also the longest lines that, being the most difficult to keep on schedule, would benefit the most from the displays. Because of the political and economic marginality of bus riders and the transit agency, the transformative possibility of a new technology is on course to fall below its potential.

## Conclusion

I have argued that policymakers and designers can shape sociotechnical change by eliminating the daily barriers to particular social practices. Providing better infrastructure for a community of practice will facilitate the growth of that community by providing a better material basis for its activities. For urban policy and design, bus-riding is an ameliorative social practice because it redresses the traffic congestion, vehicle emissions, and social stratification associated with the automobile dependence of U.S. cities (Kenworthy and Laube 1999a; Rae 2001). Yet from the perspective of daily life, bus-riding is often a poor choice because of service that is sparse, slow, or less than clean. One bus advocate explained, "I would really like to believe that there exists some level of transit subsidy which would motivate the mode shift we need. But I don't think money is the issue for the people driving the cars; if they can afford the car, they can afford to pay bus fare. The central issue is the public policy which keeps cars convenient. That 'subsidy' is the real problem." This subsidy takes on material form in public rights-of-way and adjoining land uses that, since the 1920s, have been designed to facilitate the flow of private automobiles. The consequence is an infrastructure that poorly supports bus-riding.

Bus improvements like BRT are technical means for reducing the social inequities and environmental liabilities embedded in the automobile-oriented street. Public policy is an appropriate avenue for reshaping transportation infrastructure: transit agencies are publicly chartered, streets are public rights-of-way, and both are funded with public resources. ISTEA and TEA-21 provided significant funding for intelligent transportation systems and an unprecedented level of flexibility for spending federal transportation funds on public transit. More riders would mean lower per passenger energy use and the ability to provide more frequent and extensive service. Buses are thus an inclusive public good—one that increases in value as more people use it (Best and Connolly 1982). However, an unusual opportunity for reshaping infrastructure is lost if emerging intelligent transportation systems reinforce the centrality of the automobile at the expense of its alternatives.

In contrast to Alvin Weinberg's technological fix, I have argued for shaping technical change as a means for facilitating ameliorative social change. Technologies shape social practice. By steering innovation and implementation by design, technical change will shape the relative practicality of competing social practices. The design of an infrastructure will differentially support multiple communities of practice. For example, the relative practicality of driving, bus-riding, bicycling, and walking is determined in part by how well the built environment supports each activity. This politics of infrastructure is an opportunity for affecting social change: design and policy could promote those daily activities that best embody social equity and environmental sustainability. While technology alone cannot solve the problems of urban transportation, shaping people's everyday practices through technical change is a means for decisionmakers to facilitate ameliorative social change.

## NOTES

This material is based upon work supported by the National Science Foundation under Grant No. 0115302, a dissertation improvement grant on "Design for a Pluralistic City: Multimodal Transportation in Oakland, California." Any opinions, findings, and conclusions or recommendations expressed in this material are those of the author and do not necessarily reflect the views of the National Science Foundation. The author thanks Dean Nieusma for his feedback on drafts of this chapter.

1. See Pucher (1988) and Pucher and Kurth (1995) for complementary arguments that emphasize economic and organizational considerations, respectively, for steering transportation change to facilitate walking, cycling, and transit riding.

& to make better or more tolerable

2. For an overview of intelligent transportation systems, see Whelan (1995) and OTA (1989).

3. See http://www.its.dot.gov/about.htm (accessed April 1999).

4. The comparison to European and Asian cities is increasingly dramatic: per passenger mile energy use for cars in the U.S. is 266 percent higher than for buses in Europe, 418 percent higher than for buses in wealthy Asian cities, and 532 percent higher than for buses in developing Asian cities. This comparison is based on data from forty-seven international cities (Kenworthy and Laube 1999b).

5. For overviews of bus rapid transit, see Levinson et al. (2002), GAO (2001), and Henke (2001).

6. In 2001 dollars, the report estimated the capital cost at $340 million for BRT versus $890 million for light rail. Projected operating costs in 2020 were approximately 16 percent less for bus rapid transit at $46 million and $55 million, respectively (AC Transit 2001b, 29–30). See GAO (2001, 17–25) for a national cost comparison of bus rapid transit versus light rail.

7. Neighboring cities negotiated a contract with a different advertising company that provided more favorable terms for bus riders. Under this contract, a proportion of the shelters would not have advertising. The placement of these shelters would be at the discretion of AC Transit and the municipality.

REFERENCES

AC Transit. 2001a. *AC Transit Berkeley/Oakland/San Leandro Corridor Major Investment Study*. Final Report, vol. 3, *Evaluation of Alternatives*. Oakland, CA: Alameda–Contra Costa Transit District.

———. 2001b. *AC Transit Berkeley/Oakland/San Leandro Corridor Major Investment Study, Summary Report*. Oakland, CA: Alameda–Contra Costa Transit District.

———. 2002. *Short Range Transit Plan (SRTP) 2001–2011*. Oakland, CA: Alameda–Contra Costa Transit District.

Adas, M. 1989. *Machines as the Measure of Men: Science, Technology, and Ideologies of Western Dominance*. Ithaca, NY: Cornell Univ. Press.

Barry, M. 1991. *Through the Cities: The Revolution in Light Rail*. Dublin: Frankfort Press.

Bel Geddes, N. 1940. *Magic Motorways*. New York: Random House.

Best, M. H., and W. E. Connolly. *The Politicized Economy*. 2nd ed. Lexington, MA: DC Heath & Company.

Bijker, W. E., and J. Law, eds. 1992. *Shaping Technology/Building Society: Studies in Sociotechnical Change*. Cambridge, MA: MIT Press.

Bullard, R. D., and G. S. Johnson, eds. 1997. *Just Transportation: Dismantling Race and Class Barriers to Mobility*. Gabriola Island, B.C.: New Society Publishers.

Calthorpe, P. 1993. *The Next American Metropolis: Ecology, Community, and the American Dream.* New York: Princeton Architectural Press.

City of Oakland. 2002. *City-Wide Street Furniture Program Implementation Plan and Streamlined Permitting Process.* Oakland, CA: City of Oakland.

Cockburn, C., and S. Ormrod. 1993. *Gender and Technology in the Making.* London: Sage.

Doughtery, M. 1998. Reducing transport's environmental impact: The role of intelligent transportation systems. *Proceedings of the Chartered Institute of Transport* 7 (2): 26–37.

Forty, A. 1986. *Objects of Desire.* New York: Pantheon.

Foster, M. 1981. *From Streetcar to Superhighway: American City Planners and Urban Transportation, 1900–1940.* Philadelphia: Temple Univ. Press.

Henke, C. 2001. Bus rapid transit grows up into a new mode. *Metro Magazine* (January): 43–52.

Hills, P. J. 1996. What is induced traffic? *Transportation* 23: 5–16.

Jackson, K. T. 1985. *Crabgrass Frontier: The Suburbanization of the United States.* New York: Oxford Univ. Press.

Kenworthy, J. R., and F. B. Laube. 1999a. *An International Sourcebook of Automobile Dependence in Cities, 1960–1990.* Boulder: Univ. of Colorado Press.

———. 1999b. A global review of energy use in urban transport systems and its implications for urban transport and land-use policy. *Transportation Quarterly* 53 (4): 23–48.

Lave, J., and E. Wenger. 1991. *Situated Learning: Legitimate Peripheral Participation.* Cambridge: Cambridge Univ. Press.

Levinson, H. S., S. Zimmerman, J. Clinger, and C. S. Rutherford. 2002. Bus rapid transit: An overview. *Journal of Public Transportation* 5 (2): 1–30.

Lewis, T. 1997. *Divided Highways: Building the Interstate Highways, Transforming American Life.* New York: Viking.

Marx, L. 1987. Does improved technology mean progress? *Technology Review* (January): 33–41.

McShane, C. 1994. *Down the Asphalt Path: The Automobile and the American City.* New York: Columbia Univ. Press.

Pfaffenberger, B. 1992. Social anthropology of technology. *Annual Review of Anthrolopology* 21: 491–516.

Pucher, J. 1988. Urban travel behavior as the outcome of public policy: The example of modal-split in Western Europe and North America. *Journal of the American Planning Association* 4 (4): 509–20.

Pucher, J., and S. Kurth. 1995. Making transit irresistible: Lessons from Europe. *Transportation Quarterly* 49 (1): 117–28.

Pucher, J., and J. L. Renne. 2003. Socioeconomics of urban travel: Evidence from the 2001 NHTS. *Transportation Quarterly* 57 (3): 49–77.

Rae, D. W. 2001. Viacratic America: *Plessy* on foot v. *Brown* on wheels. *Annual Review of Political Science* 4: 417–38.

Sarewitz, D. 1996. *Frontiers of Illusion: Science, Technology, and the Politics of Progress.* Philadelphia: Temple Univ. Press.

Smith, M. R., and L. Marx, eds. 1994. *Does Technology Drive History?: The Dilemma of Technological Determinism.* Cambridge, MA: MIT Press.

Smith, T. 1993. *Making the Modern: Industry, Art, and Design in America.* Chicago: Univ. of Chicago Press.

Star, S. L. 1999. The Ethnography of Infrastructure. *American Behavioral Scientist* 43 (3): 377–91.

Star, S. L., and K. Ruhleder. 1996. Steps toward an ecology of infrastructure: Design and access for large information spaces. *Information Systems Research* 7 (1): 111–34.

Transportation and Land Use Coalition. 2002. *Revolutionizing Bay Area Transit . . . on a Budget.* Oakland, CA.: Transportation and Land Use Coalition.

U.S. Department of Transportation, ITS Joint Program Office. 1996. *Operation TimeSaver: Building the Intelligent Transportation Infrastructure.* Washington, DC: U.S. Department of Transportation.

———. 1998. *You Are About to Enter the Age of Intelligent Transportation.* Washington, DC: U.S. Department of Transportation.

U.S. General Accounting Office. 2001. *Mass Transit: Bus Rapid Transit Shows Promise. A Report to Congressional Requesters.* GAO-01-984. Washington, DC: U.S. General Accounting Office.

U.S. Office of Technology Assessment. 1989. *Advanced Vehicle/Highway Systems and Urban Traffic Problems.* NTIS order no. PB94-134731. Washington, DC: U.S. Office of Technology Assessment.

Weinberg, A. M. [1966] 2000. Can technology replace social engineering? In *Technology and the Future,* 8th ed., ed. A. H. Teich. Boston: Bedford/St. Martin's.

Wenger, E. 1998. *Communities of Practice: Learning, Meaning, and Identity.* Cambridge: Cambridge Univ. Press.

Whelan, R. 1995. *Smart Highways, Smart Cars.* Boston: Artech House.

Whiteley, N. 1993. *Design For Society.* London: Reaktion Books.

Woodhouse, E. J., and J. W. Patton. 2004. Design by society: Science and technology studies and the social shaping of design. *Design Issues* 20 (3): 1–12.

Wypijewski, J. 2000. Back to the back of the bus. *Nation,* 25 December: 18–23.

# Shaping Infrastructure and Innovation on the Internet

## The End-to-End Network That Isn't

CHRISTIAN SANDVIG

## Introduction

This chapter approaches the question of how we should best reason about the design of communication infrastructures by examining a particular debate about the Internet. Specifically at issue are the benefits of the Internet for innovation. Some argue that the Internet's gift is found in an obscure design feature called the "end-to-end argument," first elaborated by Saltzer, Reed, and Clark (1984). This design feature is a network engineering strategy that promotes "stupid" networks: the center lacks intelligence and performs only a few functions, while nodes at the edge of the network—the ends—build complex applications by employing the simple building blocks of the core. The Internet, with a smarter PC and a dumber router, manifests this design strategy, while the telephone network, with its dumber handset and smarter switch, does not. Proponents of the end-to-end argument hold that with the Internet's end-to-end design, experiments can be deployed from the edges (ends) by anyone at all. Success of the Internet can be explained because experiments like the World Wide Web did just that.[1]

On the other side are commercial interests currently deploying intelligence inside the network's core. This logic speeds some traffic over others (caching), blocks traffic (firewalling, filtering), eavesdrops (snooping), and disguises some nodes as others (masquerading) (see David 2001). Some of these practices are so worrying for innovation—and for the freedom of users—that end-to-end proponents have asked the U.S. government to take action to preserve the Internet's "natural" form. Some have even argued that we might need a new Internet that retains an end-to-end design if the present one continues to erode.[2]

I take a third position. From the earliest organized communication systems, history teaches that networks tend to complexity at the core as more is asked of them. Contrary to the end-to-end argument, there is no reason to think that the Internet will reverse a three-millennium trend and evolve to be faster and more reliable than earlier communications systems while requiring fewer intermediaries and less intelligence at the center. While a more complex "middle" is already here, the important question has never been whether to preserve a simpler network structure but how and where new complexity is implemented. The key to innovation rests not on the degree of logic within intermediary nodes but in which nodes we trust. While Moors (2002) elaborates some of the technical implications of this position, this chapter addresses the implications for innovation and public policy.

I will address the future Internet by first recalling the oldest human communication networks—chiefly to remind us that although the structure of a communication network may have a technical veneer, it is a political bargain. Then, considering the Internet, I will unpack the end-to-end argument and suggest that: (a) it is not an organizing principle; (b) if it is a principle it is probably not true; and (c) even if it is true, it is probably not useful. The best outcome that normative claims premised on the end-to-end argument can offer us is to produce the right result for the wrong reasons, but we might be even better at promoting innovation if we act for the right reasons. Even worse, a dogmatic belief in end-to-end will simply retard the development of the Internet's infrastructure by limiting needed improvements in the "middle" or core. I will suggest that the right values to support are transparency and participation. I then conclude with the suggestion that underlying principles that support innovation need to be addressed explicitly, not embedded tacitly in technical arguments.

## Internet Design Is an Old Problem

To tame the policy case of the present Internet, let us recast the present problems of high technology in the relatively sedate terms of technologies long dead; they are imperfect but at least relatively settled examples. Data networks have a long history. The first things we might qualify with the term "systems" of communication were made from human couriers. In this section, I will replace the esoteric language of computer network technology with examples of older systems that involved only humans in order to bring the social relationships implicit in network topography into sharper relief.

The first courier networks were point-to-point, with most of the intelligence about the network's condition located at the network's edge — there were no "courier network commissions" or planners of overall infrastructure. Messengers relied on their own knowledge of existing paths. Once an ancient messenger left for his destination, it was by no means assured that he would arrive. There were often financial and political benefits to intercepting messages, and "courier loss"—akin to Internet "packet loss"—could also be a simple case of highway robbery. As couriers are much harder and slower to replace than Internet packets, the solution was not to re-send lost messengers, but to change the network itself.

This complexity then distinguishes between two generations of ancient courier systems. The first generation system involved only couriers using whatever existing paths or routes were at hand. Second-generation courier systems had a greatly improved capability for signaling the state of the network, improved security, better reliability, and better performance. The first second-generation courier system known arose in China early in the Chou dynasty (approximately 1000 BCE). Typically, rather than relying on the paths that already existed, second-generation networks incorporated "sponsored" roads or "post" roads that were paid for through taxation. Money spent on marking and improving the quality of the road allowed the traffic to travel faster — examples abound from the Roman *viae militares* (the military road) to the Spanish *El Camino Real* (the king's highway). But often these were not just upgrades to the same old infrastructure—beyond improvements to the old routes, the second-generation courier network added intelligence that was not at the network's edge.[3]

The Chinese system introduced repeaters: a system of "post-houses" allowed tired couriers to pass their messages to fresh, rested ones. In Chinese, Persian, Roman, and other civilizations, couriers on horseback later supplanted runner relays. Not only did the post-house or relay system increase the speed messages could travel, but post-houses were also used as a place to place guards to protect the integrity of the network from spies and robbers. They served a routing function by directing riders over alternative paths, and they maintained information on the quality of the routes. In some incarnations they billed for service, as toll stations. They filtered traffic—some post-house systems included armed guards and a capacity to inspect messengers. While Babylonian Royal Couriers were allowed to pass, the Bedouin raiders that hunted them most certainly were not (see Dvornik 1974).

One of the most ingenious additions to second-generation courier systems was the fire beacon. If the guard posts were numerous and located within sight distance, in clear weather the fire beacon could be used to send a prearranged message much faster than even a mounted courier, providing the network two modes of operation. These beacons could signal meta-information: in some systems they were used by the post-houses as a "trouble" signal warning of a problem with the network itself (a signal from the network's center to its edge, or as coordinating signal between different post-houses in the center). In special circumstances this faster mode could be quickly repurposed to carry simple information—often warning of invading armies. The integration of the beacon into courier networks (a protosemaphore) marks not just a general increase in the sophistication of the system but also the advent of segregating traffic into classes by priority, where each class has a different form and a different quality of service.[4]

## Policy Implications of Second-Generation Courier Networks

As it entailed investment in roads, the second-generation courier network required much greater standardization of traffic. Wheeled vehicles were nothing new—they predate the Chou second-generation courier network by about one thousand years—but when roads were improved, the road width had to be determined and future traffic had to conform.

Continuity was a decidedly local practice: at least eighty-four inches in width was required for a Roman road, but only fifty-five inches between shaped-stone wheel ruts sufficed for Greek sacred roads. These standards were then imposed on users by technology (the road) and also by law: the *raeda*, the fast freight cart used on Roman roads, was restricted by law to carrying 750 lbs. or less, for fear of damaging the road.[5]

Second-generation networks must have been much more effective than their predecessors and many orders of magnitude more expensive: guards, post-houses, and roads all cost money, but there were fewer dead messengers. The trend in the evolution of ancient courier networks is clear. As the demands on the networks increased, they became faster and more reliable in part by becoming more elaborate. As rulers added resources to the infrastructure, the communication network designers of the ancient world also added intelligence to the center of the network to manage and control it. They asserted this increasing control through a combination of force, law, and technology—in other words, the network's soldiers, rules, and form. As courier networks improved, the number of intermediaries must have increased at least slightly. No longer were couriers left on their own to wander. In the improved systems, the message might be normally expected to pass through many hands to be relayed, taxed, and inspected.

## The Internet and the Unexpected Reversal of History

In a hotly contested debate now occurring somewhere between network design and public policy, some argue that the optimal evolution of electronic networks will be just the opposite of these ancient courier networks—the Internet will evolve to be faster and more reliable than earlier electronic systems, and this can, in a reversal that is both paradoxical and exciting, require fewer intermediaries and less intelligence at the center of the network. In a sense, much of the early excitement about the Internet stems from this controversial design concept; it is the spring from which ideas about the Internet's "inherently decentralized nature" flow. In the remainder of this chapter I will unpack this idea, known as the "end-to-end argument," and then reassess it, employing the examples given above.[6] As additional examples will show, end-to-end has migrated from engineering circles to public policy circles. The focus of our interest here is the advocate's perception that the end-to-end argument in network design, usually evaluated by engineers on the basis of technical efficacy, may also be normatively positive: that is, that the end-to-end argument forms a kind of philosophy of network topography. Normative benefits have been claimed for two reasons: First, under some conditions end-to-end networks can dramatically increase the number and diversity of groups that can participate in the design of network applications. Second, end-to-end networks can make it more difficult for unwanted third parties to control communication on the network. In this way, the end-to-end network has been hailed as inherently more democratic, its form producing user freedom.

However, I will argue that the current technical and political debates about end-to-end are misleading, for they cloak arguments about power with appeals to a single objective technical truth where none exists. Indeed, it is not end-to-end design per se that is normatively positive, but the transparency, openness, and participatory design consultation that have come to be associated with this model of network intelligence through history and tradition (Streeter 1999). Loading the end-to-end argument with these social goals, rather than addressing normative goals directly, is a dangerous and misguided strategy because it shifts policy discourse away from normative ends in favor of technical means that may not lead where we expect.

## What Is an End-to-End Network?

As is common in many areas of technological development, computer system and computer network development have exhibited since their inception a trend toward modularization. What was once a single piece of technology ("the computer") becomes an assemblage of different parts, some of which are standardized and produced by different parties. The innovations of the Internet reside in software, but modularization has still proceeded in a manner similar to the standardization of parts for a car. Today's programmers do not start by writing binary codes to control the hardware of a computer. Instead they assemble software by relying on a variety of standard components such as code libraries and features of the operating system. The modularization of software has enabled the growth of computing as a sociotechnical system: it facilitates innovation by allowing developers to incorporate standardized components in new and larger projects, increases reliability in software by distributing what is effectively pre-packaged expertise, reduces software development time by changing the process from one of construction to one of assembly, and drives down software cost by allowing competition in development and provision of these modular subcomponents. The clearest expression of this modularization trend in the subdomain of computer *networking* has been the development of a standard model—called the seven-layer model—for representing communication between computers.[7] This model effectively divides the task of communicating into subtasks (layers), making the development of new network applications such as Web browsers and e-mail clients much simpler because the application programmer can leave most of the job of communicating to other subsystems. The entire means for sending messages does not require reinvention each time a new kind of message needs to be sent.

The seven-layer model is then useful for reasons of efficiency because it saves development cost. But as networking and the idea of network layers became widespread in the 1980s, so too did disputes about the technically correct level (that is, module or part) where any given functionality should be incorporated into a communication system. To translate this development to the ancient world, should the post-house count the couriers, or the king? Because different actors worked at different layers, these disputes could also take the form of the question: "Should this software company be responsible for a given task, or should that one?" In a now-classic paper in network engineering, Saltzer, Reed, and Clark (1984) elaborated a design strategy for computer systems called the "end-to-end argument," which take the position that:

> (1) If a particular function requires the participation of the endpoints of the system, it should not be implemented in any other location in the system.

This statement is the first of six principles that I extract from arguments made on behalf of end-to-end networks. It can be explained for the functional goal of reliability with a somewhat oversimplified scenario of mailing postcards:[8] If person A mails a very important postcard to person B, how does person A know that the postcard has arrived safely? One option would be to apply a variety of strategies at points in the postal system where loss of the postcard is likely to occur: the location of each item of mail in the system could be constantly tracked, duplicate postcards could be made at several points and delivered as insurance, or postcards could be fabricated out of a durable metal instead of flimsy paper. The end-to-end argument, however, notes that while these strategies would increase the reliability of the postal system as a whole, they are expensive and might tend to slow the delivery of all mail. Indeed, regardless of any of these costly improvements to the postal system, person A could never be absolutely certain that person B did in fact receive a particular postcard. To attain peace of mind, person A would still have to ask person B if the postcard had arrived. It is likely more efficient, then, to skip the improvements to the postal system as a whole and for person A to plan on asking person B. As long as the use of the post is cheap and simple, if some of the postcards do not arrive, person B can ask for them to be sent again—by, for instance, using another postcard. In this scenario, person A and B are the "ends" and thus the repositories of all the intelligence necessary to confirm receipt of the message. No complexity was added to the middle (or core) of the postal network; instead it was the behavior of the "devices" connected to the edge of the network—the users—that changed to support the goal of reliability. In the eyes of the end-to-end proponents, simpler networks and smarter ends are always the answer.

## A Competing Strategy: The End-System Model

The early telephone network, in contrast, was initially developed with a network design philosophy that has been called "end-system" (Kruse, Yurcik, and Lessig 2001).[9] Most of the functionality of the network was located in the telephone switch, and the "end" (the telephone) was a fairly simple device that could do little except relay simple commands to the switch (off-hook, on-hook, 1234567890*#). To expand slightly on the postcard example above, we see that the end-to-end argument is clearly about who does what. If the communication problem is to insure that a postcard sent between person A and person B arrives, some end-to-end strategies might be that person A keeps very careful track of the postcard, A makes a duplicate of the postcard in case it is lost, or A chooses to make the important postcard out of a very durable material. In comparison, the end-system strategies would be that the post office implements a tracking system to keep very careful track of all postcards, the post office makes duplicates of all postcards when they are mailed, and the post office requires that all postcards be made more durable. And so in computer system design these questions about the location of functionality can seem like purely technical concerns, but the questions drawn from this example are also clearly "who is in control?" and "who pays?"

End-to-end arguments have waxed and waned. When Internet pioneers first proposed end-to-end, it was controversial—end-system designs ruled the day in computer communication. In the related sphere of telecommunications, the dominant thinking was later called the "intelligent network" movement (Mansell 1994). In the late 1990s, backlash against this centralization led to a rediscovery (or independent discovery) of end-to-end in telecommunications under the catchphrase "the dumb network" or "the stupid network" (Isenberg 1997). Note that the categories of "end-to-end" and "end-system" are useful but not exhaustive. They are not a dichotomy, and indeed may not even be on a continuum. It is possible to characterize the structure of some networks as predominantly end-to-end or end-system, but other networks defy this characterization—they are confusing hybrids.[10]

Nevertheless, and despite the danger of over-generalization, table 11.1 attempts a rough categorization by way of example. As "end-to-end" and "end-system" networks are design strategies and ideal types, no network can clearly be labeled one or the other, but in this table I attempt to match the ideal types with a few examples that may come closest.

The earliest ad hoc courier networks were end-to-end in that the network was so simple that no complexity existed, except at the edge. All applications (diplomacy, commerce) were bargains between end-points. Although the earliest networks of paths were probably not, strictly speaking, "designed," still no intermediaries at all were planned. (Bandits are an example of an unplanned intermediary.) The earliest, nonswitched "party-line" telephone systems could also be conceptualized as end-to-end. Really they are broadcast networks, as all terminals received all transmissions, and there was no complexity at all in the middle of the network. This example brings us to a point of some confusion: the implied relation between end-to-end and switching. Note that the distinction between end-to-end and end-system networks is not related to switching. Switching is just one function of a network that might be implemented with an end-to-end or end-system strategy. For instance, the Internet is a packet-switched network that end-to-end proponents worry is becoming less end-to-end—but it will still be packet-switched. A nonswitched, that is, broadcast network could implement many functions in a complicated network core that are unrelated to switching. Last in the table is the network that gave birth to the concept

Table 11.1. Networks characterized by design strategy

| Ideal types: End-to-end networks | End-system networks |
|---|---|
| The earliest ad-hoc courier networks | The Chou Dynasty "post-house" courier network |
| Early nonswitched ("party-line") telephone networks | Early switched telephone networks |
| The Internet before widespread content caching | |

of end-to-end: the recent Internet before the advent of technologies that allegedly "break" end-to-end.

In the right column, the Chinese "post-house" system of the Chou dynasty could be considered end-system if we take some liberties with the analogy between courier systems and data networks. The post-house system also implemented routing at the core, but in addition it probably supported spying, taxation, and some other functions. It is a difficult example for this ideal type.[11] We also have the network where the concept "end-system" originated, telephone service before the advent of technologies that increase intelligence at the terminal device, or end (such as ISDN), and muddle the example. While the Internet's core handles routing—an example of some intelligence at the center—the Internet is still considered an end-to-end network because little other function was implemented in routers at the core.

This broad distinction between two ideal types in overall strategy for communication network design has consequences far removed from the practice of computing. The design of any technology that allows humans to communicate must have social and political consequences, but more specifically the perceived benefits of end-to-end design have been put forward in the case of the Internet in three areas: (1) user-driven innovation (2) protection from unwanted intermediaries, and (3) technical correctness.

## The Implications of End-to-End for User-Driven Innovation

Advocates of end-to-end point out that it reduces the complexity of the "core" network, and that the resulting generality of the network fosters innovation, because new and unanticipated services can employ the basic building blocks of a simpler network core (Blumenthal and Clark 2001). The innovation principle could be embodied succinctly as:

(2) The lowest layers of a system should provide the greatest flexibility possible, so as to permit applications that cannot be anticipated.

The Internet is a manifestation of this principle: it provides a fairly general set of facilities to allow the transfer of any data. A variety of different applications have emerged (e.g., remote login, electronic mail, the World Wide Web) that all share the same Internet. If certain functionality is required to allow only one of these applications to work, it is located in software (the electronic mail client, the Web browser) on computers attached to the network's "edge" (Blumenthal and Clark 2001, 92).

It is a key feature of end-to-end design that *users* of the network are able to create new applications—applications that eventually influence the broader technological system (cf. von Hippel 1988; Bar et al. 2000; Bar and Riis 2000).[12] This process of "user-driven innovation" occurs with other technologies, but because the control mechanisms for the Internet reside in software, user-driven innovation has a much more central role. Consider that the two most popular Internet applications were developed not by a centralized network authority or the owner of the network but by users: Ray Tomilson, who developed electronic mail for his group at BBN in 1973; Tim Berners-Lee, who developed the World Wide Web for his group at CERN in 1990.

By contrast, in an end-system model the device at the edge of the network—the telephone, for example—is simple and inexpensive to manufacture. The network equipment to which the device connects provides all of the intelligence and functionality of the system. In the early telephone network, adding new technology at the end was expressly forbidden—adding an answering machine or even a piece of cardboard was a violation of the system's design principles, and in some cases was illegal in the United States as late as 1976.[13] Both the design philosophy and the difficulty in configuring hardware act against user-driven innovation on the telephone network. Prior to electronic switching, an innovation such as three-way calling would have required a user to find a screwdriver and link three wires. Even since electronic switching and software control, if a user wishes to implement a new service such as voicemail on the telephone network, no facility exists to allow it. Telephone companies control the system, and the user may not tamper with it—or even discover how it works—without permission.

## End-to-End as Protection from Intermediaries

Those with technological determinist leanings have pointed out that the structure of the end-to-end Internet is itself a protection from anyone who would limit the freedom of communication. By this logic, adopting an end-to-end Internet design will produce freedom irrespective of legal or social arrangements. For much the same reasons that would-be innovators can deploy any new application that suits their fancy, would-be communicators do not need permission to speak and need not subject their speech to anyone's review. This argument says that if the network has no functionality to examine the messages it carries, communication is then more free.

## The Argument from Technical Correctness

Some have sought to offer end-to-end design principles as true in some objective sense. Principle 2 is less overtly concerned with objective correctness, but the next arguments considered are related to the idea that if functionality is removed from the network core, the core itself likely becomes cheaper, faster, and easier to administer. They argue that a simpler core is necessarily more transparent and easier to model, and that acceptance of this argument is a step toward a more rational, scientific, and rule-based network engineering (for an account and critique of this specific point, see Moors 2002). These "correctness" arguments have been made and remade in the landmark papers on end-to-end, in fora like the end-to-end mailing list and, one imagines, in engineering meetings around the globe with increasing frequency over the last twenty years. It has been claimed that end-to-end is technically correct because:

> (3) Any function implemented in the core network may be redundant because this functionality has already been implemented at the end-point.
> (4) Any function implemented in the core network may be redundant because some applications will never need it.

Let me return again to the postcard example to explain these claims of end-to-end proponents. To illustrate principle 3, consider that if both the end-to-end and the end-system strategies in table 11.1 are implemented, there will be a significant duplication of effort. Furthermore, to illustrate point 4, imagine person C who does not need to verify if her postcards are delivered or not (or does not wish to pay extra for it). If person C pays for the postal system (through stamps or taxes), part of her contribution to the postal system through taxes will go to support improvements she does not need or want. The network will be more expensive to operate, with no benefit to her. A stronger form of statement 4 is that no function should be implemented in the network unless all clients of the network (or that network layer) will need it, because:

> (5) The end-point tends to have more information about what it needs than the network, and
> (6) Any function implemented in the core network adds cost and complexity that is borne by all network users, even if they do not use the function.

While principle 5 seems to be an argument for end-to-end design, it presupposes an end-to-end network where there is no central authority dictating what applications should be used, and where the technology and form of communication remain unsettled. If the Internet will continue to be a place where new applications arise continually (e.g., peer-to-peer file sharing in the form of Napster) there is some merit to this point. Statement 6 is merely a rearticulation of the conclusions arrived at earlier with the postcard example.


## The Perceived Challenge to End-to-End

Attempts to promote new Internet applications have proponents of end-to-end design feeling as if they were under siege. Proponents have warned that developments in software, policy, and use are "compromising the Internet's original design principles" (Blumenthal and Clark 2001). Blumenthal and Clark note four examples of emerging requirements for Internet applications that challenge end-to-end design principles: (1) the need to manage untrustworthy end-points (2) demands for better throughput required by streaming audio and video (3) differentiation of service between competing Internet Service Providers (ISPs), and (4) the rise of increasing third-party involvement in communication. That is, each of these developments might imply a need to add intelligence to the network, for example (1) the "firewall" to block "hostile" network traffic (2) the addition of proprietary distributed cache systems for multimedia content (e.g., Akamai) (3) the distinction between different kinds of content by the ISP to provide different levels of service quality, and (4) the use of filters to block unwanted or illegal traffic, or the use of traffic analyzers to eavesdrop on suspect traffic.

The perceived danger of these changes is that they each constitute new intermediary points in the path of network traffic where some third party may exert control. Those concerned about the censorship of content or the leveraging of the control of the Internet's wires into the control of its content have pointed out that "end-to-end was initially chosen as a technical principle. But it didn't take long before another aspect of end-to-end became obvious: It enforced a kind of competitive neutrality. The network did not discriminate against new applications or content because it was incapable of doing so" (Lessig 2000). Indeed, one of the original authors of the seminal end-to-end paper equates end-to-end with "the default situation [where] a new service among willing endpoints does not require permission for deployment." Abrogation of end-to-end design principles leads to the case where "new chokepoints are being deployed so that anything new not explicitly permitted in advance is systematically blocked" (Reed 2000, 4).

The strength of the arguments made to ensure future end-to-end design relies on the presumptions that end-to-end is the current design scheme, and that the current design scheme has been an effective one. That is, if it ain't broke, don't fix it. As Lemley and Lessig (2000, 4) state, "We do not yet know enough about the relationship between these architectural principles and the innovation of the Internet. But we should know enough to be skeptical of changes in its design. The strong presumption should be in favor of preserving the architectural features that have produced this extraordinary innovation." These appeals have been made in policy fora and seek to constrain those who would depart from end-to-end design principles. These threatening parties are private actors, such as ISPs and telecommunications companies, free to attach their new software and equipment to the Internet as they see fit. In other words, end-to-end proponents wish to make the case that while the Internet has a "fundamentally" decentralized and distributed nature, it now requires policy action to prevent private actors from departing from its technical design principles.

The most visible recent manifestation of this conflict has been the U.S. cable industry "open access" debate. Cable providers in the U.S.

have deployed extensive intelligence in the network's middle in the form of a technology called a "caching gateway" at the junction of their subscriber network and the slower Internet. Because cable providers already own government-sanctioned monopoly franchises, if they are allowed to require that users of broadband cable modem service also use an ISP that they own, it is likely that they will be tempted to leverage their monopoly power into control of Internet content. This leveraging could be accomplished by hosting "strategic partners" or in-house content on these caching gateways that they alone control and that are close to cable modem subscribers. This arrangement would provide subscribers with faster access to content that generates profits for the cable company, and indeed also provides the cable company with a temptation to slow traffic from competing entities. The user in this scenario would never know why some parts of the Internet were faster and some slower. The U.S. Department of Justice, the Federal Trade Commission, and the Federal Communications Commission were lobbied with this rationale to require open access to competing ISPs in merger proceedings involving large cable providers (Bar et al. 2000).

## End-to-End: Already Over or Never Was?

In this section I critique end-to-end approaches in order to show that the end-to-end debate is not one of technical correctness but of sociopolitical control. First, I should point out that what has been presented in the section above is a classic case of a technology in the early period of its development, when its inner workings are subject to interpretive flexibility (Pinch and Bijker 1987). Several relevant social groups have been identified in this controversy, for example, private firms seeking to profit from the Internet in some way, and the "old school" of Internet pioneers, designers, and computer scientists. The Internet is a technology that they each seek to shape, and in order to assert control over it they are engaging in a definitional controversy about what features are essential to the object "Internet." They attempt to control the emerging technology by painting the encroachments of other groups as antithetical to the natural form of the thing that we call "Internet."

As the authors of the original end-to-end paper note, end-to-end is an "argument" rather than a rule, law, or even principle. Determining what is meant by "ends" in any given engineering problem is extremely difficult, and the resulting debates are open to much interpretation. Even the strongest case made for end-to-end, the case for the promotion of innovation, is problematic. The end-to-end approach only facilitates innovation of the kind where the new application services can be built with the low-level building blocks provided by the network. Since network engineers do not agree on a kind of "periodic table" of low-level network building blocks with which every possible service can be built, the end-to-end goal of a simple and easy-to-build-from network is actually an easy-to-build-from network where it is easy to build things that are made from the blocks or functions that are present. If the service you want to build cannot be built from these functions, an end-to-end design strategy will not make your new idea easier to build.

For instance, those who wish to introduce robust multimedia streaming services object that it is not possible to build quality of service guarantees when starting from the building blocks that are available. To work at all, the development of streaming for multimedia content seems to require changes to the core functionality of the network. It is hopelessly impractical to treat broadcasting as a point-to-point operation in the present Internet. With any large number of users wanting to view a broadcast stream simultaneously, the load on the provider of the stream quickly becomes unmanageable, and the network near the provider is subject to congestion because the provider must send multiple identical copies of the stream, one for each user that desires it—an ugly and inefficient approach. The "stupid" network does not notice that each stream is the same thing and cheerfully sends a thousand copies when one would be best.

The cooperative, open approach to redesigning the Internet's core to support this semibroadcasting of multimedia content is termed "multicasting," and it involves modifying the Internet's basic protocols to reduce the duplicate transmission of multiple streams when one stream would do. End-to-end proponents have described these core network modifications as a necessary departure from an end-to-end strategy for reasons of performance.

However, providers of privately distributed stream-caching services now offer reliable broadcasting on the Internet without modification of existing protocols or the network's core. These providers, Akamai and formerly Inktomi, for example, locate servers in many data centers around the world. The content stream is sent first to these data centers, where it is then duplicated for users that are nearby. In other words, users are directed to obtain the content from the point that is nearest them. This approach has been decried by end-to-end proponents as a violation of end-to-end.

Clearly the approach used by Akamai reduces the transparency of the network. Akamai is a private company, and it has introduced a proprietary facility for providing content that is available only to its subscribers. In contrast, if multicasting were fully realized within the Internet's protocols, it would provide a facility inside the network for providing such content that could be used by anyone. In effect, a key difference between these two approaches is cost distribution. Modifying the core network, as Akamai does not do, distributes the cost of broadcasting Internet content across all users: those who want the content, those who provide it, and those who never use it. The Akamai approach requires that multimedia content providers and providers of popular (nonmultimedia) content pay for this infrastructure as an added service, in advance. The decision of where to locate the functionality determines how Internet broadcasting is funded and who has access to it.

## Conclusion

In considering policy decisions related to the design of the Internet, it is easy to despair when a decision must be made without much precedent. But the history of communication networks abounds with useful comparisons in the distribution of intelligence within networks. Indeed, the present debate can be usefully informed by the history of courier networks and the difficulties of asserting control through network protocols. The word "Internet" evolved from what ARPANET protocol designers called "the Inter-network problem," that is, the between-network problem. "Networks represent administrative boundaries of control, and it was an ambition of [the ARPANET] to come to grips with the problem of integrating a number of separately administrated entities into a common utility" (Clark 1988, 107). Coordination of the second-generation courier network of the Chou dynasty, then, needs to be extended further still, to ask what happens when the couriers reach the edge of a kingdom and must pass onto courier networks controlled by other kings. To borrow the language of computer networking, the network designers of ancient China did not consider technical means for ensuring the integrity of their communication across these administrative boundaries of control, because any protocol they might devise would depend entirely upon the cooperation of rulers of neighboring kingdoms. If it was in the strategic interests of other rulers to intercept or otherwise compromise messengers, the problem is not a technical one but one of politics. The only way to make sure that adjacent interconnected networks behave the way you want them to is to assert control over them in some fashion.

The use of end-to-end as a design principle has the effect of pushing intelligence to the borders of the network, creating what end-to-end advocates argue are "application insensitive" networks that are optimized to deliver few "low-level" services at the core. They also claim that these networks offer not only building blocks upon which many types of use can be built but also an environment where there need be few intermediaries. I hope to have demonstrated that this "low-level" of service is better described as one form of service among many possibilities, for the Internet is application-insensitive as long as the application is similar to something its first designers envisioned. The Internet has always had intermediaries, but end-to-end proponents were happiest with the intermediaries they knew—service providers who initially were academic computer scientists and then universities. The end-to-end argument is an effort to stop the new intermediaries by arguing that they are technically incorrect—a definitional debate about the form of the technological artifact "Internet," which has not yet stabilized.

I do not intend to underestimate engineers. My argument does not imply that those who advance end-to-end arguments as technical solutions to their political problems are in any way simpleminded. Many of the participants in these debates are well aware of the dangers of advancing normative principles as technical principles, or conflating the technical and the normative. But the technical argument is alluring because it offers the promise of objective correctness to trump messy compromises.[14] Instead, I suggest that these clever engineers are themselves underestimating policymakers and corporate opponents. To stand on expertise instead of principle will work only if no other experts contradict you. When those who wish to modify the network in "undesirable" ways do not like your engineering, they will simply buy new and better engineers. The use of technical arguments as proxies for normative arguments produces a strange debate. If one argues that "the end-to-end principle renders the Internet an innovation commons" (Lessig 2001, 40), one then cedes the debate to whichever specialists are successful at defining the historical, or true Internet. In arguments from tradition, those who can define the past get to define the future.

It is seductively easy to conceptualize technical, social, and legal mechanisms of control as different sorts of levers one can pull to steer the Internet (Clark et al. 2002). Writing in the end-to-end debate has pointed out the prevalence of technical control, and it has also called for more social and even legal control as a better "balance." In fact, what seems like "technical" control is nothing new. The technical, social, and legal always interpenetrate, and there is no purely technical way to guarantee that the Internet will evolve or be used in one way and not another without some broader assertion of control, by owners or by governments.

The process of network design should continue to include considerations of transparency, participation, and flexibility, but these should be explicit goals and not pursued under the rubric of technical correctness or the end-to-end argument. Furthermore, the legitimate public policy role for governments lies not in protecting the Internet against those who would "break" it. Such a policy would merely grant authority to whoever is designated to interpret the Internet's fundamental nature and write its history. Reflecting on the Internet's boon to innovation provides a logical rationale for regulating transparency and participation. This role is not a new one for government, even with respect to the Internet.

## NOTES

The author would like to thank Helen Nissenbaum, Stephen Barley, Paul David, Ian "Gus" Hosein, William Drake, and Dieter Zinnbauer for their helpful suggestions. This research was kindly supported by the Markle Foundation Information Policy Fellowship at the Programme in Comparative Media Law and Policy at Oxford University, and by a Visiting Research Fellowship at the Oxford Internet Institute. An earlier version of this chapter that did not contain the arguments related to innovation was presented at "Computer Ethics: Philosophical Enquiries," 15 December 2001, Lancaster, UK.

1. The most visible popularizer of this notion may be Lessig (2001), the most precise Blumenthal and Clark (2001).

2. This intriguing suggestion was made by David (2001) and also deemed impractical by him.

3. The historical details in this introduction are taken from Holtzmann and Pehrson (1995).

4. The simple beacon of light (fire) or smoke may predate the organized courier network, but the point here is the integration of the two.

5. The details in this paragraph are taken from Lay (1992) and Forbes (1954).

6. In telecommunications engineering, the notion of a single carrier owning all segments of a connection between two parties is also called "end-to-end," but this use of the term is not relevant to this paper. This paper's use of "end-to-end" refers to the definition used in computer networking from the 1970s forward.

7. The seven-layer model was developed as part of the Open Systems Interconnect (OSI) initiative of the International Organization for Standardization (ISO).

8. The original example given in the article was the problem of "careful file transfer" (p. 510).

9. Meaning that as an approach to designing the system, the design of the end is oriented to communication with the system (in this case, the switch) and not another endpoint. To simplify the term it might be easier to conceptualize it as "from end to system."

10. For instance, Integrated Services Digital Network (ISDN), a way for computers to communicate via telephone deployed in the 1980s, was an attempt to move some intelligence from the center of the (end-system) telephone network to the edge, and to implement some features in smarter terminals (called ISDN terminal adapters) than the traditionally dumb telephone.

11. Stations in military courier networks like the post-house system were usually also distribution points for news and gossip and accepted nonmilitary traffic when extra capacity was available. We can imagine, if we adopt today's terms, a richer topography with nonuniform nodes, some broadcasting, some caching, and some filtering. All of this was, if not planned, at least accepted as normal.

12. Note that "users" may not necessarily mean novices—some users have considerable skill with the technology they are using.

13. See, for instance, a review of the Hush-a-Phone and the Carterfone (Neuman, McKnight, and Solomon 1998, 176–78).

14. In this section I take issue with those who cover for normative goals with technical conclusions, but it is doubtful that the distinction between "technical" and "normative" is ever very useful. Every technological decision includes normative assumptions, even if these are so generally accepted or unexamined that they are ignored.

## REFERENCES

Bar, F., and A. Munk Riis. 2000. Tapping user-driven innovation: A new rationale for universal service. *Information Society* 16 (2): 99–108.

Bar, F., S. Cohen, P. Cowhey, J. B. DeLong, M. Kleeman, and J. Zysman. 2000. Access and innovation policy for the third-generation Internet. *Telecommunications Policy* 24 (6/7): 489–518.

Blumenthal, M. S., and D. D. Clark. 2001. Rethinking the design of the inter-
   net: The end-to-end arguments vs. the brave new world. In *Communications
   Policy in Transition: The Internet and Beyond*, ed. B. M. Compaine and S.
   Greenstein. Cambridge, MA: MIT Press.
Clark, D. D. 1988. The design philosophy of the DARPA Internet Protocols.
   *Computer Communication Review* 18 (4): 106–14.
Clark, D. D., J. Wroclawski, K. R. Sollins, and R. Braden. 2002. Tussle in
   cyberspace: Defining tomorrow's internet. Paper presented at the Annual
   Meeting of the ACM Special Internet Group on Data Communications
   (SIGCOMM), in Pittsburgh, PA. 19 August.
David, P. A. 2001. The evolving accidental information super-highway. *Oxford
   Review of Economic Policy* 17 (2): 159–87.
Dvornik, F. 1974. *Origins of Intelligence Services: The Ancient Near East, Persia, Greece,
   Rome, Byzantium, the Arab Muslim Empires, the Mongol Empire, China, Muscovy.*
   New Brunswick, NJ: Rutgers Univ. Press.
Forbes, R. J. 1954. *Roads to c. 1900.* Vol. 4 of *A History of Technology*, ed. C. Singer.
   Oxford: Clarendon Press.
Holzmann, G. J., and B. Pehrson. 1995. *The Early History of Data Networks.* Los
   Alamitos, CA: IEEE Computer Society Press.
Isenberg, D. S. 1997. The rise of the stupid network. *Computer Telephony* (Au-
   gust): 16–26.
Kruse, H., W. Yurcik, and L. Lessig. 2001. The InterNAT: Policy implications
   of the Internet architecture debate. In *Communications Policy in Transition: The
   Internet and Beyond*, ed. B. M. Compaine and S. Greenstein. Cambridge,
   MA: MIT Press.
Lay, M. G. 1992. *Ways of the World: A History of the World's Roads and the Vehicles
   That Used Them.* New Brunswick, NJ: Rutgers Univ. Press.
Lemley, M. A., and L. Lessig. 2001. The end of end-to-end: Preserving the
   architecture of the Internet in the broadband era. *UCLA Law Review* 48:
   925–72.
Lessig, L. 2000. Innovation, regulation, and the Internet. *American Prospect* 11 (27
   March). Available at http://www.prospect.org/print/V11/10/lessig-1
   .html.
———. 2001. *The Future of Ideas: The Fate of the Commons in a Connected World.* New
   York: Random House.
Mansell, R. 1994. *The New Telecommunications: A Political Economy of Network Evolution.* Thousand Oaks, CA: Sage.
Moors, T. 2002. A critical review of "End-to-End Arguments in System De-
   sign." In *Proceedings of the IEEE International Conference on Communications (ICC).*
   Vol. 5, , 1214–19. New York: IEEE.
Neuman, R. W., L. McKnight, and R. J. Solomon. 1998. *The Gordian Knot: Po-
   litical Gridlock on the Information Highway.* Cambridge, MA: MIT Press.

Pinch, T. J., and W. E. Bijker. 1987. The social construction of facts and arti-facts: Or, how the sociology of science and the sociology of technology might benefit each other. In *The Social Construction of Technological Systems: New Directions in the Sociology and History of Technology*, ed. W. E. Bijker, T. P. Hughes, and T. Pinch. Cambridge, MA: MIT Press.

Reed, D. P. 2000. *The End of the End-to-End Argument* [online]. April 2000 [cited 17 September 2004]. Available at http://www.reed.com/Papers/endofendtoend.html.

Saltzer, J. H., D. P. Reed, and D. D. Clark. (1984). End-to-end arguments in system design. *ACM Transactions on Computer Systems* 2 (4): 277–88.

Streeter, T. 1999. "That deep romantic chasm": Libertarianism, Neoliberal-ism, and the computer culture. In *Communication, Citizenship, and Social Policy: Re-Thinking the Limits of the Welfare State*, ed. A. Calabrese and J. C. Burgel-man. New York: Rowman & Littlefield.

von Hippel, E. 1988. *The Sources of Innovation*. Oxford: Oxford Univ. Press.

# 12

# Technology Policy by Default

## Shaping Communications Technology through Regulatory Policy

CAROLYN GIDEON

## Introduction

The relationship between regulatory policy and technology is difficult to trace, especially when much of the policy that influences technology has a different purpose. For example, policies designed to influence industry structure will often create incentives that influence investment in technology development and deployment. Communications in particular has a long history of policy that predetermines industry structure and thus technology development. Regulation designed to manipulate industry structure has often brought about technologies that favor certain market structures. Currently, the debate on mandatory unbundling of telecommunications networks to promote competition in local telephone and broadband access features the question of whether such policy encourages or discourages investment in technology. This chapter addresses this question by exploring how regulatory policies targeting industry structure shape technology development in telecommunications and Internet.

To clarify the relationship between regulation and technology, I first analyze select policies in the history of telephony and the Internet in the

United States to see how they may have influenced investment in technology. I then discuss the network pricing game, a model that shows when a market is more likely to be a natural or inevitable monopoly and when it is more likely to sustain competition, as a counterfactual for how technology investment might be different had network firms continued to compete. These discussions are then applied to the debate on mandatory network unbundling, considering three types of innovation: cost reduction, facilities deployment, and development of new products and services. The chapter shows that the incentive for investing in cost-reducing innovation does not change with mandatory unbundling, while investment in deployment of broadband facilities and new products and services are likely to benefit from these requirements. The chapter concludes with policy implications about the shaping of technology through regulatory policy.

## Regulation and Communications Technology: A Historical Perspective

Industry structure has long been considered critical to investments in technology. Most prominently, Schumpeter ([1942] 1975) claimed monopoly returns are necessary for investment in new technology. It is tempting to simply reject this idea when we think of the many industries—including computer equipment and semiconductors—that are not monopolies and are characterized by significant investment in technology. In fact, the history of telecommunications regulation and investment in technology shows that competition may better favor technology advancement. This section provides a brief historical analysis of the relationship between policies targeting industry structure and the development of technology for telecommunications and the Internet. While anecdotal—and not meant as a comprehensive history of the policy or the technology—the analysis suggests a strong positive relationship between policies promoting competition and the advancement of technology.

### Predivestiture

When the initial Bell patents for telephones and local network service expired in 1894, there was a rush of entry. Competitors entered, building their own facilities, reaching toward markets neglected by Bell. In particular, they built facilities to provide residential service. Bell had never intended to provide residential telephones and service, considering telephones instruments for business. Predictably, Bell responded by quickly building facilities in new markets, including residential (Noll 2002). The result was a high level of intense competition, with Bell's market share as low as 56 percent in some markets, and many advances in telephone technology, often developed by Bell's rivals (Faulhaber 1987; Brock 1994; Noll 2002). This is an example of technology benefiting from competition without the sharing of facilities. The additional firms developed markets and technologies ignored by the incumbent.

The telephone market's evolution from this competitive state to monopoly was not the natural evolution of a natural monopoly.[1] In fact, early telephone system technology had decreasing returns to scale (Mueller 1997). The Bell Company used its exclusive control of the developing long distance technology and facilities to regain monopoly, refusing to connect non-Bell service providers until threatened with antitrust action. Bell Chairman Theodore Vail also adopted a strategy of embracing regulation through advocating the social goal of universal service, then defined as a single network for all telephone service in the country. Vail successfully promoted a single national monopoly telephone network owned by Bell. The result was a protected national monopoly. Once entry was prohibited, no firms remained to invest in technology that might change the industry's cost structure. The incentive for the monopolist was to improve its cost position given its monopoly structure, thus investing in cost improvements that increased economies of scale. "It is improbable that a firm would be interested in directing technological change and lowering costs at scales of output where it does not operate" (Wenders 1992, 14). By this logic, the capital intensity of telecommunications was likely the result, not the cause, of natural monopoly classification. Thus the monopoly in telephony may be more the result of policy choices than economic conditions. The technology that resulted in natural monopoly cost structure was the result of monopoly created by policy. Cost structure is not exogenous to market structure, and natural monopoly was predetermined by policy.[2]

The long-prevailing monopoly structure appears to have suppressed the development of technology for new products and services. AT&T as a monopolist did not neglect innovation. Through Bell Labs, AT&T invested heavily in research, creating many significant innovations in technology, economics, and other fields (Noll 2002). Despite its substantial capabilities, however, AT&T repeatedly failed to introduce digital and advanced services until pressured by competition to do so. One likely explanation for this failure is that as a monopolist, AT&T tried to limit the uses of the network to those that supported its monopoly status.

The long-distance telephone market provides examples of competition promoting technology advancement when there is sharing of facilities.[3] Long-distance competition began in 1959 when the FCC permitted private use of microwave by large businesses to establish their own networks (Crandall and Waverman 1995). With entry came new technologies not pursued by the monopolist. Many value-added services were first introduced by these firms. The long-distance entrants also brought new technology to the equipment segment, creating demand for new kinds of equipment substantially different from that used by AT&T, and opportunities for entry of new equipment suppliers to develop the equipment needed for the new services offered by the service entrants (Crandall 1991).[4] Competition also resulted in increased development of cost-reducing technology compared to monopoly. For example, long-distance competition brought enormous technological changes in switches, resulting in a dramatic fall in switching costs and an increase in efficiency (Crandall 1991).[5] Thus, the policy decisions that promoted competition spurred significant advances in the development of new technology.

Divestiture

AT&T's divestiture of the regional Bells in 1984 raised concerns that telecommunications research and development (R&D), and Bell Labs in particular, would suffer from the subsequent onslaught of competition (Noll 1987). However, real telecommunications R&D increased substantially during the 1980s—though some amount of this increase was for defense-related research contracted by the government (Crandall 1991). Capital expenditures in telephony tripled from 1970 to 1988, with more than half of this growth contributed by the new competitors (Crandall 1991). After the divestiture, R&D shifted from being centrally controlled by Bell Labs to being distributed across multiple firms, including non-Bell entrants. This, and the greater level of competition in telecommunications services, seems to have resulted in increased investment in technology. The new competitors also invested in new technologies neglected by the AT&T monopoly, namely digital and other advanced services. The result was a "revolution in electronics and digital compression technology [that reduced] the cost of providing a wide array of telecommunications services over copper wires, coaxial cables, and radio circuits" (Crandall and Waverman 1995, 5). In the ten years following divestiture, central office technology became more advanced.

Entry in local telephony also appears to have benefited the development and use of new technology. The entrants were the first to adopt and deploy fiber in local telephone infrastructure, followed by incumbent imitation. Early local telephone competition was fueled by the entrants' advantage in offering higher reliability and greater capability for data and video transmission on their fiber networks than the incumbents could provide on their copper-based infrastructure. The incumbents began to increase their investments in fiber a few years later. After the Telecommunications Act of 1996, competitive local exchange companies (CLECs) entered local service markets largely to provide Internet access. The incumbents did not provide residential broadband service until 2000, and then only under increasing pressure from CLEC DSL offerings (Ferguson 2002).[6] While the technology to provide digital and advanced services in the local markets existed, firms did not invest in these services until there was competition. Similar to the long-distance experience described earlier, competition was the catalyst to providing the new technology.

## Internet

Unlike in the telephone sector, early decisions regarding the Internet targeted technology directly. The public origins of the Internet contrast to the entrepreneurial beginnings of telephony. There was no Internet industry for many years after its development and significant proliferation. The issues of technology policy as a second-order result of regulatory policy are only beginning to emerge. Economic concerns of the market and the firms providing service on different layers of the Internet are beginning to overshadow the technological concerns.

In 1965, when the Defense Advanced Research Projects Agency's first wide-area network experiment showed both how well time-sharing could work and how inadequate the circuit-switched telephone system was for supporting networked computers (Leiner et al. 2000), the

National Science Foundation (NSF) instituted policies to give the private sector incentives to invest in new technologies to improve networking. Later, NSF encouraged its regional networks to extend beyond their academic communities and expand their facilities to seek and serve commercial customers. The economies of scale of the expanded networks could then be exploited to lower subscription costs (Leiner et al. 2000). At the same time, NSF denied commercial use of its national backbone. This exclusion encouraged the commercial development of competitive long-haul networks (Leiner et al. 2000). Unlike the examples from telephony, such policies directly targeted the development of technology, successfully encouraging the capabilities of interest.

This and other design and policy choices allowed for a vibrant and competitive Internet Service Provider (ISP) industry when the Internet was privatized and markets were formed (Downes and Greenstein 1999). While there was less competition in Internet backbones than ISPs, it was not a monopoly. This industry structure supported rapid, decentralized innovation by maintaining end-to-end architecture and philosophy (Blumenthal and Clarke 2001).[7] Lemley and Lessig (2001) claim that this design was critical to the tremendous innovation experienced thus far on the Internet. Another decision made by the engineers and scientists at the earliest stages of Internet design that contributed greatly to subsequent technological development was to base the Internet design on the technical principle of open architecture networking. Concerns have been raised about the evolution of Internet architecture now that it is in private hands. Some advocate regulation to prevent network owners from deviating from these open architecture and end-to-end characteristics in order to allow for the development of a greater variety of applications (Lemley and Lessig 2001; Leiner et al. 2000; but see Sandvig, chapter 11 in this volume, for a contrasting view). Innovation flourished in the Internet industry with multiple providers at all levels, supported by reasoned technology policy.

As policies relating to industry structure, such as unbundling, are now debated, potential implications for the shaping of technology abound, and the above-mentioned cases from telephony and the Internet can be instructive. Three important overriding lessons emerge. First, anecdotal evidence strongly suggests there is greater innovation and investment in technology when there is competition. Second, investments in technology reinforced policies directed at industry structure. Third, policies directly addressing industry structure and technology can be successful.

## An Alternative Evolution: Competing Networks

What if the government had not intervened and created a telecommunications monopoly? Given the cost structure of telephony at the time the monopoly was formed and protected by law, competition may have persisted. An analysis of how network firms compete can provide one counterfactual to how the industry might have evolved without regulatory distortions. The network pricing game (NPG), an analytic framework for understanding how network firms compete in the absence of price regulation, considers pricing and exit decisions to show when competition is sustainable and when monopoly is the inevitable or more likely outcome.[8] Once we understand the factors that make it possible for the small firm to survive, or enable the large firm to induce its rival's exit, we can surmise how the firms may invest in technology in order to improve their positions in the game.

The NPG represents a stylized network communications market in its simplest form, where firms simply maximize profits without price regulation. The game consists of two network firms competing in price in a market with a fixed number of subscribers; thus firms are not competing for new business but simply fighting over each other's existing subscribers, reflecting a more mature market. Networks are fully interconnected, eliminating any network externality differences between them.

The NPG is a two-stage game. In the first stage, the players, Firm 1 (larger) and Firm 2 (smaller), learn their initial market-share allocations, determined exogenously.[9] Firm 1 then sets its price. In stage 2, Firm 2 will exit or stay and set its price. Firm 2 exits when it cannot earn positive profits. The larger firm (Firm 1) does not exit. Subscribers have a preference for lower price, tempered by a hesitancy to switch from their current provider. They will choose to stay or switch networks based on the price difference and their inherent propensity to switch networks for a given price difference, and so reallocate themselves between the firms. Subscribers then purchase service and firms earn profits. Regulators are not players in this game. Firms are assumed to be profit maximizers with identical average fixed and marginal costs.[10] This cost assumption reflects facilities-based competition, though relaxing these assumptions to reflect alternative facilities arrangements does not change the nature of the results.

For each market there is some critical market share, the minimum size Firm 2 must be at the beginning of the game in order to earn profits and stay in the market. If this smaller firm meets this critical market share condition, it will stay in the market and grow.[11] If not, it will exit and leave a monopoly. The critical market share increases when fixed costs and subscriber propensity to switch increase. Critical market share becomes a means of categorizing markets based on these underlying characteristics. Monopoly is inevitable when critical market share is greater than half the market, as only one firm can have more than half the market. Duopoly is always sustainable in markets with critical initial market share of zero.[12] The remaining markets are indeterminate, with both monopoly and duopoly possible. In this third type of market, the smaller firm's initial market share allocation will determine if it can stay in the market or if it must exit.[13] Duopoly is sustainable when the smaller firm meets the critical market share level, which is between zero and half the market.

A market is more likely to be a monopoly when the fixed cost of serving the market is high and when the propensity of subscribers to switch networks for a lower price is high, as this increases the critical market share, or the size a firm must be to stay in the market. In the indeterminate markets, where either monopoly or duopoly is possible depending on firm characteristics, monopoly is more likely when the firms' initial market-share allocations have greater disparity and when the maximum price that can be charged by a monopolist without subscribers beginning to drop service is higher. Even when Firm 2 has a cost advantage, it cannot stay in the market unless it meets the critical market-share condition.

The results of the NPG help us to identify where the firms might direct their investment to improve their profits. Both firms will attempt to gain more market share sooner, by investing in new services or better quality, or by changing those underlying market parameters that can be changed. The large firm will choose, whenever possible, investments to favor monopoly: increasing fixed costs, increasing subscriber propensity to switch, increasing its market share at maturity, and increasing the monopolist price. The rival firm, trying to stay in the market and grow, should target its investment on technologies that would reduce the fixed cost of serving the market or reduce the propensity of subscribers to switch networks, thereby lowering the critical market share.[14]

This analysis suggests that competition benefits technology development, because the two firms have conflicting technology objectives and thus will invest in different technologies. Also, the dominant firm invests in different technologies than it would as a monopolist. The NPG also illustrates the potentially overwhelming importance of market share, which helps explain the current difficulty in bringing competition to local telephone markets. Many competitors may be unable to survive because, even with advanced infrastructure offering higher quality and increased functionality in the form of new services, they could not get enough subscribers to switch from the incumbent to counter the power of the incumbents' installed base.

## Unbundling and the Incentive for Technology Investments

To promote telecommunications competition, the Telecommunications Act of 1996 requires incumbent local telephone companies to share their facilities by leasing unbundled network elements (UNEs) to their competitors. Similarly, open access of cable-provided broadband service is a topic of current debate. Industry incumbents argue against such resource sharing, claiming that sharing eliminates their incentive to innovate, especially as rivals will free ride on their investments and their risk (e.g., Jorde, Sidak, and Teece 2000; Speta 2000; Lopatka and Page 2001). They claim this argument holds for technology for new services, as well as for technology to improve the cost for existing services and deploy advanced facilities. This section applies the relationships found between industry structure and technology innovation in the historical discussion and theoretical results of the NPG to address these arguments for each of these types of technology investment.[15]

### New Products and Services

Some argue that monopoly returns are needed to induce firms to invest in developing technology to provide new broadband services (Jorde, Sidak, and Teece 2000). Incumbents, they argue, will not want to share the means of providing new innovative services with rivals. Any revenues the incumbent would expect from offering a unique or superior product would be decreased if it is forced to share these capabilities with its rivals, who did not make their own investments in this technology. Unbundling allows the rival to provide the same new service without investment in imitating the innovation.

This effect may be mitigated by the effect of competition, however, if facilities sharing is the only way to achieve competition in these industries, as seems to be the case. In the history of telephony, new products and services were introduced largely in times of competition, often by the competitor. Monopolists with large amounts of capital with long depreciation lives, as in telecommunications, are particularly unlikely to invest in new technologies if they might threaten existing products and their revenues.[16] Thus unbundling, by enabling competition, may increase incentives to invest in new products and services. While the incumbent's incentive to invest in innovation when unbundling is required may be less than optimal, the monopolist's incentive is also less than optimal. It is not clear that the incentive to innovate with unbundling would be lower, or even different, than it was for the firm as a monopolist. The decrease in revenues from a potential new product or service due to competition is the direct effect of unbundling on the incumbent's incentive to invest. There is also the indirect effect of unbundling—the incentive to invest to compete with the rival—as well as the rival's incentive to invest in new technology. The policy question is whether the direct effect, which reduces investment, outweighs the indirect effect and the competitor's incentive, which increase investment. Nielsen (2002) shows that in most cases investment is higher in duopoly than in monopoly, a conclusion supported by the foregoing historical analysis of both long distance and local telephone service.

Jorde, Sidak, and Teece (2000) argue that the value of being the first-mover in developing new technology for new services will erode if the incumbents are forced to share the infrastructure for providing these services with their rivals. Although first-mover advantage may be decreased, it is not eliminated. The NPG results show that the advantage of offering the service first and building initial market share is significant. There will always be a lag before rivals have the opportunity to observe the incumbent's new service, request the UNEs (unbundled network elements) needed to provide the service, await the incumbent's setting of prices for these elements, actually obtain the elements from the incumbent, integrate the elements into their own facilities, and market the service and provide it to customers.[17] This delay gives the incumbent opportunity to build market share.

Even if unbundling did chill investment by the incumbent network owners, Lemley and Lessig (2001) point out that there is a bigger pool of potential investors in innovation who would be chilled if networks are not kept open. Lemley and Lessig (2001) argue that open access to competing Internet service providers (ISPs) is necessary to preserve the end-to-end architecture that enabled vast amounts of innovation at the higher layers of the Internet. If broadband network providers are not required to share their networks with competing ISPs, ISP monopolies will cut off potentially infinite unknown opportunities for innovation. Network owners may also redesign the physical layer architecture to be more specifically optimized for the owner's purpose, giving them control to limit new uses of the network.[18] There is evidence that the owners of the networks are already attempting to prevent use of their networks for applications that threaten their legacy businesses (Ferguson 2002). Telephone companies provide asynchronous service for broadband, where upstream speed is significantly lower than downstream, making the network less conducive to voice service that would threaten their traditional telephony markets; cable broadband providers similarly have set limits to the size and duration of broadband video streaming, preventing substitution away from cable television.

Lessons from the historical analysis and NPG indicate that, on balance, unbundling is likely to benefit investment in new products and services. Unbundling enables competition, both bringing the possibility of new product introduction from competitors and increasing the incumbent's incentive for investment. Unbundling also increases the opportunity for investment in new products and services in layers above the physical network.

## Cost Reduction

Incumbents argue that unbundling eliminates their incentive to invest in cost-reducing technology for existing facilities if its "competitors can achieve the identical cost savings by regulatory fiat" (Jorde, Sidak, and Teece 2000, 8). While this argument is intuitively appealing, a closer look suggests it is deficient.

The incumbents' incentive to invest will remain as long as profitable returns are expected. Even when leasing facilities to rivals, an incumbent maintains its full incentive for cost-reducing innovation as it still receives the full benefit of the difference between the old cost and the new cost. The incumbent's benefit from the cost reduction does not change when it must provide the network element to its rival at some arbitrary wholesale price. Regulatory pricing is most often based on incentive regulation, using benchmarks other than a provider's actual network costs to eliminate any disincentive to improve network costs. Incentive

regulation preserves incentives to improve costs, since cost improvements are not passed through to the retail or wholesale purchasers. Thus
cost-reducing innovations improve the incumbent's margin, regardless
of the actual price. Since neither retail nor wholesale prices are based
on costs (retail prices are based on price caps, and wholesale prices are
based on hypothetical networks), revenues do not change with the cost
improvement from innovation. The benefit of the innovation is based
solely on cost savings, which do not change if there is unbundling.[19] The
incumbent's revenues are changed by the unbundling requirement, but
the benefit of a cost-reducing innovation is not. Unbundling changes
the incumbent's revenues whether the investment in innovation is made
or not. If this innovation is profitable when there is no unbundling, then
it is profitable when there is unbundling. Thus unbundling does not
change the incentive for the ILEC (Incumbent Local Exchange Carrier)
to invest in technology that improves its costs. This result is the same
when the innovation lowers fixed costs, or both marginal and fixed
costs, and it would also apply to a cable broadband access provider sharing its network with a competing ISP.

## Deployment

Many argue that requiring broadband owners to share their physical
networks with rivals will eliminate the returns necessary for them to invest in further deployment of broadband infrastructure (e.g., Jorde,
Sidak, and Teece 2000; Speta 2000). Yet historically, deployment activity seems to increase with the existence or threat of competition, both
with and without sharing rules. Prior to the Telecommunications Act of
1996, several states that experienced local telephone entry with established interconnection rules also experienced deployment of fiber by
the local telephone incumbents, imitating the entrants' infrastructure
investments. This is similar to the experience in long distance, when rivals who began as resellers invested in deploying fiber, followed by
AT&T's fiber deployment in response. The persistence of the relationship between competition and increased deployment activity suggests
that competition with unbundling is not likely to harm incentives to invest in broadband deployment.

    With unbundling, network elements are sold in a wholesale market
with price regulation, as is the case for retail business. Like retail prices
(under price cap regulation), network element prices are derived using

incentive regulation, based on efficient costs and not the incumbent's actual costs. Greenstein, McMaster, and Spiller (1995) show that incentive regulation increases the incumbent's incentives to deploy digital infrastructure.[20] In this sense, we could expect that just as incentive regulation of retail prices encourages advanced infrastructure deployment, so would such wholesale price regulation. However, if it is true that regulated wholesale prices for network elements are damagingly below actual costs, the effect could be different as the incumbent's ability to fund such investments would be damaged. If it is found that prices are not appropriate, the argument that network sharing with price regulation does not harm deployment incentives holds. It would simply indicate that a new price algorithm is needed.

## Conclusion

Observations from the history of telephony and the Internet provide some indication of how policies that target industry structure can shape technology development. A brief historical overview shows a strong positive relationship between policies that promote competition and technological advance in telecommunications. Similarly, it appears from this history that policies favoring monopoly will tend to result in capital-intensive technology that reinforces dominance by increasing barriers to entry. The development of the Internet shows the potential for open networks designed for multiple purposes to enable both competition and innovation. Policy properly directed can effectively promote technology and competition.

The network pricing game shows an alternative way network firms may compete without the legacy of regulation. It shows that some markets, based on their underlying characteristics, are inevitable monopolies. Other markets will always support two network service providers, and still others will support a second firm only if that firm can maintain sufficient market share. It also shows that firms in this situation would target their investment in technology to change the underlying parameters of the markets to their benefit.

Applying these theoretical and historical findings to unbundling suggests how unbundling might shape technology. On balance, unbundling appears to increase investment in technology by enabling greater competition. Thus recent decisions to eliminate the line-sharing requirement for DSL service and not impose a similar requirement on cable-modem networks may have been mistakes. Without line sharing and open access requirements, innovation at the physical network layer will be in the control of monopolists with incentives to invest less than duopolists and thus direct their investment to technologies that reinforce their dominance.

It is encouraging, however, that the effects of regulation on technology in communications currently are being considered more explicitly in such decisions. Innovation was given a prominent role in the rationale behind the recent proceedings regarding the sharing of unbundled network resources for broadband service (FCC 2003). Yet it seems to be used in the debate almost as a basis for obfuscation. Both sides of the debate evince equal concern for the development of technology, indicating that we need a better understanding of how regulatory policy shapes technology development. This chapter takes a step in this direction by providing a framework for understanding how regulatory policy shapes technology, and by raising many questions about this relationship that remain to be answered with empirical analysis.

Technological change and innovations that alter the underlying parameters of specific markets can overcome monopolies. Investment in technology is part of the feedback loop that enables monopoly to reinforce itself, not only with market dominance that gives it power in the pricing game but also by making the type of innovation that could change the cost structure or make real disruption more difficult to achieve. The historical evidence and theoretical findings presented here show that competition increases advancement in technology. If this finding is true, policies that promote competition will generally promote innovation. Deregulation will not necessarily have this effect. As the network pricing game shows, some markets are inevitable monopolies and require intervention to sustain competition. Much of the argument against line-sharing arrangements have emphasized that such policies would sacrifice investment. However, policies that sacrifice competition to promote investment in technology are likely to be suboptimal on both accounts, resulting in less innovation after all.

NOTES

1. A natural monopoly occurs when the total cost of serving the entire market is lowest when the market is served by a single firm. Note that this condition is based on cost structure, not the actual number of firms in a market.

2. Regulation and policy are also not truly exogenous, influenced by lobbying, and particularly vulnerable to a single monopolist or dominant firm with deep pockets.

3. MCI began providing service as a reseller, later building its own facilities.

4. For example, Northern Telecom introduced the first digital, time-division local-office switch in 1976. AT&T did not have an equivalent offering until 1982 (Crandall 1991).

5. Efficiency was measured in terms of total factor productivity (TFP), which is the ratio of output to capital and labor inputs.

6. There is also some anecdotal evidence that AT&T resisted the Internet's development and interconnection of other networks to its own because of the threat to its business model. See Lemley and Lessig (2001).

7. The end-to-end design of the Internet requires that the lower, physical layer of the network be designed as a general purpose network, with all specific application-level functions built into the higher layers, creating an open platform that maximizes innovation opportunities by enabling competition in innovation.

8. This section briefly summarizes the network pricing game developed in Gideon (2003). A complete description of the model and its findings can be found in that dissertation.

9. The exogenous process includes past circumstances that occur before the game begins.

10. In reality, with different types of firms crossing traditional industry boundaries to compete in providing the same type of networked service, though very likely with a different technology or infrastructure, we would expect to see different costs for different firms. The assumption of identical costs is made for simplicity and to isolate the effect of asymmetric initial market shares. Relaxing this assumption does not change the conclusions.

11. Fixed costs are assumed to be the costs of business that are not sensitive to subscribers, market share, or volume. Fixed costs may include maintenance of infrastructure, depreciation, administrative costs, and the like. Fixed costs occur in each period. Fixed costs do not include the capital investment already made to enter the market, except in terms of depreciation and potentially high maintenance costs.

The larger firm always charges a higher price, so some of its subscribers will switch to the smaller network. For proofs and detailed explanations of these results, see Gideon (2003).

12. In such markets, a firm with zero market share gains enough subscribers in the switching process to earn profits. This occurs when fixed costs are low, reducing the amount of contribution needed above marginal costs, and also when propensity to switch is low, enabling the firms to set higher prices.

13. There is also a dynamic version of this game in which firms have one

period to invest in market share before the small firm makes its exit decision in the second period. The results are qualitatively the same. In the dynamic game, Firm 2's ability to stay in markets where either monopoly or duopoly is possible depends also on the maximum price that can be charged by a monopolist without inducing subscribers to drop their service.

14. While an increase in subscriber propensity to switch increases the critical market share, making it harder for the smaller firm to stay in the market, the role of propensity to switch in determining the potential for sustainable competition can change in certain cases. For a monopoly where competition will come only from entry, propensity to switch must be sufficiently high to allow entry if there is to be any competition.

15. The historical analysis provides parallels to the current situation for both facilities-based and reselling competition, at different times and different stages of market maturity.

16. See Scherer's (1996) account of IBM's delaying the introduction of new products to prevent cannibalism of existing products.

17. Additional delays will be experienced if wholesale price disputes must be resolved through regulatory or legal channels.

18. According to Lemley and Lessig (2001), in the end-to-end network, layers at the lower level of the network should provide a broad range of resources that are not particular to, or optimized for, any single application, even if that means sacrificing a physical layer design that is more efficient for at least some applications. They later contrast this with the physical layers of the networks constructed by monopolists in telephony and cable television, which are optimized for their single applications.

19. Even if the regulated price was based on the incumbent's actual cost, there is still incentive to innovate as there is a lag time before the cost improvement is reflected in a reduction in the regulated prices.

20. This study analyzed digital infrastructure deployment after the 1984 breakup of AT&T using data from 1986 to 1996. Note that not all markets had competitors at this time. They also found that when price-cap regulation was combined with earning sharing schemes, the incentive for investment in infrastructure deployment was reduced.

### REFERENCES

Blumenthal, M. S., and D. D. Clark. 2001. Rethinking the design of the Internet. In *Communications Policy in Transition*, ed. B. M. Compaine and S. Greenstein. Cambridge, MA: MIT Press.

Brock, G. W. 1994. *Telecommunications Policy for the Information Age*. Cambridge, MA: Harvard Univ. Press.

Crandall, R. W. 1991. *After the Breakup*. Washington, DC: The Brookings Institution.

Crandall, R. W., and L. Waverman. 1995. *Talk Is Cheap*. Washington, DC: The Brookings Institution.

Downes, T. A., and S. M. Greenstein. 1999. Do commercial ISPs provide universal access? In *Competition, Regulation and Convergence*, ed. S. E. Gillett and I. Vogelsang. Mahwah, NJ: Lawrence Erlbaum Associates.

Faulhaber, G. R. 1987. *Telecommunications in Turmoil*. Cambridge, MA: Ballinger Publishing Company.

Federal Communications Commission (FCC). 2003. FCC adopts new rules for network unbundling obligations of incumbent local phone carriers. Press release, Feb. 20.

Ferguson, Charles H. 2002. The U.S. broadband problem. Policy brief number 105. Washington, DC: The Brookings Institution.

Gideon, C. 2003. Sustainable competition or inevitable monopoly? The potential for competition in network communications industries. PhD diss., Harvard Univ.

Greenstein, S., S. McMaster, and P. Spiller. 1995. The effect of incentive regulation on infrastructure modernization. *Journal of Economics & Management Strategy* 4 (2): 187–236.

Jorde, T. M., J. G. Sidak, and D. J. Teece. 2000. Innovation, investment and unbundling. *Yale Journal of Regulation* 17 (1): 1–37.

Leiner, B. M., V. G. Cerf, D. D. Clark, R. E. Kahn, L. Kleinrock, D. C. Lynch, J. Postel, L. G. Roberts, and S. Wolff. 2000. A brief history of the Internet. www.isoc.org/internet/history/brief.shmtl (August).

Lemley, M. A., and L. Lessig. 2001. The end of end-to-end. *UCLA Law Review* 48 (4): 925–72.

Lopatka, J. E., and W. H. Page. 2001. Internet regulation and consumer welfare. *Hastings Law Journal* 52: 891–928.

Mueller, M. 1997. *Universal Service*. Cambridge, MA: MIT Press.

Nielsen, M. J. 2002. Competition and irreversible investments. *International Journal of Industrial Organization* 20: 731–43.

Noll, A. M. 2002. Telecommunication basic research: An uncertain future for the Bell legacy. Paper presented at 30th Research Conference on Information, Communication, and Internet Policy, TPRC.

———. 1987. Bell System R&D activities. *Telecommunications Policy* 11 (2): 161–78.

Scherer, F. M. 1996. Computers. Chap. 7 in *Industry Structure, Strategy and Public Policy*. New York: HarperCollins College Publishers.

Schumpeter, J. [1942] 1975. *Capitalism, Socialism, and Democracy*. Rev. ed. New York: Harper Perennial.

Speta, J. 2000. Handicapping the race for the last mile? *Yale Journal of Regulation* 17 (1): 39–91.

Wenders, J. T. 1992. Unnatural monopoly in telecommunications. *Telecommunications Policy* 16 (1): 13–15.

# PART 4

# SHAPING LIFE

S cience and technology policy has, like many areas of study and action, had to shift its foci to keep pace with the changing scientific and political arenas that it engages. This section acknowledges that contemporary advances in the life sciences, particularly genomics and reproductive technologies, and their associated research programs and therapeutic techniques, are a central scholarly and practical concern across the globe.

Although the strands of DNA are often portrayed as things that unite us as humans (and indeed, even reveal our commonalities with chimpanzees, worms, and yeast!), the authors here examine how social conditions including culture, ethics, race, and ethnicity create differences in ethical perceptions of DNA technologies, ethical practices, and the distribution of benefits. For these authors, the cultural and political inputs into policy are just as critical in shaping the human outcomes of the technologies of life as are the scientific tools and the therapeutic techniques.

The Human Genome Project promises myriad and beneficial genetic technologies. The question remains, however: Which segments of society will benefit from these technologies? Majority communities have and continue to benefit a great deal more than minorities. Tené Hamilton Franklin, a genetics counselor, describes in her chapter the Communities of Color and Genetics Policy Project, which used a model of civic mobilization developed during the civil rights movement to engage African Americans and Latinos of diverse socioeconomic levels in dialogues about genetic research and technology. From these dialogues, the project developed recommendations for laws, professional standards, and institutional policies regarding the use and application of genomics. The project conducted focus groups and multiple dialogue sessions with fifteen community-based organizations in Alabama and Michigan, with discussions that addressed not only genomics but also concerns about access to health care, minority inclusion in human subjects research, and safeguards to protect against research abuses. Replicating such efforts will facilitate public education about genomics and enable the public to collaborate with ethicists, researchers, medical practitioners, and lawmakers to shape science, technology, and policies in ways that provide more equal opportunities for underrepresented and underserved communities.

The next chapter, by philosopher Michael Barr, confronts a different aspect of community in the age of genomics. Barr investigates the ethical aspects of the linkage of large-scale, population-based DNA samples with personal medical information, a much-anticipated method for studying the combined effect of genes, environment, and lifestyle factors on diseases. The United Kingdom, for example, approved £60 million ($90 million) for a national genetic database to link personal medical information contained in National Health Service records, with blood samples collected from 500,000 middle-aged volunteers. Known as BioBank UK, the project is organized by the Wellcome Trust, the Medical Research Council, and the Department of Health. However, BioBank UK has called into question long-standing ethical principles, foremost among them informed consent. Traditional notions of consent were not designed for research on the scope or scale of population-based genetic studies. Moreover, protocols often differ based upon whether research will reveal clinically relevant information. These complications raise questions including the possibility and desirability of community consent, the thinness of information when the future use of samples is unknown, and the opportunity for re-consent. The chapter draws upon British and North American sources to evaluate options for the shaping of genetic epidemiology in an ethical way.

Charlotte Augst, a genetic information project manager, also uses a comparative framework to explore the legal shaping of life science and technology in Britain and Germany. Both countries began debating embryo research and infertility treatments in the mid-1980s and regulating it in 1990. Although similar concerns about alienation, eugenics, the value of life, and women's reproductive decisionmaking surfaced in both national debates, the legislatures produced different conclusions. The British Human Fertilisation and Embryology Act permits embryo research, the treatment of unmarried women, and gamete donation and surrogacy. The German Embryo Protection Act outlaws these practices. Augst's close reading of the parliamentary discourses suggests a struggle with the effects of modernization, understood as the occasionally conflicting and always ambivalent importance of processes of individualization and of scientific and technological progress. She concludes that this struggle cannot be resolved without contradictions, and yet both national "solutions" are based on a denial of the complex and contradictory effects of reproductive technologies as instances of modernization. Augst's reading of parliamentary records allows us to understand how politics comes to terms with a controversial new technology and shapes the lives that women will lead and give rise to.

Many scholars of globalization point to the uniformity of contemporary science and technology to explain increasing transnational linkages. Shobita Parthasarathy, a scholar in science and technology studies, questions the relationship between contemporary genomic medicine and the new global order. Our genetic codes are 99.9 percent identical to one another, and multinational corporations fund much research and development in this area. Parthasarathy's chapter examines the attempt by the American genomics company Myriad Genetics to globalize its service for genetic testing for breast cancer. In the mid-1990s, Myriad acquired patents for BRCA1 and BRCA2, the breast cancer genes, to become the sole provider of BRCA testing in the United States. As it tried to develop an international network of testing services across Europe and Asia, however, Myriad faced considerable opposition to both its technology and its understanding of the relationship among health care professionals, patients, and testing laboratories. Parthasarathy describes in detail how scientists, clinicians, and governments in Britain reasserted a national approach to genetic testing technology, forestalling the expansion of Myriad's services. The chapter provides an empirically rich demonstration of how national priorities and norms can shape the technologies of life, even within the processes of globalization.

Together, these authors expand the purview of science and technology policy to one that is more global in vision and more sensitive to ethnic, cultural, and political contexts. They directly confront both the great and the troubling potential that life science and technology offer, and they argue for the importance of asserting community, culture, and politics in sharing a role in the shaping of human life.

# 13

# Engaging Diverse Communities in Shaping Genetics Policy

*Who Gets to Shape the New Biotechnology*

TENÉ HAMILTON FRANKLIN

## Introduction

The Human Genome Project brings promise of new gene therapies and possible cures for many medical conditions. But this technology includes the risk of employer or insurance discrimination based on personal genetic information; the fear of unethical research practices; and the challenge of providing equitable access to the benefits of this technology. These concerns particularly affect ethnic minorities such as African Americans and Latinos because of the past injustices that African Americans historically have suffered in our society, including those at the hands of health-care professionals and scientists.

As an effort to confront this history of mistreatment, inequity, and distrust, the *Communities of Color and Genetics Policy Project* developed a process for engaging African Americans and Latinos of diverse socio-economic levels in dialogues relating to genome research and its resulting technology, and from these dialogues developed recommendations for laws, professional standards, and institutional policies regarding the use and application of genome research and technology. In this paper I first discuss the process we implemented for cultivating community discourse, and then present the results of our dialogue process and its implications for public policy.

## Background

The Human Genome Project (HGP) is an international, collaborative research program whose goal is the complete mapping and understanding of all the genes of human beings—that is, the human genome. The HGP began formally in 1990 as a thirteen-year effort coordinated by the U.S. Department of Energy and the National Institutes of Health. The project originally was planned to last fifteen years, but rapid technological advances accelerated the completion date, and in April 2003, the completion of the HGP was announced. The HGP has successfully sequenced and mapped 99.9 percent of our human DNA. After the genes are identified, the sequences of the three billion chemical bases that make up human DNA are determined. This information is stored in databases where it will be used to develop genetic technologies. The existing and ultimate products of the Human Genome Project will give the world a resource of detailed information about the structure, organization, and function of the complete set of human genes. This information can be thought of as that basic set of inheritable "instructions" for the development and function of a human being. The Human Genome Project promises great advances in medicine and research. However, this technology will continue to raise significant ethical, legal, and social implications that may require societal response. For this reason, approximately five percent of federal funding for the National Human Genome Research Institute has been allocated to address the ethical, legal, and social implications that might arise from the project (NHGRI 2002).

Which segments of society will benefit from the myriad technologies likely to derive from the knowledge created by the Human Genome Project? Traditionally, majority communities have benefited a great deal more from genetics research than ethnic minorities. Will this pattern continue to hold?

### Distrust

In the African American community, human subjects research, including research in genetics, is often accompanied by distrust of the researchers. The distrust reflects the history of research abuses that have been suffered by people of color. In a particularly notorious example of human subjects research abuse, the *U.S. Public Health Service Study of Syphilis at Tuskegee*, African American men of Tuskegee, Alabama, and the surrounding Macon County unknowingly participated in an experiment to examine the affects of untreated syphilis in the black male. These men were not informed about the research risks associated with this study and therefore were not able to provide informed consent or their permission to participate in the study (Jones 1981).

Today, research in genetic medicine has progressed to a point where human subjects participation is imperative. Cures and treatments for many medical and genetic conditions depend on clinical trials and other tests involving informed human participation. But the legacy of the *Tuskegee* study magnifies the need for human subjects research to be monitored to protect people from abuses. This need is particularly acute for African Americans, who, from slavery, have been subjected to abusive research practices that have created a legacy of distrust and a need to take precautions to make sure that such abuses do not occur again (Savitt 1978).

Barriers to entry into a clinical research study include distrust, economic factors, lack of awareness about clinical research, and poor communication. It is not uncommon for some researchers explaining the research study to the participant to do a poor job conveying the risks and benefits. This creates a situation in which the participant may not completely understand the limitations of the study, thereby not being able to grant true informed consent.

These barriers can be overcome by providing education about the research, and fostering trust that can develop from open communication. Trust can be developed by using culturally sensitive study staff with good communication skills, by treating the patient with respect, and by taking the time to explain the research to potential subjects in understandable terms (Gorelick 1998).

## Underserved and Underrepresented

African Americans familiar with the Human Genome Project recognized at an early stage that it was more than an exercise in listing the locations of genes; it was from the onset a project with profound social, medical, and political implications, particularly for groups not systematically included in the research but subject to its health and policy implications. Furthermore, African Americans are underrepresented among genome scientists and are likely to be highly vulnerable to any genome-based technologies and policies that emerge from research efforts but do not reflect the breadth and depth of African American genetic diversity (Jackson 1999).

Health disparities for ethnic minorities are rooted in the sociopolitical and economic history of the American society (see, e.g., Fiscella 2000; Clayton and Byrd 2000). For example, unequal access to employment opportunities leaves a large percentage of people of color such as Mexican Americans uninsured or underinsured. Therefore, they do not have a regular source of medical care; they often endure pain and illness until symptoms become severe, and thus more often seek medical care from hospital emergency rooms (Castro et al. 1995). Moreover, the perception of the occurrence and extent of ethnic disparities in health and health care differs between white and ethnic minority populations in the United States (*Perceptions* 1999).

## Community Organizing

Engaging the community to develop public policy is a historical concept that was particularly important during the civil rights movement of the twentieth century. The civil rights movement was largely successful because it was able to draw upon grassroots organizations, preexisting networks, and communication infrastructures, through which experienced leaders were able to provide direction to the movement (Morris 1984). In the 1930s and 1940s, African Americans joined neighborhood relief committees and attempted to shape public policy, particularly in relation to poor people. They fought for welfare relief and public health services, and against evictions and foreclosures (Kelley 1990). Given the challenges raised by the Human Genome Project for African Americans, community-based action offers an appropriate and well-tested approach for confronting the HGP's implications.

# The Importance of Community-Based Organizations

The *Communities of Color and Genetics Policy Project* was conducted by a consortium of the University of Michigan, Michigan State University, and

Tuskegee University, working closely with fifteen community-based organizations in Alabama and Michigan over a period of three years (1999–2003). Focus groups and multiple dialogue sessions were held with each community-based organization over a period of approximately one year to identify and discuss issues of concern and ultimately make policy recommendations about genome science and technology. The discussions were not limited to the genome technology but also included concerns about access to health care, minority inclusion in human subjects research, and safeguards to protect against research abuses. Using such discussions as a model in working with the public, the project was able to effectively engage ethnic minorities, a segment of our society that is currently underrepresented and underserved in the fields of science, technology, and medicine.

Community-based organizations (CBOs) participated in the *Communities of Color and Genetic Policy Project* by hosting dialogue sessions and by recruiting dialogue participants. Five African American dialogue groups were held in Tuskegee, Alabama, and five were held in various Michigan communities. All five of the Latino dialogue groups were in Michigan. Each of the community-based organizations hosted a set of dialogues that met for a total of six sessions. The composition of the Michigan African American groups varied from 31 percent to 65 percent male; the Tuskegee groups were all predominantly female, with the proportion of women varying from 64 percent to 100 percent.

The CBOs in Tuskegee were identified by word of mouth. Since the project team was new to the area, we elicited the assistance of individuals who were familiar with the community in identifying possible organizations to participate in the project. Three churches were recommended because of their relationships and previous commitment to university-related projects. The housing authority was recommended as a source of identifying grassroots individuals to help balance out our sample of participants. A support group for individuals with disabilities was identified because it was thought that this group could provide a unique perspective since they had firsthand experience about genetic conditions.

CBOs in Michigan were chosen in a similar fashion. An effort was made to ensure that both men and women were represented. In addition, efforts were made to include both African Americans and Latinos with a range of educational backgrounds and life experiences.

Community-based organizations played a significant role in the *Communities of Color and Genetic Policy Project* in three ways. First, they were knowledgeable about the constituents and the components that made up their community. Second, the CBOs hosted the dialogue sessions in their office space. Some organizations met in churches, community centers, and in public housing developments. Others met in the homes of the members. The project team conducted the dialogue sessions at the locations where the CBOs held their general meetings. Finally, the CBOs took on an advocacy role for their local communities (Tuskegee and Michigan), as well as to the academic project team responsible for leading the investigation, and later to policymakers.

## Engaging CBOs

Community-based organizations are viewed as the centerpiece of the community, and they served as effective vehicles for engaging local populations in dialogues about genetics. Having a researcher or the project team go into a particular community to recruit dialogue participants by making general announcements and passing out flyers would not have been as effective as using the community-based organizations to perform these tasks. We consciously adopted a model for recruitment and organizing dialogue groups that was similar to that utilized by grassroots organizations and activists in the southern civil rights groups. The primary concern of the project team, similar to the concern of civil rights activists, was building a relationship with people in Alabama and Michigan. We were able to do this by working with CBOs that were respected through the communities, and then relying upon their guidance and direction to approach the larger communities and their constituents.

Because genetics is strongly tied to issues of health care and health-care practice, the project and its political implications resonated among the participating CBOs. Generally, participatory initiatives—whether they involve larger organizing campaigns or research activities—will have greater success if they are viewed as beneficial and appear to meet the needs of the communities that are the subject of the study. CBOs then are useful for raising the awareness of their constituents and the larger community to public policy issues. The academic team and project directors consciously sought to assume a role secondary to the CBOs. We did this by continually seeking the advice and consultation of the

CBOs on how this investigation should take place throughout the multiple phases of the project. This was important particularly because the CBOs were knowledgeable about their constituents, and thus were able to suggest methods for informing the broader community in order to increase participation in the dialogue groups. More importantly, the CBOs had some insight about the dangers of unfair health-care practices in their communities. In Tuskegee, for example, the CBO members, regardless of their socioeconomic background, had some knowledge about previous health-care practices, especially the infamous *Tuskegee* study. They knew how to effectively inform the broader community about the project, thus leading to increased participation. Because community-based organizations are an integral part of the community, it followed naturally that they would be able to serve as advocates to the community, to the project team, and later to the policymakers.

In identifying community-based organizations to participate in the project, CBOs that were well known in the African American and Latino communities of Tuskegee and the areas near the University of Michigan and Michigan State University were sought out by the project team. The CBOs that agreed to participate in the investigation also agreed to take on the role of community partners to the academic project team. They proposed effective ways to inform individuals about the project, ultimately leading to involvement of a diverse cross-section of participants. For example, the CBOs in Tuskegee uniformly proposed that the project team first hold informal sessions with members of the organizations. This allowed a member of the project team (usually the project director) to describe the project and propose opportunities for collaboration. More importantly, it allowed the CBO members to have a face-to-face meeting with one of the principal investigators of the project before the official start of the focus groups and dialogue sessions. Again, it is important to point out that relationship-building between the investigators and dialogue participants was instrumental to the success of the overall investigation. This important step allowed the community to feel comfortable about the potential of being a part of the research project. In Tuskegee, for example, the project director met several times with various members of each organization, starting with an introductory meeting with one person affiliated with the CBO to gauge potential interest. Next, the project director met with the leaders of each organization to explain the project in detail. Upon approval of the leadership of the organization, the project director offered an additional orientation of the project to members of the entire organization. After this series of meetings, the members of each organization then decided if they wanted their groups to participate in the project. All of the groups approached in Tuskegee elected to participate in the project. Finally, the project director and the representatives collaborated on finding effective ways to engage community participants.

Dialogue sessions were the primary method that the project team used to gather data. The long-standing presence of the CBOs in the communities where we anticipated implementing the project offered a great advantage for organizing these sessions. CBOs linked their constituents, families, and community and social institutions to the project team. In fact, some organization members invited their own family members, friends, and neighbors to participate in the dialogues. The church organizations invited people from that community, who were not members of that particular church, to participate.

## Community Advisory Board

Each of the fifteen community-based organizations chose a representative to serve on the Community Advisory Board (CAB) and act as a liaison between their organization and the academic project team. The representatives to the CAB consisted of rank-and-file community-based organization members, as well as individuals who held high standing in their communities or were considered prominent spokespersons or leaders.

The CAB members were modestly compensated for their involvement. Most held full-time jobs and were also active in their respective organizations. Nonetheless, they were required to carry out a number of time-consuming tasks in order to guarantee that the project was successful. For example, they participated in project team meetings and relayed information between the project team and their organizations. They assisted in the recruitment activities, such as placing phone calls to their organization members, posting announcements in area churches, and issuing personal invitations to participate in the dialogues. Furthermore, CAB representatives helped to facilitate appropriate times and locations in order to accommodate the schedules and demands of the participants, particularly those who were low-income, had no transportation, and had difficulties finding childcare. This was important since most of the participants were women who had children to care for and because the majority of the meetings occurred in the early evenings.

Dialogue Sessions

Each community-based organization participated in one focus group and later met for a total of six dialogue sessions. The majority of the organizations met on a weekly basis while a few organizations met when time allowed over a period of two months. To encourage continued participation at each session, the CAB members sent out weekly letters and followed up with phone calls to the participants reminding them of the upcoming dialogue sessions. In addition to this, the project director also took on an organizing role by assisting the CAB members in following up with the project participants.

Each dialogue session was audiotaped and included a trained facilitator and observer. Both the facilitator and observer took notes on the session, and each dialogue session lasted approximately two hours. By the time the discussions came to a close, the groups were each able to provide consensus recommendations.

The topics of discussion for the dialogue sessions were identified during the focus groups. The most requested topics across all of the focus groups were chosen and a set of materials developed to help focus each topic of discussion. Not surprisingly, the participants brought diverse viewpoints about genetic research and technologies to the table, and opposing views and opinions were discussed. Each person called upon a set of experiences that helped form their opinions about the sometimes controversial issues of genetic technologies. However, after the participants thoroughly discussed a given topic, a general consensus was reached, and on this basis a set of recommendations was proposed.

Reports of recommended actions were developed from each dialogue session. These reports were taken back to the groups to verify that the recommendations accurately reflected each group's intent. After each group verified the accuracy of the recommendations, a summary report was developed that included all sides of the issues discussed among the participants. Even though the various groups often arrived at different results for a given topic, the project team carefully merged the individual topic reports from each organization into one summary report.

From the report, members of the CAB and project team developed position papers. The position papers and recommendations were disseminated to the Alabama and Michigan state legislatures as well as the U.S. Senate and House of Representatives. Members of the community of various CBOs presented the project findings at policy briefings for the state and federal governments. The position papers are also being distributed to federal and state agencies, advocacy organizations, professional organizations, and health-care providers.

## Recommendations from the Dialogue Process

The *Communities of Color and Genetics Policy Project* engaged African Americans and Latinos of diverse socioeconomic levels in dialogues relating to genome research and its resulting technology, and from these dialogues developed recommendations for laws, professional standards, and institutional policies regarding the use and application of genome research and technology in three areas: genetic research, education about genome technologies, and issues surrounding trust and distrust in utilizing this technology.

### Genetic Research

The project addressed the issue of how the African American community can continue to embrace the promise of genetic research and yet ensure that the proper safeguards are in place. The dialogues yielded the following recommendations:

1. In order to ensure equity and equality in the conduct of research, steps must be taken to adequately represent people of color among researchers and research participants.
2. There should be careful assessment of the risks associated with all human research, and this information needs to be widely publicized. There also needs to be ample opportunity for debate and community input on these issues via community forums.
3. There must be a strong commitment to the principles and practices of voluntary informed consent for all research involving human subjects.

Gov focus on possible benefit + no analysis of possible risk?

4. Human embryonic stem cell research should be approached cautiously. If public funds were to support such research, then human embryos that are left over from in vitro fertilization efforts should be used only if the couples give their explicit consent for the specific use of their embryos. However, public funding should not go to the creation of human embryos for the purpose of stem cell research. There must be policies in place that would prevent human embryonic stem cells from becoming a market commodity. Therefore, payment for sperm and egg donors in connection to human embryonic stem cell research should not be allowed.

5. There should be policies encouraging private companies doing genetic research to hold community forums during which they would inform the public about the type of research that they are engaged in (*Communities* 2002).

 Education

The project considered the issue of who should be responsible for the education of the community about genetics and the emerging technologies. The dialogues yielded the following recommendations:

1. The federal government should take the lead in funding broad public educational efforts regarding emerging genetic technologies, current genetic research, and current policy issues.

2. A diverse group of educators, parents, community leaders, and citizens should jointly decide the precise form of genetics education in the school system. This education should include both science issues and the ethical, legal, and social implications of genetics research and technology.

3. The health-care industry should take measures to include people of color when engaging in genetic education programs (*Communities* 2002).

Trust and Distrust

Issues of trust and distrust run through each of the six position papers. All dialogue groups voiced distrust of the ability of government, research institutions, and private sector organizations to conduct research and apply genetics technology to benefit communities of color and thus avoid discrimination against and exploitation of these communities. In view of this distrust, the dialogues offered the following recommendations:

1. Advisory committees to the principal governmental agencies carrying out genetic research and regulating such research must be diverse in their lay and expert membership.

2. Private industry, including research organizations and health-services providers, should be urged to create advisory bodies reviewing their genetics research and the provision of genetic services, to assure the protection of research subjects and service recipients. These groups should be diverse in their lay and expert membership.

3. Advocacy organizations should network with the advisory groups referred to in recommendations 1 and 2, to share their findings and recommendations, to strengthen the impact of their work, and where appropriate, to recommend and advocate for government policies furthering their goals.

4. Federal funding for genetics research must be maintained at a high enough level relative to private funding of research, to ensure that research is guided by social needs as well as to maintain openness of the research and minimize the impact of profit motivation on research goals.

5. Institutional Review Board procedures should be amended to address the protection of groups as well as individuals, with particular emphasis on minorities.

6. More people of color should assume ownership and executive positions in private industry. Increasing the number of people of color who receive genetics education is one strategy to promote such involvement (*Communities* 2002).

## The Discussion beyond Genetics

These recommendations were not always easily extracted. Among the participants, the point was often made that discussions about genetic technologies were moot until basic health-care needs were being met in the minority communities. Sometimes the topic of conversation digressed from debating the pros and cons of genetic research to discuss the reality in American society that ethnic minorities as a whole are less likely to receive adequate health care compared to that of white Americans. The facilitators made a decision to engage the groups in conversations outside the realm of genetic technologies because the participants felt that the discussions about health-care disparities were inherent in critical thinking about genetics.

The participants felt that perhaps the discussions should revolve around the topic of access to health care as opposed to the topic of access to genetic services. The reasoning that the participants provided can be summarized by a Tuskegee, Alabama, participant: "What good will talking about genetic services do when we do not have adequate health care in our communities? Perhaps we should talk about ways to ensure better health-care services for our [African American] communities."

## Conclusion

The *Communities of Color and Genetic Policy Project* yielded over two hundred hours of discussion from African Americans and Latinos about genome technologies and the Human Genome project. The recommendations that I have summarized do not do justice to the richness of the discussion and the diversity of reasons that lie behind the recommendations. In particular, the recommendations reflect but cannot capture the tone of the dialogues and the way in which they reflect the experiences of the African American and Latino cultures here in the United States. For example, the discussions reflected on the health-care disparities that are invariably experienced in the African American and Latino communities. Discussions of the promise genetic technologies were thus accompanied by consideration of how such technologies might make a difference in combating health-care disparities in these communities.

In January 1994, a small group of African American biological and social scientists and community activists met to discuss the options available for African Americans in major genomic studies. This discussion resulted in the creation of a *Manifesto on Genomic Studies among African-Americans*. Five aspects of the manifesto mirror the discussions of the *Communities of Color and Genetic Policy Project*.

1. African Americans expect full inclusion in any world survey of human genomic diversity.
2. It is imperative that systematic sampling, either model-based or design-based (probability) be used to identify the broad range of variation existing among African

Americans and that this diversity be linked to other relevant social, cultural, historical, and ecological features of the African American existence.

3. Establishment of a National Review Panel for ongoing evaluations of genomic studies among African Americans.
4. African Americans must participate in the research design, research implementation, data collection, data analysis, data interpretation, and dissemination of research results.
5. Genomic sampling of African Americans will be linked to improvements in the provision of health and educational services to the African American community. The largest proportions of health problems in the African American community are due to disorders that are preventable. These preventable environmental and gene-factor diseases should be addressed in conjunction with genomic studies. Indeed, the target of both disease prevention and genomic studies among African Americans must be the improved health and well-being of the population and its enhanced survival into the coming centuries. Groups and individuals wishing to conduct genomic sampling among African Americans are expected to concurrently provide meaningful educational and training opportunities for African Americans (Jackson 1999).

The Human Genome Project has the potential of helping to find cures or therapies for conditions that are commonly found in the African American community, including sickle cell disease, heart disease, and certain cancers. People of all ethnicities—including African Americans—must participate in the research needed for these medical advances to occur. However, along with the promise of research comes the fear of research abuses. In order to protect against research abuses, potential subjects must be fully educated and informed about potential risks and benefits.

Our project demonstrated that the assumption that African Americans and Latinos do not want to participate in genetic medical research is false. On the contrary, minority groups want to do everything in their power to help such advances come to fruition. But the African American and Latino communities that participated in the *Communities of Color and Genetic Policy Project* feel that it is imperative to bring their voices to the table and be part of the discussions that affect policy issues of genetic research and technologies.

Replicating these efforts to engage diverse communities in dialogue about genetic technologies will facilitate public education about science and technology. It will also enable the public to come together with ethicists, researchers, medical practitioners, and lawmakers in a forum to develop policy relating to science, technology, and medicine. These fora may ultimately provide opportunities for underrepresented and underserved communities to benefit more fully from science and technology, as well as gain improved access to health care.

## REFERENCES

Castro, F. G., et al. 1995. Mobilizing churches for health promotion in Latino communities: Compañeros en la salud. *Journal of the National Cancer Institute Monographs* 18.

Clayton, L. A., and W. M. Byrd. 2000. *An American Health Dilemma: A Medical History of African-Americans and the Problem of Race, Beginnings to 1900.* Vol. 1. New York: Routledge.

*Communities of Color and Genetic Policy Project.* 2002. 22 February 2002 [cited 1 October 2004]. Available at http://www.sph.umich.edu/genpolicy/current/index.html.

Fiscella, K. 2000. Inequality in quality: Addressing socioeconomic, racial, and ethnic disparities in health care. *Journal of the American Medical Association* 283 (19): 2579–83.

Gorelick, P. B. 1998. The recruitment triangle: Reasons why African-Americans enroll, refuse to enroll, or voluntarily withdraw from a clinical trial. *Journal of the National Medical Association* 90 (3): 141–45.

Jackson, F. 1999. African-Americans' responses to the Human Genome Project. *Public Understanding of Science* 8:181–91.

Jones, J. H. 1981. *Bad Blood: The Tuskegee Syphilis Experiment.* Vol. 12. New York: Free Press.

Kelley, R. 1990. *Hammer and Hoe: Alabama Communists during the Great Depression.* Chapel Hill: Univ. of North Carolina Press.

Morris, A. 1984. *The Origins of the Civil Rights Movement: Black Communities Organizing for a Change.* New York: Free Press

National Human Genome Research Institute. 2002. *Introduction to the Human Genome Project.* Bethesda, MD: NHGRI.

*Perceptions of How Race and Ethnic Background Affect Medical Care.* 1999. Menlo Park, CA: The Henry J. Kaiser Family Foundation.

Savitt, T. L. 1978. *Medicine and Slavery: The Diseases and Health Care of Blacks in Antebellum Virginia.* Urbana: Univ. of Illinois Press.

# 14

# Informed Consent and the Shaping of British and U.S. Population-Based Genetic Research

MICHAEL BARR

## Introduction

The much heralded genetic revolution promises a new era of predictive medicine and better treatment. Researchers are increasingly looking for gene-disease associations in order to translate advances in genetic technology into reliable information with clinical, etiological, and public health relevance. A key question for policymakers and practitioners is how to create a climate in which potential health benefits can be realized without harming the interests of donors. Throughout the latter half of the twentieth century, the interests of human research subjects have been protected through the ethical principle of patient autonomy and the requirement of informed consent. Scholars and practitioners increasingly recognize, however, that traditional notions of autonomy and consent are highly problematic in population-based genetic research.

In this chapter, I evaluate the issue of consent by focusing on the shape of genetic epidemiology in Britain and the United States. After a review of population-based genetics in both countries, I discuss the challenges that genetic information poses for consent requirements and how each country has addressed such challenges. I argue that ethicists, policymakers, and practitioners may benefit from understanding what motivates people to donate and what beliefs potential donors hold about medicine and technology. In addressing consent, it seems important to acknowledge both the sociological context of the act of donation and the historical context of traditional consent regulations. I conclude that it may be necessary to reshape current ethical frameworks to encompass greater appreciation of the reciprocity that exists between the benefits of medical research and the obligation to participate in such research.

## Population-Based Genetic Research

One way of using the large amount of genetic data produced from the Human Genome Project is through population-based genetic research, often referred to as genetic epidemiology or "biobanking." Such research aims to better understand the relative contributions of genetic, environmental, and lifestyle factors on common conditions such as cancer, diabetes, asthma, and heart disease. To be successful, biobanking requires linking a large collection of DNA (taken through blood or other tissue samples) with donors' personal medical information contained in clinical health records. By locating reliable and quantifiable relationships between environmental exposure (defined as infectious, chemical, nutritional, and social factors), genetic information, and health outcomes, researchers hope to improve diagnostic methods and develop drugs that are tailor-made to an individual's genotype, thus increasing efficacy and decreasing unwanted side effects (House of Lords 2001).

In April 2002, the British House of Lords approved plans for the world's largest biobank, UK Biobank (UKBB). UKBB is organized and funded by the Wellcome Trust, the Medical Research Council, and the Department of Health, and is estimated to cost upwards of £60 million, or $90 million (Draft Protocol 2002). Starting in 2003, UKBB plans to collect DNA samples (via blood) from 500,000 British volunteers. Recruits will include men and women between the ages of forty-five and sixty-nine, the usual age of onset for many common diseases. Apart from the initial donation, patients will be asked to fill out extensive lifestyle and medical history questionnaires. In addition, researchers will periodically re-contact donors for at least ten years beyond the original donation in order to inquire about the volunteer's health status. UKBB researchers will also monitor general practice, hospitalization, and prescription records, as well as disease and morbidity registers.

All data will be stored and used in an anonymous format. Samples will be numerically coded and genetic information will be encrypted so that a special key will be needed to decrypt and make sense of the stored data. Access by third parties such as insurers or employers will be prohibited. Investigators will require data linkage in order to add new information (taken from records or follow-up questionnaires) to the cohort. Crucially, there will be extensive collaboration with the pharmaceutical industry in order to translate research findings into products of direct benefit to the general population. Organizers argue that "due to the unique combination of a large heterogeneous population and a centralized National Health Service" the UK is an ideal place to conduct genetic epidemiological studies (Draft Protocol 2002, 6).

While UK Biobank is unique in size, it is not so in kind. Other smaller biobanks, both public and private, have operated in Britain since the early to mid 1990s. For instance, the North Cumbria Community Genetics Project (NCCGP) collects blood and tissue samples from the umbilical cord of newborn babies, maternal blood samples, and personal health information derived from questionnaires (Chase et al. 1998). Local midwives inform pregnant women in the West Cumbrian region of England of the project. Midwives also request consent from the women to take their afterbirth at the time of delivery. The NCCGP raises a number of ethical questions that are unique to its recruitment procedures. These include requesting samples from a "captive audience" of women receiving prenatal care (albeit with the assurance that refusal does not compromise treatment) and requesting mothers to give consent on behalf of their babies (although the child can withdraw their sample at the age of sixteen). Nonetheless, the NCCGP enjoys a participation rate of over 80 percent and holds nearly ten thousand samples, which provide a large source of DNA for future genetic studies. To date, researchers have used NCCGP data to conduct studies on heart disease, cancer, and neural tube defects.

Apart from such biobanks, there are also large collections of human tissue samples from which DNA could be extracted. For instance, hospitals collect neonatal bloodspots (known as "Guthrie cards") from all newborn babies in the UK for diagnosis of phenylketonuria. Although these samples were not collected for the purpose of genetic research, they could be used for such work since they provide large sources of

DNA from a normal population (House of Lords 2001). That the wide-spread existence of such specimens and their possible use in genetic epidemiological research was not anticipated at the time of collection seems to strengthen calls for sound and uniform consent requirements.

The United States also maintains a number of tissue collections from which DNA could be extracted. However, unlike Britain, the United States does not have a government sponsored and nationally coordinated biobank. Instead, genetic epidemiological studies are conducted across many private, academic, and public sector activities. Many sample collections, usually derived from whole blood or buccal cells (taken from cheek swabs), are ongoing projects. The National Health and Nutrition Examination Survey, for instance, is a continuous study that began in 1999 and collects specimens from approximately 5,000 people each year. More recently, Howard University announced plans for a genetics database that would collect DNA profiles from patients at Howard University Hospital, which serves a predominantly black and medically underserved population in Washington, DC (Goldstein and Weiss 2003). Organizers at Howard plan to collect samples from 25,000 patients in order to study diseases, such as diabetes and prostate cancer, which afflict African Americans more than whites. Finally, to cite another, albeit unrepresentative example, Gene Trust, run by the company DNA Sciences, aims to create a "huge database of information about people" in order to "drastically speed up the rate of medical advances" (Gene Trust 2002). Organizers claim to have registered online over 10,000 donors in its first year of operation. While such private schemes are more frequent in the United States than the UK, the ultimate target of population-based genetic studies in both countries focuses on the two biggest killers in the developed world: heart disease and cancer (Steinberg et al. 2001).

Despite the differences between UK and U.S. biobanks, the ethical issues raised by genetic epidemiology are similar on both sides of the Atlantic. Foremost among these is the informed consent and the challenges posed by the unique nature of genetic information.

## Informed Consent and Genetic Information

The principle of patient autonomy refers to the idea that an "individual of adult years and sound mind has a right to choose what shall happen to his or her body" (Cardozo 1914). In medicine, autonomy is largely protected through the process of informed consent, which stems from the Nuremberg trials and the impact of so-called whistle-blowers such as Henry Beecher in the United States and Maurice Pappworth in the UK, both of whom exposed unethical practices in medical research during 1950s and 1960s (Rothman 1991). More recently, in the United States, the National Bioethics Advisory Commission and Federal Regulations on the Protection of Human Subjects in Medical Research have detailed consent requirements (National Commission 1979; Title 45 1998). In the UK, Medical Research Council guidelines also detail the requirements of obtaining fully informed consent (MRC 2001). However, biobanking and the information that derives from it seems to challenge traditional consent requirements on at least three accounts (Chadwick 2001).[1]

The first challenge concerns the disclosure element of consent, which requires researchers to inform subjects of all relevant facts of the proposed study. Generally, this requirement covers areas such as the potential benefits and risks to the donor, the researcher's personal interest in the project (e.g., financial connections), the aims and methods of the research, and the subject's right to withdraw (Beauchamp and Childress 2001). In the case of biobanking, however, it is seemingly impossible for the subject to be truly informed of the use of the sample when the researchers themselves often do not know the full range of studies that might be done. Samples are stored in liquid nitrogen tanks and retain their scientific value for well beyond the duration of the project for which consent was obtained. Therefore, it is possible that a person who donates their DNA today for a study on breast cancer may not be informed twenty years hence when their sample has been used to study genetic influences on, say, sexual orientation.

Second, population genetics poses problems for the voluntariness requirement of consent, which states that persons must make their decision without being under the control or undue influence of others (Beauchamp and Childress 2001). Many believe that genetic information is exceptional in nature compared to other forms of medical data (House of Lords 2001). Whether one accepts such arguments or not, it is certainly the case that individual donors share a portion of their genetic profile with family members. Thus, if research ever led to clinically relevant results, it could have enormous consequences for the donor's relatives (assuming that results could be traced back, of course). In other words, the predictive value of genetic information could create a situation where relatives of a biobank donor involuntarily learn of their genetic attributes, such as predisposition to a certain condition.

Genetic information is exceptional in other ways as well. Some argue that DNA has become the secular equivalent of the human soul (Nelkin and Lindee 1995). In addition, the popular press and popular science books often depict the "gene" as the secret or blueprint to life (Condit 1999). Given this representation of DNA, what are researchers really asking people to donate—a mere piece of their biological tissue, or an intimate and unique part of their personal identity?

Third, genetic epidemiology is problematic for the understanding requirement of consent, which holds that subjects must be able to justify the beliefs leading to the nature and consequence of their decision (Beauchamp and Childress 2001). There is debate about the extent to which the public is, or can be aware of, genetic science and how that may affect their ability to give consent. While this lack of understanding may be true of many kinds of research, the potential for misunderstanding may be greater in genetics. Not only is genetic science new, but even clinicians have difficulty keeping abreast of the latest research due to the sheer speed of developments. To make matters worse, some segments of the media (sometimes encouraged by researchers) cloud the issue with sensationalist claims that go beyond what is known and currently achievable.

Despite the difficulty of understanding genetics, people do seem to have their own coherent lay understandings (Kerr, Cunningham-Burley, and Amos 1998). However, lay understandings may not translate well in clinical or research settings (and vice versa). Many people cannot define a Mendelian pattern of inheritance nor differentiate between dominant and recessive disorders (Richards 1996). In addition, people have different levels of risk-aversion or tolerance. Lay understandings of the difference between genetically determined and genetically predisposed may not match clinical definitions of the same terms (which may themselves evolve). Such challenges have led observers to reevaluate informed consent procedures in hopes of achieving consent while not sacrificing research priorities.

## Approaches to Achieving Informed Consent

Although an increasing number of writers have addressed the problem of consent in biobanking, the issue is far from settled. Numerous options have been put forth, few of which are mutually exclusive. A key distinction in these discussions is between broad and narrow versions of consent (Berg 2001).

Broad consent would allow investigators to conduct a range of studies, not all of which would be spelled out explicitly at the time of taking consent. The UK's Human Genetics Commission (HGC) advocates broad consent so long as participants are given a "clear explanation of the potential scope of the research" (HGC 2000, 93). The commission argues that broad consent is "practically necessary in a fast-moving field with constantly developing new technology" (94). As part of its call for written evidence, the HGC claims to have received significant support for broad consent—not only from researchers but from patient advocacy groups as well. Part of my argument is to support broad consent on the basis that it best secures donors' expectations of the benefits of medical research.

In contrast, narrow consent would restrict sample use to studies explicitly mentioned at the outset of a research project. If investigators wanted to conduct further studies, they would have to re-contact donors. While narrow consent may help subjects stay better informed, the HGC has claimed that re-contacting people to seek further consent would impose an unnecessary burden on donors and, more importantly, increase risks of a security breach since sensitive medical data would have to be decoded (HGC 2002). In addition, narrow consent could be an obstacle to research. Not only would it increase costs for investigators but it would complicate matters in cases where the donor had since died.

Whether consent is broad or narrow, however, questions still remain as to who is consulted and how the process of consent ought to unfold. Answers offered to these questions have included a reevaluation of the language and substance of consent forms, some level of group or community consent, and public consultation exercises.

## A Matter of Form?

One approach, more prominent in the United States, has focused on the content and language of actual consent forms (Beskow et al. 2001; Deschenes et al. 2001). Recently the Centers for Disease Control and Prevention (CDC) formed a multidisciplinary working group to address the issue of consent. Its conclusions were not fundamentally different from existing guidelines in public health research, but rather an extension of those guidelines (Khoury 2001). The CDC panel concluded that the best way of protecting subjects' interests and autonomy is by making forms as thorough as possible regarding the details of a study. The panel suggested that forms contain information on why the study is being done, what it will involve, how information will be kept private, the study's risks and costs, the possibility of receiving results, and the status of the sample once the study is complete. An approach that focuses on detailed consent forms is designed to exhaust the disclosure element of consent requirements.

It is worth noting that the CDC recommendations directly connect the probability of harm and benefits to the meaning of the results for the health of the participant. In other words, consent ought to be based from the start on an assessment of whether the results would generate information that could lead directly to an evidence-based intervention, such as drug treatment or lifestyle alteration. The CDC panel concluded that family-based research guidelines were not suited to biobank research since they do not distinguish between studies that are likely to yield clinically relevant results and ones that may have significant public health implications but carry few physical, psychological, or social risks to individuals—as in the case of population-based genetic research. By trying to ensure that no one learns of a meaningful result without having consented, CDC has sought to meet the voluntariness element of consent requirements.

Focusing on the language of forms has severe drawbacks, however. One criticism is that this strategy seems to be more for the protection of the interests of researchers and Institutional Review Boards than those of the subject (Annas 2001). Skeptics argue that forms end up looking more like legal contracts than documents of explanation and education and that focusing on the language of consent is a legalistic and bureaucratic approach.

Concerns with the content of forms, particularly in the United States, has drawn frequent criticisms in the UK as well. The British medical establishment has long sought to avoid an American-style litigation culture. Thus, critics in the UK argue that by focusing on forms, researchers are too heavily preoccupied with the so-called audit society, in which institutions are monitored for accountability by external review (Power 1996). Such a society seeks accountability through observing and ultimately judging the activities of professionals and institutions

(O'Neill 2002). Another criticism is that an emphasis on forms is the product of a consumer-driven culture. "In a world where medicine has become a good to be consumed, where patients are customers to be wooed, informed consent becomes the disclosure of the contents on the back of the box" (Wolpe 1998, 49).

The danger is that concentrating on the language of forms runs the risk of turning consent into a sufficient ethical justification, which it is not. I shall return to this point later in the chapter.

## Group Consent

Another way of meeting the challenge posed by the exceptional (or familial) nature of genetic information is by asking for group or community consent (Greely 2001). This system has received more attention in the United States than in Britain, perhaps because of the presence of the Native American population and the work of the North American Regional Committee of the Human Genome Diversity Project (HGDP).

Proponents argue that group consent is necessary since genetic research could have implications beyond the individual donor or even local community. Research done on a Chinese community in California, for instance, could have consequences for all Chinese, no matter where they live or if they consented. As mentioned above, the fear is that if research indicates that a population is predisposed toward developing a certain condition or behavioral characteristic, this information could have adverse effects in terms of insurance or employment, or, more broadly, could even stigmatize the population under investigation.

However, achieving group consent is fraught with so many difficulties that it may be impossible to achieve. Advocates of group consent assume that populations are stable objects that can be used to define consenting subjects. In reality, though, group consent creates groups that have a "questionable status both in nature and in society" (Reardon 2001, 372). For instance, how does one define the group or community? Should it include only the subject community, or should it include all who may be influenced by the research? Should the group be limited only to families? Should it include the so-called disease organizations, such as the American Lung Association? How would the group actually express consent? Should researchers work with "culturally appropriate authorities," as the HGDP suggests (Greely 2001)? Or should a vote be held, no doubt at great cost to the research project?

Another factor against group consent is that it may end up re-enforcing controversial stereotypes not grounded in reality. It is open to debate whether there are any genetic differences among ethnic groups. Some scientists believe that the distinctions are miniscule and that race has no real genetic basis. Asking for group consent from a specific eth-nicity, however, implies that race does have a genetic underpinning (Juengst 1998).

In fairness, proponents argue for community consent as an extra layer of ethical protection rather than as a full alternative to traditional consent requirements (Weijer and Emanuel 2000). Despite the commu-nal nature of genetic information, most observers do not take group consent seriously for the reasons highlighted above. Even in the United States, group consent seems to have only lukewarm support. Its most vocal advocates in the HGDP have reservations about implementing the idea and recognize that in some situations, where communities do not have a culturally appropriate authority, for example, group consent would be impossible (Greely 2001).

## Public Consultation

An alternative position, recently adopted in the UK, is to hold pub-lic consultations rather than seek direct community consent. But public consultation raises problems as well. For example, how can researchers adequately consult fifty-eight million people? As one of the first bio-banks in Britain, the NCCGP held public consultations in an effort to inform the local community of its plans to collect DNA samples and to win support for its initiative (Chase et al. 1998). However, accounts of these meetings varied. One community group conducted a street poll after the consultations and found that 90 percent of those surveyed had never heard of the project (CORE 1995). If it is difficult to achieve ade-quate consultation with a small and stable community in the northwest of England, then the difficulties of doing so on the level of national pop-ulations may be insurmountable.[2]

Furthermore, the term "consultation" implies a one-way relationship between two distinct parties (Haimes and Whong-Barr 2003); but consul-tation will always be a process of talking with representatives of the com-munity. In other words, some of the problems encountered in obtaining group consent resurface in community consultation exercises. How rep-resentatives are selected and how consultation occurs is underpinned by political processes, which in turn underpin the ways in which ethical is-sues are handled. Therefore, even if a more evenly balanced term such as "dialogue" replaces consultation, the problem of who has a voice in that dialogue remains.

All of these approaches—individual and group consent and consultation—rely upon a bottom-up approach, that is, they seek au-thority for research from the subjects involved. An alternative (or per-haps parallel) approach, which I discuss later, is to develop a system of effective governance whereby a publicly recognized and accepted body would be authorized to decide appropriate types of research and limit sample uses.

## Consent and the Act of Donation

Curiously, what is missing from most debates about consent is a discus-sion of the wider context. Such a discussion would address, for example, why people donate their biological tissue in the first place. What moti-vates them? What expectations, if any, do they maintain in exchange for their donation? What do donors want from informed consent requirements?

I argue that the process of informed consent is embedded in a whole series of social processes—including, crucially, beliefs about the role of medicine, health, and technology in every day life. In other words, by focusing exclusively on who gives consent and how it is given, and on the language of consent forms, we may be in danger of "losing the plot"—of mistakenly seeing the act of consent as a singular event some-how abstract from other considerations of why and wherefore. Perhaps it would prove helpful to pay closer attention the act of donation itself and the belief patterns underpinning it.

Recently in the UK, the Medical Research Council (MRC) has the-orized tissue donation for genetic research by referring to the "gift rela-tionship" and the altruism of donors (MRC 2001). MRC's guidelines on the collection, storage, and use of tissue samples cite the work of Rich-ard Titmuss (1970), who believed that people donated blood out of a sense of altruism. Titmuss argued that a U.S.-style market-based system of blood donation would lead to more problems than the British vol-untary system, including administrative inefficiency, greater costs, and higher amounts of contaminated blood.

MRC (2001, 8) guidelines state that the gift approach is "prefer-able from a moral and ethical point of view, as it promotes the 'gift relationship' between participants and researchers, and underlines the altruistic motivation for participation in research." The MRC (2001, 8) goes on to say, "Gifts may be conditional (that is, a donor may specify what the recipient can do with a gift), and it is very important that the donor understands and agrees to the proposed uses of the donated ma-terial. The assumption by the donor is that nothing will be done that would be detrimental to his or her interests, or bring harm to him or her." The organizers of one of the UKBB public consultations also cite altruism as a reason for donation. Their final report claims, "the one primary motivating force that stimulates people to volunteer . . . [is] al-truism" (People, Science and Policy 2002, 11)

Attempts to understand the significance of donation are highly im-portant. Yet the original meaning of the gift relationship, as developed by anthropologist Marcel Mauss's work on the indigenous cultures of Polynesia and among Native Americans, should not be neglected or for-gotten. According to Mauss, the idea of gift giving is based on reciproc-ity. A "gift relationship" stresses that exchanges may appear voluntary and motivated by altruism, but they are in fact based on interlocking ob-ligations. To refuse to give "is to reject the bonds of alliance and com-monality" (Mauss 1997, 13). In other words, altruism is rarely, if ever, un-calculated. In the words of Mary Douglas (1997, vii–xviii), "there are no free gifts. . . . A gift that does nothing to enhance solidarity is a contradic-tion." People are willing to give, it seems, not as unilateral acts of kind-ness but as part of a interdependent system of giving and receiving—of sharing.[3]

The MRC has several reasons for shaping medical donations as "gifts." First, and most obviously, medical research depends on peoples' willingness to donate. Second, characterizing genetic donations as "gifts" enables the MRC to avoid legal uncertainties over ownership. In the UK it is not legally possible to own a human body per se. However, the law is not clear whether one can own samples of human tissue or whether donors can have property rights over their samples. According to the MRC, for human research, the important consideration is not legal ownership but "who has the right to control the use of samples" (MRC 2001, 8). The term "custodianship" is used instead of ownership to imply responsibility of safe storage and control of tissue samples. Thus, by referring to donations as gifts, "any property rights that the donor might have in their donated sample would be transferred, to-gether with the control of the use of the sample, to the recipient of the gift" (MRC 2001, 8). In other words, theorizing donations as gifts pro-vides the MRC with a "practical way" of avoiding the possibility that donors' may later claim legal rights to their samples.

Returning to epidemiology, the gift concept raises the question of whether the ethical framework within which consent is theorized itself needs to be reconsidered. Increasingly, ethicists are reevaluating an em-phasis on individualism and autonomy in favor of ethical frameworks based on solidarity and equity. This reevaluation raises the question of whether "one has a duty to facilitate research progress and to provide knowledge that could be crucial to the health of others" (Chadwick and Berg 2001, 320). Such a principle, if adopted, would seem to contradict the view that research should not be conducted unless it benefits those participating in the study (Annas 2001), yet it draws attention to the interdependent nature of medical research—a point often forgotten in discussions that center on rights and autonomy.[4]

## Reasons for Donation: Examples from the NCCGP

Considerations of the gift relationship and solidarity are not mere aca-demic exercises. A study of participation in the NCCGP biobank con-firms the view that people are motivated by the expectation that they or those close to them may someday benefit from advances in medical re-search. When asked about their overall reasons for donating, interview-ees responded that they gave samples out of an obligation to help. Many held the view that by donating they had "done their bit" to facilitate re-search, in case "there was something wrong with somebody in the fam-ily" (M018; M052). For example, donors explained:

> If it's going to be best for the future, I mean, as a child I suffered right from the age of three from asthma and I didn't have a very nice child-hood and I haven't been able to do the same things that every other, like some of my friends have been able to do, and if I think that is going to affect [my son] the same way, so me helping, it's my way of, I've done my bit (M018).
>
> I mean if we had anything wrong with any of our children or our-selves and they needed a major organ or something like that, if some-one donated . . . that is the reason why you donate yourself, you

> know . . . you will like to think that you would help someone else out,
> if you found yourself in that situation, and somebody would do the
> same for you. (M006)

Several respondents were aware that they personally had benefited from fertility research done previously, and this had influenced their own decisions: "Because we had had the IVF treatment, you think, 'well, if they hadn't done a lot of research about that then,' you know . . . I think that was the main reason why we agreed that we would donate" (M013). Similarly, another woman that had benefited from IVF said, "Well, I think we had help so should try and help" (M031).

These statements show how vital a component reciprocity is in donation. Although the respondents did not use academic phrases as "gift relationship" or "solidarity," their responses indicate that biobank donors expect a return on their donation. The anticipated return is future health benefits—either for themselves, their families, or the wider community.[5] Notwithstanding the exceptional nature of genetic information highlighted above, donors seem most concerned with helping to facilitate medical research that may improve both diagnosis and therapeutic intervention. They also saw themselves in historical continuity with other real and potential donors.

## Implications for Informed Consent

In discussions on biobanking, there is a growing recognition that there "might be reasons to question the transferability of rules and principles developed in one context, to the problems of today and tomorrow" (Chadwick and Berg 2001, 320). Current consent requirements are direct products of World War II and were not designed for research on the scale or with the scope of genetic epidemiology. This is not to suggest that consent, as codified in the Declaration of Helsinki, is no longer valid or important. Rather, I wish to recover a more balanced notion of consent and the idea that underpins it, the right to autonomy. In other words, the problem with consent is, in large measure, a problem with autonomy.

As mentioned earlier, individual patient autonomy has been the most dominant idea in twentieth-century medical ethics. In part, the dominance of autonomy stems from the rights movements of the 1960s. Yet by focusing so heavily on a discourse of rights, bioethics has run the risk of missing the other half of the equation.⸢In order to claim a right to health, perhaps there is a duty to help.⸥ Such claims may be more common in Britain where the national discourse around health has been premised on notions of community and public service. Yet even in the United States, the reciprocal nature of donation and health benefits is fundamental to the success of medical science.

In other words, I claim that expectations to receive the best possible care ought to be premised on the recognition of an obligation to contribute to the process by which effective treatments are created (Harris and Woods 2000). There are at least two reasons for this claim. First, donation is in the interests of both the volunteer and the public. By participating in genetics research, of course, there may be a degree of personal sacrifice or inconvenience (e.g., time away from work or family, the discomfort of fasting or giving blood). One useful way of thinking about medical related duties is to place them in a category with compulsory military service or jury service, both of which are obligations that people perform for the public good. Jury service also requires a sacrifice on the part of the citizen, yet most people recognize the need for such a system. In other words, there are clearly jurisdictions where self-sacrifice is expected and is not unreasonable. My argument is that medical research is one such jurisdiction.

In addition to the public good, volunteering as a participant in medical research is, arguably, a matter of fairness. To profit from the contribution of others while refusing reciprocity is akin to accepting a free ride on the sacrifice of others. Most people, I think, would find it unfair if people benefited from medical research but refused to participate in it (Harris and Woods 2000).

Recognizing a modest duty to participate in medical research need not undermine the gift relationship. As I have tried to emphasize, the critical point of gift giving is the element of reciprocity and expected return. As Mauss writes, "exchanges and contracts take place in the form of presents; in theory these are voluntary, in reality they are given and reciprocated obligatorily" (Mauss 1997, 3). Thus, while the languages of gifts and duties may seem at odds, sociological and anthropological insights (based on actual practice) clearly show that this is not the case.

The reciprocity of rights and obligations may have important implications for informed consent, at least in DNA banking. Indeed, data from the NCCGP study suggests that subjects are well aware of these interlocking obligations. Thus, it remains to be seen if subjects desire consent procedures requiring their approval for each different study. To put it another way, rules for population based research that demand narrow consent seem oddly paternalistic in that they may run counter to donors' wishes. Narrow consent is needed in a clinical setting or in research that may have a direct impact on the health of the participant. But genetic epidemiology is unique in the sense that not even researchers know what future uses of the samples might be. If ethics committees have approved studies, then geneticists should be free to conduct the research that they think will most likely lead to medical advances.

Allowing for broader consent neither diminishes the importance of a researcher's duty of care to donors, nor legitimates coercion as a means of recruiting research participants. The principle of nonmalfeasance (do no harm) still applies, as does the need for strict scrutiny of research aims, methods, and processes. Indeed, a critical aspect of the arrangement I propose is a trustworthy and effective governance mechanism. In order for a governance system to work, a key factor would be the involvement of lay participants to help ensure that ethics committee decisions are taken in the interests of the public good and not just for the benefit of the scientific establishment or for trivial or market-driven research (such as improved cosmetics). Laying out a detailed governance framework is beyond the confines of this chapter, but such a system would, in effect, rely upon models of democratic and representative government to entrust empowered authorities to make the "right" decisions.

However, consideration of the social context of donation helps illustrate the limits of consent. Informed consent, after all, is only one part of a wider set of ethical requirements. If consent were a sufficient ethical requirement, then there could be no objection to developing a free market system in human tissue or body parts, so long as donors agreed (O'Neill 2002). Yet many find that idea repugnant and for good reason.

People donate because they realize that they and those near them are vulnerable to disease. It is not clear, though, that donors are looking for in-depth forms where they can check off boxes for the types of research that they do and do not approve of. Are exhaustive descriptions of the proposed research what donors want? Or what investigators feel they need in order to protect themselves from external audit and possible litigation? In biobanking, at least, perhaps the perspective of the investigator is not that of the donor. Perhaps, in the case of consent forms, less is more.[7]

# Conclusion

Genetic research promises a new era of preventive medicine through improved diagnostics and therapeutics. It also presents novel ethical and legal challenges, to which "solutions" are not easily reached.

This chapter has shown the different shape of population-based genetic research and informed consent in the United States, where there has been much focus on the nature and language of consent forms and the possibilities of group consent, and in Britain, where ethics commissions have advocated broad consent and characterized donation as a "gift relationship." I have argued that theorizing donation as a gift and allowing for broad consent seems to reflect donors' expectations and wishes regarding medical science, that is, future benefits to themselves, families, and communities. I have also claimed that there may be an obligation to participate in medical research since donation is in the interest of both the individual and the public.

There are several implications to the claim that there is an obligation to donate to genetics research. First, consent requirements for genetic epidemiological studies ought to be broad-based enough to allow researchers to progress with their work without seeking additional consent when each new study arises. Narrow consent, as the HGC states, seems not only impractical (when donors may be untraceable or dead) and potentially insensitive, but, crucially, it also risks wasting limited resources better spent on what donors want most: improved medical services. Second, consent forms that are laden with technical language and legal jargon miss the point of informed consent requirements. Donors and relatives, as O'Neil (2002, 157) argues, may find "that being confronted with the full detail of research protocols provides excess, unassimilable information, to which they can hardly hope to give genuinely informed consent."

In short, subjecting volunteers to the demands of the audit agenda seems ethically questionable and not necessarily in accord with donors' wishes. Reshaping the ethical framework for population-based genetic research to encompass the concepts of donation and obligation may provide a morally sturdier opportunity to further inquiry about human life.

NOTES

The author would like to thank Erica Haimes and two anonymous reviewers for their comments on an earlier draft of this chapter. I am also indebted to the Wellcome Trust for helping to fund both my attendance at the Next Generation Symposium and research into the NCCGP biobank.

1.  My discussion assumes a competent subject, although I do not wish to minimize the issues involved in achieving informed consent with children or the mentally impaired.

2.  In the planning stages of UKBB, organizers held three public consultations with special interest groups, members of the general public, and health professionals. The findings suggested that people, practitioners included, had little knowledge of the methods involved but overall support for the aims. Respondents viewed consent as "crucial" but "potentially problematic." See Cragg Ross Dawson (2000); Hapgood et al. (2001); and People, Science and Policy (2002).

3.  Admittedly, the application of these ideas to an industrial or service society is not unproblematic. No one can say for certain that the act of giving in modernity springs from a foundation of solidarity rather than individual competition. The point I wish to make, however, is that the gift relationship brings to light a forgotten aspect of ethics and donation—that of duties and reciprocity.

4.  I have not here addressed the philosophy that underpins such a view. However, support for my position can be found in some elements of feminist ethics and in the recent work of Onora O'Neill (2002), who revives a Kantian notion of "principled autonomy," which seeks to ground human rights in human obligations.

5.  Perhaps it is not surprising that "solidarity" is a term often employed in the UK, given Britain's publicly funded National Health Service (as opposed to the American private-based health-care system).

6.  It can be difficult to argue that rights necessarily imply duties. I recognize, for instance, that if X has a right, then it does not necessarily follow that X has a duty. It is plausible to assert so, but not conceptually true, as in the case of an infant who clearly has rights but cannot have responsibilities.

7.  I thank an anonymous reviewer for suggesting a system whereby interested donors would have the right to monitor the use of their sample and prohibit its inclusion in studies they find disagreeable. One of several problems with such a arrangement, however, is that in order to meet donors' requests, researchers would have to access donor names by decoding and decrypting personal details. Arguably, the more often this happens, the greater likelihood

of a breach in confidentiality. However, my research suggests that biobank researchers may find that donors appreciate greater discussion on benefit-sharing, such as profit sharing, greater equity in drug treatments, and community projects.

REFERENCES

Annas, G. 2001. Reforming informed consent to genetic research. *Journal of the American Medical Association* 286:2326–28.

Beauchamp, T., and J. Childress. 2001. *Principles of Biomedical Ethics.* 5th ed. Oxford: Oxford Univ. Press.

Berg, K., 2001. DNA sampling and banking in clinical genetics and genetic research. *New Genetics and Society* 20: 59–68.

Beskow, L. M., W. Burke, J. F. Merz. 2001. Informed consent for population-based research involving genetics. *Journal of the American Medical Association* 286: 2315–21.

Cardozo, B. 1914. *Schloendorff v. New York Hospital* 211 NY 127, 129, 105 N. E. 92, 93.

Centers for Disease Control and Prevention. 2001. Supplemental brochure for population-based research involving genetics: Informed consent: Taking part in population-based genetic research. Available at http:www.cdc.gov.

Chadwick. R. 2001. Informed consent and genetic research. In *Informed Consent in Medical Research,* ed. L. Doyal and J. Tobias, 203–10. London: BMJ Publishing Group.

Chadwick, R., and K. Berg. 2001. Solidarity and equity: New ethical frameworks for genetic databases. *Nature Review Genetics* 2: 318–21

Chase, D., J. Tawn, L. Parker, J. Burn, J., and P. Jonas. 1998. The North Cumbria Community Genetics Project. *Journal of Medical Genetics* 35: 413.

Condit, C. 1999. *The Meanings of the Gene.* Madison: Univ. of Wisconsin Press.

CORE. 1995. Findings of street-poll for NCCGP public consultations. CORE pamphlet. Barrow-in-Furness, Cumbria.

Cragg Ross Dawson. 2000. *Public Perceptions of the Collection of Human Biological Samples.* Report prepared for the Wellcome Trust and Medical Research Council.

Dedschenes, M., G. Cardinal, , B. M. Knoppers, and K. C. Glass, 2001. Human genetic research, DNA banking and consent: A question of form? *Clinical Genetics* 59: 221–39.

Douglas, M. 1997. No free gifts. Foreword to M. Mauss, *The Gift, the Form, and Reason for Exchange in Archaic Societies,* vii–xviii. Trans. W. D. Halls. London: Routledge.

Draft Protocol for Biobank UK: A study of genes, environment and health. 2002. Available at http://wellcome.ac.uk.

Gene Trust. 2002. See: http://www.dna.com.

Goldstein, A., and R. Weiss. 2003. Howard Univ. plans genetics database. *Washington Post Online,* 27 May. Available at http://www.washingtonpost.com.

Greely, H. 2001. Informed consent and other ethical issues in human population genetics. *Annual Review of Genetics* 35: 785–800.

Haimes, E., and M. Whong-Barr. 2001–2003. A comparative study of participation and non-participation in the North Cumbria Community Genetics Project. Project Grant funded by the Wellcome Trust.

———. 2003. Competing perspectives on reasons for participation and non-participation in the North Cumbria Community Genetics Project. In *DNA Sampling: Ethical, Legal and Social Issues.* ed. B. M. Knoppers, 199–216. Leiden: Brill Publishers.

Hapgood, R., D. Schickle, and A. Kent. 2001. Consultation with primary health care professionals on the proposed UK population biomedical collection. Report prepared for the Wellcome Trust and the Medical Research Council.

Harris, J., and S. Woods. 2000. Rights and responsibilities of individuals participating in medical research. In *Informed Consent Medical Research,* ed. J. Tobias and L. Doyal, 276–82. London: BMJ Books.

House of Lords. 2001. Select Committee on Science and Technology Fourth Report: *Human Genetic Databases: Challenges and Opportunities.*

Human Genetics Commission. 2002. *Inside Information: Balancing Interests in the Use of Personal Genetic Data.*

Juengst, E. T. 1998. Groups as gatekeepers to genomic research: Conceptually confusing, morally hazardous, and practically useless. *Journal of the Kennedy Institute of Ethics* 8:183–200.

Kerr, A., S. Cunningham-Burley, and A. Amos. 1998. The new genetics and health: Mobilizing lay expertise. *Public Understanding of Science* 7: 41–60.

Khoury, M. 2001. Informed consent for population research involving genetics: A public health perspective. Available at http://www.cdc.gov.

Medical Research Council. 2001. Human tissue and biological samples for use in research: Operational and ethical guidelines. Available at http://www.mrc.ac.uk/pdf-tissue_guide_fin.pdf.

National Commission for the Protection of Human Subjects of Biomedical and Behavioral Research. 1979. *The Belmont Report: Ethical Principles and Guidelines for the Protection of Human Subjects of Research.* Washington, DC: National Commission for the Protection of Human Subjects of Biomedical and Behavioral Research.

Nelkin, D. and M. S. Lindee. 1995. *The DNA Mystique: The Gene as Cultural Icon.* New York: W. H. Freeman.

O'Neill, O. 2002 *Autonomy and Trust in Bioethics.* Cambridge: Cambridge Univ. Press.

People, Science and Policy Ltd. 2002. *Biobank UK: A Question of Trust: A Consultation Exploring and Addressing Questions of Public Trust.*

Power, M. 1997. *The Audit Society.* Oxford: Oxford Univ. Press.

Reardon, J., 2001 The Human Genome Diversity Project: A case study in co-production. *Social Studies of Science* 31: 357–88.

Richards, M. 1996. Lay and Professional Knowledge of Genetics and Inheritance. *Public Understanding of Science* 5: 217–30.

Rothman, D. 1991. *Strangers at the Bedside: A History of How Law and Bioethics Transformed Medical Decision Making.* New York: Basic Books.

Steinberg, K., J. Beck, D. Nickerson, M. Garcia-Closas, M. Gallager, M. Caggana, Y. Reid, et al. 2002. DNA Banking for Epidemiological Studies: A Review of Current Practices. *Epidemiology,* 13: 259–64.

Title 45 CFR Part 46: Protection of Human Subjects, 1998. http://206.102.88.10/ohsrsite.

Titmuss, R.M. 1970. *The Gift Relationship: From Human Blood to Social Policy.* London: Allen & Unwin.

Weijer, C., and E. J. Emanuel. 2000. Protecting Communities in Biomedical Research. *Science* 289: 1142–44.

Wolpe, P. 1998. The Triumph of Autonomy in American Bioethics: A Sociological View. In *Bioethics and Society: Constructing the Ethical Enterprise,* ed. R. DeVries and J. Subedi, 38–59. Englewood Cliffs, NJ: Prentice-Hall.

# 15

# Embryos, Legislation, and Modernization

## *Shaping Life in the UK and German Parliaments*

CHARLOTTE AUGST

## Introduction

This chapter investigates the legislative shaping of new reproductive technologies. It focuses on the debates about embryo research, in vitro fertilization, gamete donation, and eugenics between 1988 and 1990 in the British and German parliaments. Both sets of parliamentary debates, running more or less in parallel, led to legislation: the Human Fertilisation and Embryology Act (HFEA) of 1990 and the Embryonenschutzgesetz (ESchG, Embryo Protection Act), of the same year. Despite overt similarities, however, the legislative outcomes were quite different. The British HFEA allows egg and sperm donation, embryo research, the treatment of unmarried women, and surrogacy; the German ESchG outlaws all of these practices.

I argue in this chapter that debates about reproductive technologies are debates about human life under the conditions of modernization; indeed, I see new reproductive technologies as instances of modernization. In order to disentangle the many concerns expressed in the two sets of debates about this modernization of human life, I first outline how the modern attitude toward human life is crucially shaped by ambivalence—about reason and rationality on the one hand and about individual freedom and the human condition on the other. Both these strands of modern thought shape the development, use, and understanding of reproductive technologies. Second, I show how this ambivalence characterizes both sets of national debates and how it is ultimately overcome. Each set of parliamentarians chose different strategies for dealing with the ambivalence of modernization, which led to different discursive constructions of human life, reason, nature, and science. Crucially, both debates are based on an unwillingness to accept that the concerns triggered by modernization cannot be neatly dissolved by distinguishing good medicine from dangerous science (as in Germany); or modern reason from unenlightened irrationality (as in Britain). I conclude that the risky nature of modern science (Beck 1986) and the precariousness of human life in modernity cannot be rendered harmless through denying the complexity of modernization.

All boundary drawing, including lawmaking, necessarily denies complexity (Gieryn 1995). This denial enables laws to be passed and issues to be settled. However, whether this denial enables the two laws in question to respond effectively to the challenges posed by new reproductive technologies remains an open question. This chapter suggests that some of the dangerous potential of modern science could be more effectively managed through a legislative framework that acknowledged a higher degree of ambivalence.

## Modernity's Ambivalence

This chapter treats modernity as a perspective, that is, as a certain way to make sense of and change the world. My concern here is not with modernity as a precisely dated historic period or as a term in art history. I use "modernity rather as an attitude than as a period of history . . . , a mode of relating to contemporary reality" (Foucault 1984, 39). This understanding of modernity is closely linked to the idea of the Enlightenment (see Douzinas et al 1991, 6), of which I emphasize two strands.

The first emphasis is the Enlightenment idea of setting humanity center stage. Humanity and, more importantly, the individual is the central figure of the enlightened universe. The individual is to be emancipated from God, religion, and tradition, and from tribe, village, and

monarch. The individual is the central character of Enlightenment discourses of emancipation, equality, and rights; it epitomizes human life
in modernity. Human life in modernity is "individualized," "detraditionalized," and "disembedded" (see Franck 2000; Giddens 1994; Beck
and Beck-Gernsheim 1992).

The second strand of the modern attitude I want to emphasize has
to do with the centrality of knowledge: Human rationality can know the
world. It can understand the world and look at it objectively, with the
eyes of an outsider (Douzinas et al. 1991, 9; Adorno and Horkheimer
1999, 9). Modern persons, in order to understand the world, have to
stop thinking from their own subjective standpoint. People must look at
the earth and at human life as if they were placed in the universe, elevated to a position of objectivity, and detached from their individual,
limited viewpoint. This perspective, of course, leads to great prominence for the natural sciences in modernity's worldview. The natural
sciences represent everything that modern knowledge is about. They
are objective; they are instrumental; they are concerned with the universal and not the particular, with the rule and not the exception, and
with order rather than ambivalence (also see Giddens 1990, 40; Bauman
1998; 2001). Because of all of these characteristics, modern knowledge
allows for the control of certain phenomena in the natural or, by means
of transfer, the social world.

In this chapter, these two strands of the modern perspective—the
tendency for individualization and the use of science to shape the
world—stand for my use of the terms "modern," "modernity," and
"modernization." This designation is less than a coherent definition of
modernity as such and more a pointing out of some trajectories that, in
this view, cast the shape human life can take in modernity.[1] I argue that
these two strands of the modernization project conflict to a certain extent, and that this conflict is particularly visible in the field of reproductive technology.

## A First Glimpse of the Problem

The tension inherent in the modern project becomes apparent when
thinking about human nature, or the biological side of human life. The
typically modern way of dealing with nature is to treat it as the object of
human action, a material resource to be manipulated and managed.
However, proclaiming that humans are always to be treated as subject

and nature as object creates an immediate tension when thinking about *human* nature (see Freud 1985, 274–77). Accordingly, human materials such as tissues and organs become very contentious: Are they to be used as resources for manipulation and exploitation? Or is there something inherently and already human about any tissue derived from the human body that means it must be thought of as subject rather than object (see Santos 1995, 28–29)? These tensions arise very clearly in controversies about organ donation and the Human Genome Project as well as embryo research. The claims to rationality and to individualism, to objectivity and subjectivity, which shape the modern project and the concept of human life, are difficult to reconcile (Cooke 1990, viii; McGuigan 1999, 37; Connolly 1987). This difficulty keeps parliamentarians in Germany and the UK busy.

## New Reproductive Technologies as Instances of Modernization

The working definition of modernity outlined above is based on, first, the modern claim to a distinct, objective, scientific rationality, and second, the modern tendency to individualization and its erosion of limits imposed by nature or tradition. These two strands of modern thought shape the development, use, and understanding of reproductive technologies. On the one hand, reproductive technologies instantiate the science of human life and human reproduction. Just like any other natural process—including the reproduction of animals or the growth of cells—the phenomenon of human life can be analyzed and dissected. What was hidden can be made visible; what seemed mysterious can be explained. Elements of it can be altered or manipulated, replicated, and controlled. The creation of human life can be understood as the succession of distinct steps of development, and each can be the object not just of scrutiny but also of engineering.

On the other hand, there are those who use the technologies. Sometimes they are called patients, and mostly they are women. For them, technological intervention into reproduction offers liberations from limits imposed by nature or tradition. Even if your Fallopian tubes are blocked, you could conceive. Even if your partner has a fertility problem, you could conceive with him. Even if you do not want to have sex with a man, you could achieve pregnancy. Even if you are postmenopausal— lesbian or single—with the help of new reproductive technologies you do not have to forfeit your wish for a child. You can indeed reproduce.[2]

These two potentials of new reproductive technologies are expressions and effects of processes of modernization. Reproductive technologies are based both on the logic of individualization and on rational planning and engineering. Humans are enabled as possibly powerful agents, living their lives according to their own wishes, yet they may also be vulnerable victims, objects of someone else's master plan. Parliamentarians talking about new reproductive technologies debate the promises of modernization: the relief of suffering, the control of hitherto uncontrollable processes, and the creation of happy families and healthy children. But they also get caught up in the contradictions created by fears of ambitious scientists engineering the human race and reckless single women selfishly insisting on having a child outside of the limits circumscribed by tradition or nature. That is, parliamentarians who set out to legislate on new reproductive technologies, shaping the reproduction of human life, are dealing with the ambivalence of modernization.

## Deciding about Ambivalence: Creating Boundaries for Science

The parliamentarians in the UK and German legislatures are worried about a whole range of issues. They speak about the engineering and the individualization potential of reproductive technologies. They are concerned about women's reckless use of reproductive technologies, and the implications such use might have for the "traditional" family.[3] They are concerned about "men in white coats" manipulating the embryo und humankind.[4] They express anxieties about the commodification of children[5] and about the risks for women[6] who are desperate enough to try everything in order to conceive.[7] And yet, they are not simply rejecting technologies of reproduction, because technologies of reproduction also stand for the treatment and cure of the suffering, for controlling the gamble of reproduction (as expressed through the risks of infertility[8] or disability[9]), and for personal autonomy and freedom in reproductive matters.[10]

It is possible to dissect the parliamentary discourses and show in detail how this ambivalence about individualization and science is played out. Here, I focus in on one aspect of this ambivalence in action, the nature of science and scientific progress. Science and its progress form the basis of modernity's promising future: we will indeed be able to do things that we cannot yet do. We might be able to understand and cure more illnesses. We could take the risks out of reproduction, and we might avoid the uncontrollable occurrence of disability. Yet science, in its quest for ever more knowledge, ever more control, might also lead to the manipulation of humans. It might treat humans simply as material and the world as a big laboratory, violating people's integrity and dignity. Scientists might hold "the promise of a brighter future" (Durant 1998), yet they might also "be up to no good, and must not be allowed to proceed without scrutiny" (Warnock 1985, xiii).[11]

Both German and British parliamentarians express this ambivalence about the promising and the dangerous potential of science in legislative debates about technologies of reproduction. Science, it is argued, "can both create and destroy" (Kevin Brown, 2 April 1990, HC, vol. 170, col. 962).[12]

How do the parliamentarians in Germany and Britain attempt to draw a boundary between the safe and the dangerous side of science in their attempts to tackle the inescapable ambiguity of modern scientific progress and its implications for human life? We will see that national discourses have radically different points of departure, reflected in the outcomes of a relatively science-friendly HFEA and a relatively cautious German ESchG. However, neither boundaries drawn by these distinct discourses are logically sustainable. The parliamentarians are involved in the construction of partly fictional distinctions. Both pieces of legislation are ultimately based on a denial of ambivalence and undecidability.

## Germany: Embryo Research as Everything That Is Wrong about Science

In the German parliamentary debates, research on human embryos stands for everything that is wrong about modern science in general and about technologies of reproduction in particular. Embryo research is the epitome of dangerous science—the objectification, manipulation, and use of human life without regard for its inherent dignity. All members of parliament (MPs) of all parties agreed that research with embryos must be outlawed and that embryos must not be manipulated or destroyed for research because "human life" must not be used for "any reason" (Government Bill, 25 October 1989, *BD* 11/5460, 10).[13]

The real political disagreement lay in the question of whether any fertility treatment that does not directly lead to the destruction of embryos (e.g., IVF) should be outlawed as well (the Green position); whether it should be very strictly regulated (the Social Democratic position); or whether Parliament should leave all aspects of "treating" women to the medical professionals' body, the Bundesärztekammer (the position of the Conservatives, who ran the government at the time).[14]

## The Natural Order

Central to understanding the German discourse is a set of considerations about the natural order and the place humans occupy in it:

> The human power over nature has now reached humanity itself. . . . Man can become the creator of himself. We have to doubt that we would be able to come up to the role of creator. Our best intentions, according to experience, tend not to suffice for that role. Not only can they be terribly abused, they also are not immune against ignorance and negligence. . . . [There is a possibility] for profound sacrilege, sacrilege against what constitutes man in the creational order. On the basis of my Christian beliefs I confess that I reject the opinion that man should take the responsibility for all of human life in the world. . . . This power does not become us. (Dr. Albrecht, 25 November 1988, *BP* 595, 428/429)[15]

Not surprisingly for a speaker from the conservative Christian Democratic Union (CDU), Dr. Albrecht refers to Christian values to underpin his position. However, this perspective is shared by MPs who do not argue from a Christian standpoint: "It does not matter where one has one's roots. Whether they are of a general philosophical or moral nature, or whether they are based on Christian beliefs and ethics. It is all important that in many questions we are moving on shared ground" (Einert, ibid., 441).

So what is this shared ground the MPs refer to? Dr. Albrecht's speech shows that it is a very distinct conceptualization of a natural order, and of the space that humans and their science must occupy in this order. Nature is a complex system. Whether it is held to be God's creation or not, humanity is part of this complex system, not its master. Humans must think of themselves as creations, not as creators. To go ahead with research on embryos would mean to "meddle with nature's or God's business" (Dr. Seesing, 8 December 1989, *PP 11/183*, 14171).

The natural order must therefore be thought of as essentially good, right, and meaningful. It becomes humans, and scientists in particular, to appreciate this essential rightness of the natural order, rather than to embark on short-sighted projects of manipulation and control. All speakers share a concern that things can go horribly wrong if science assumes control over the uncontrollable: "We know now that nuclear power cannot be mastered. . . . Our children and their children will still have to live with the consequences of Chernobyl. In the face of this experience we have now made, we must not repeat the same mistakes" (Dr. Peters, 25 November 1988, *BP* 595, 443).

## Ambivalence Reintroduced

Keeping this general agreement in mind, it may be surprising that the ESchG does not simply outlaw all aspects of reproductive technologies. In order to understand how the cautious conceptualization of science and scientists can be reconciled with the ESchG, which gives much freedom to doctors who work in the field of reproductive medicine, it is necessary to look at another typical contribution to the German debate:

> There are some who have a distinct position, because they see the dangers that come with the new possibilities. They say: alright, childlessness just is fate. . . . This might well be the case, Ladies and Gentlemen, and I do understand this standpoint. However, I do not share the consequences that are drawn from this, namely that none of the methods of artificial insemination should be allowed. I am of the opinion that there might be cases, for example if a woman cannot have children of her own with her husband or long-time partner due to an accident or an illness, cases in which one can help, if the new insights allow it. I think this is right and justifiable. (Dr. Däubler-Gmelin, 8 December 1989, *PP 11/183*, 14168)[16]

Here Dr. Däubler-Gmelin concedes that one needs to be critical and cautious about science and claims to understand the position that advocates nature as a limit to human freedom. She admits, however, that she herself does not entirely subscribe to this logic. There are cases where one can and should help. It is the logic of helping in circumstances where new technologies or medical treatments can alleviate suffering that allows the German parliamentarians to overcome their generally negative assessment of science. In a way, one could argue that the morality of the natural order is supplemented by a different morality: the morality of care for those who are suffering. With the introduction of this ethic of care, we can see that nature, as a discursive resource, provokes ambivalence.

## Serving or Controlling Human Life?

This ambivalent use of nature as a discursive resource—the limit for scientific ambitions and a tragedy to be alleviated—leads to a certain way of talking about scientists and their actions. The good and moral scientific behavior of doctors depends on their intentions: "That the doctor can take part in the creation of an individual does not at all imply that therefore he [*sic*] may exercise control over that individual" (*Bundesminister für Forschung und Technologie* 1985, 2). Another speaker explained that the line between terrible abuse and appropriate use of the technologies has to be drawn where the doctor is no longer "the servant of nature," but claims "the role of co-creator" (Dr. Berghofer-Weichner, 25 November 1988, *BP* 595, 431).

It seems that doctors act responsibly only if they take part or serve in an otherwise natural process. Doctors must not claim authorship, control, or the role of "co-creators" of human life. It further seems that the  difference between these two understandings of the medical profession is based less on what doctors actually do than on the spirit in which they do it. If a doctor fertilizes a woman's egg with her husband's sperm and transfers it back to her, then the doctor is performing a good and responsible act only if she or he thinks of that work as helping nature or helping "the whole person, the whole couple" (Government Paper, 23 February 1989, *BD 11/1856*, 2). The work is abusive, however, if it is done in the spirit of "playing God."

## Intentions as Boundary Marker

The conceptual basis of the ESchG should be clearer now. Research on embryos is forbidden. However, the ESchG permits the treatment of individuals. The ESchG makes it a crime "to fertilise an egg *for any other reason* than to bring about a pregnancy in the woman whose egg it is"; or "to acquire, use . . . or keep" an embryo "*for any other reason* than to bring about a pregnancy" (§ 1 (1) 2, and § 2 (1) and (2) ESchG, emphasis added). It is thus illegal to fertilize an egg, or to do anything with an embryo, that does not aim to treat an individual woman. The same practices can thus be criminalized or permitted, depending on what the scientist intends. [This judgment reflects the earlier argument that science, even medical science, must be pursued in a spirit of modesty and service and not the assertive spirit of the control or manipulation of nature.] ←

About the treatment side of new reproductive technologies, the ESchG says hardly anything at all. Unlike the HFEA, it does not contain provisions, for example, about the duties of clinics to offer information, advice, and counseling to their patients, to record their treatment and its success, or to establish data about children born after IVF or sperm donation. It does not contain provisions about which women should receive treatments, or how good practice is to be established or ensured. It leaves the entire regulation of practice to the medical profession itself (see Bundesärztekammer 1988; 1998). Legislation in Germany therefore combines a strict criminalization of some scientific practices (embryo manipulation for research) with a lenient provision for self-regulation of other medical practices (embryo manipulation for treatment). It expresses a deep mistrust of scientists, yet at the same time it delegates unsupervised responsibility to professional bodies who apply science to alleviate suffering (on this also see Betta 1995; Augst 2001; Waldschmidt 1993).

Some scientists and some representatives on the political left attack this boundary drawn between research and treatment as unsustainable, albeit for different reasons. The scientists want both to be legalized[17] and the critics want both to be outlawed.[18] Effectively, both groups of opponents to the government proposal argue that the ambivalence born of scientific progress is not resolved by the government bill but rather is simply eclipsed. To them, drawing a boundary between dangerous research and safe treatments makes no sense. The scientists argue that as treatments for fertility problems are beneficial, one must logically also support the research that makes these treatments possible. The Green Party argues that, as everyone agrees about the dehumanizing potential of these technologies, one must simply avoid the arena of technological interventions into human reproduction altogether.[19]

Both groups of critics argue that the risky nature of scientific progress cannot be resolved in the way the majority suggests. However, the ESchG is premised on isolating the medical benefits from the risks of embryo research. Drawing a boundary between good medical uses and bad scientific abuses, the lawmakers in Germany construct a miraculous boundary: it incorporates the benefits of the excluded scientific insights firmly on the side of the included medical practice. Apparently, the law-makers can have their cake and eat it, too.

## Britain: Similar Anxieties, Different Solutions

Ambivalence about the nature of scientific progress and its shaping of human life occurs in the London debates, too. This ambivalence and the anxieties it triggers, however, are resolved in very different ways in Britain. Here, the majority of MPs wanted to continue embryo research but were still confronted with anxieties about the destructive and dangerous potential of science.[20] They thus needed to find ways of distinguishing the embryo research they wanted to continue from the horrors envisioned if science proceeded without restrictions. They did so by drawing boundaries between safe science and science fiction and between reasonable confidence and irrational fear. This cartography also led to a certain conceptualization of the minority opposition to embryo research as irrational and uninformed.

### The Majority Construction of Science and Scientists

In the British debates we find positive images of scientists that are entirely absent from the German discussions. British MPs consistently praise the decency, courage, and trustworthiness of scientists by describing them as "wonderful, dedicated and brilliant people, [whose] work is carried out with much love, with a great deal of service to humanity" (Lord Ennals, 7 December 1989, HL, vol. 513, col. 1013).[21] In contrast, German MPs often talk about corporate power and about science's being driven by the financial interests of a few powerful agents.[22] The contrast could not be greater with, for example, Lord Ennal's praise of scientists, which he concludes with the remark that they do all this wonderful work "at not very high wages" (ibid., col. 1013). In England, those in favor of embryo research do not distinguish between scientists and doctors or research and treatment; both are good. However, those in favor of embryo research still have to deal with an opposition that draws upon examples of science gone wrong in the past and that talks about the dangerous potential of embryo research or genetic engineering. The other side of science cannot simply be ignored.

## Science vs. Fiction, and Reason vs. Irrationality

One strategy employed by the majority to dismiss anxieties about science, and thus about embryo research, is to draw a boundary between the science they are really talking about and science fiction. The majority of speakers held that certain scientific practices such as cloning and hybridization, about which there was much concern, are not what the bill in Parliament really is about: "We have heard much scare talk about hybrids, clones and designer babies, but *such talk comes from people who do not understand* the limitations of the work. . . . They remain in the science fiction arena" (Mr. Turnham, 23 April 1990, HC, vol. 171, col. 64, emphasis added).[23] As can be seen from this quote, the distinction between science and science fiction implies a further distinction between those who are able to distinguish the two, and those who are not. This distinction allows the majority discourse to construct itself as rational and enlightened and its opposition as irrational and unenlightened.

The majority speakers argue frequently that those opposing embryo research are informed only by unworldly religious doctrine.[24] A rational discourse with them is thus nearly impossible. They frequently assure the minority that they are welcome to keep their beliefs (after all, we live in a "multicultural, pluralist and free society"), but that they must not impose such beliefs on others.[25] The majority discourse constructs contrasting identities for itself and the opposition to embryo research and demands that "semantics and scruples" be replaced with "science and sense" (Dr. Goddson-Wickes, 2 April 1990, HC, vol. 170, col. 962).

The identification of the minority position with religious belief also allows the majority to employ another discursive strategy. Many majority speakers tell a story that begins with papal intolerance and the conviction of Galileo,[26] and leads via the prohibition of anatomy lessons,[27] to the burning of books,[28] and now to the controversies around research on the human embryo. This narrative has a well-known conclusion: the church has lost every battle since it tried to tell science what not to do. Those resisting scientific progress on the basis of religious convictions will most probably lose this battle, too. The opponents of embryo research therefore find themselves cast as losers from the outset of the narrative.

The majority's discursive strategy also makes bold statements about itself. Identifying with Galileo, those in the majority tell the story of the

Enlightenment's battle with the "forces of darkness." Lady Warnock, for example, claimed:[29]

> We are now in the twentieth century and irrevocably as it is, part of an ~~age where we must be allowed to take the possible risks of our own knowledge.~~ We are not in the same position as people were in the seventeenth century. We must be able to take risks and take them into account when we pursue knowledge. . . . We cannot undo the Enlightenment. In my view it would be morally wrong to place obstacles derived from beliefs that are not very widely shared in the path of science. (7 December 1989, HL, vol. 513, col. 1036)

In Warnock's Enlightenment discourse, the pursuit of knowledge is the way forward, and any attempt to stop science is a futile effort to undo progress. The "we" subjects of her discourse seem a rightly self-confident bunch of people. We know that it is right to take risks. We can control possible negative effects of scientific progress, and our ambition must be to overcome irrational anxieties and lack of knowledge. The difference with the German view on controlling nature could not be greater. In Bonn, the speakers warn against the delusion that we could successfully assess and control all the relevant risks. Warnock and other majority speakers in London instead see no reason to be overly modest about human abilities. It would indeed be "paradoxical" to assume anything but human grandeur "at a time when freedom is expanding in a glorious way in so many fields" (Lord Ennals, 7 December 1989, HL, vol. 513, col. 1015).

We can observe the majority's attempts to draw a boundary between themselves and the antiresearch minority. This boundary separates progress from regression, the scientific mind-set of openness and rationality from religious intolerance and superstition, and universal reason from particularism. The German case showed how difficult it is to decide whether it is reasonable to fear or to trust science. In London the majority of speakers successfully erect a boundary that neatly distinguishes between these sentiments. To be concerned about embryo research is to be superstitious, old-fashioned, and irrational; to trust science is modern, responsible, and reasonable.

## Progress and Its Other

Taking this boundary seriously would mean that there was no superstition, irrationality, or religious sentiment to be found on the majority side of the fence. However, pro-research speakers who claimed to be informed solely by facts and reason were regularly carried away into mystic, enthusiastic descriptions of a brighter future. They speak of "fantastic advances" in science, of "magic substances" (Earl of Halsbury, 7 December 1989, HL, vol. 513, col. 1046) and of the "scientific miracle of gene-mapping" (Lord Glenarthur, ibid., col. 1042).[30] The claim of the pro-research majority to be nothing but rational and cool-headed when making judgments about scientific progress is itself fictitious. The pro-research speakers revert to the language of religion, wonder, and fairy tale when they describe the work of scientists. Beck (1986, 344–45) has called this attitude modern *Fortschrittsglaube*, that is, an almost religious belief in progress that "is the self-confidence of modernity based on its own creational power in the form of technology." This "othering," displayed in the UK discourse about progress is a strategy to overcome ambivalence by ascribing to a distinct "other" one set of characteristics (irrational, religious, particular, narrow-minded) and assuming the self to be anything the other is not (Bauman 1991, 53).

Modern belief in "the forces of progress—with a capital 'P' if one wishes" (Lord Ennals, 20 March 1990, HL, vol. 517, col. 234) is per se not essentially different from Christian belief in God (see Freud 1985, 280). Neither can claim to be intrinsically rational or to be based only on facts. The difference between the majority and the minority in British parliament is not that one is rational and the other irrational, but that the beliefs of the majority, supporting embryo research, are constructed not as beliefs but as sober and rational judgments. The majority's trust and dependence on science is held not to be as desperate and blind as religious faith, because it is held to be based on reason and knowledge.

## Conclusion

In order to legitimize the boundaries they draw, German and British parliamentarians turn to specific constructions of science, nature, knowledge, progress, and human life. German lawmakers placed human life firmly within notions of the "natural order," whereas British parliamentarians saw human life emerge in a miraculous world of heroic scientist battling against human suffering.

The British HFEA can be read as a paradigmatically modern law, taking up modern ideals of progress and rationality. It grounds itself in the rejection of anything that can be constructed as "countermodern"—irrationality, religious beliefs, or the anxieties of the antiresearch minority. In this model, human life is best served by permitting scientific and medical progress. The German ESchG, on the other hand, does not whole-heartedly embrace the rationality and spirit of science. It is modeled on an order perceived to be natural and thus timeless. In this perspective, law exists to resist changes to this natural order. Human life needs to be protected from the erosive forces of permanent progress. Yet, the rejection implied in either of these positions cannot be upheld without contradictions. The British rejection of religion is pursued quasi-religiously, and the German dismissal of modern science breaks down where the ESchG denies that modern medicine necessarily relies on it.

Both laws therefore contain a crucial element of denial that what is rejected and what is embraced are not fundamentally different: Belief in progress is not fundamentally different from irrationality, neither is modern medicine from modern science. Through making laws that claim to regulate reproductive technologies, parliamentarians attempt to assert control over the inescapable ambivalence provoked by the nature of science, progress, rationality and reason, and individualism. In short, they attempt to give shape to human life under the conditions of modernization, while they feel deeply ambivalent about the risks and opportunities generated by new technologies. In order to do this, both laws have to deny "the ambiguity that modern mentality finds difficult to tolerate, and modern institutions set out to annihilate" (Bauman 1991, 52).

Taking this "notorious duality of the modern tendency" (Bauman 1991, 52) seriously, it is clear that a degree of denial is always necessary to achieve discursive closure, draw a boundary, pass a law. The neatness of demarcating clearly what is inside these boundaries and what is out can never reflect the inescapable ambiguity of reality. In this respect, the denial of ambivalence is a necessary element of law. Through a partly fictional distinction between what it embraces and what it excludes the law in both countries manages to cast an—albeit fragile—shape to human life in modernity. "Through the fiction, law can assert what is clean contrary to recognized truth. But what law is thus asserting of itself is hollow. . . . The legal fiction, then, reveals law as the most dependent yet the most independent thing" (Fitzpatrick 2001, 88).

NOTES

I would like to thank Dr. Ian Gibson, MP, Chair of the HC Science and Technology Select Committee, for giving me time off from his busy office to attend the Next Generation Conference, and Munizha Ahmad for holding the fort in my absence.

The following abbreviations are used consistently through the references for legislative documents in the United Kingdom and Germany: House of Lords (HL); House of Commons (HC); *Bundestagsdrucksache (BD); Plenarprotokol (PP);* and *Bundesratsprotokoll (BP).*

1. Santos (1995, 2) describes two "pillars" of modernity, regulation and emancipation, each of which "tends to develop a maximalist vocation" that leads to tensions. There is obviously an overlap between Santos's and my conceptualization of modernity, but I prefer to think of what he calls pillars as dynamics or trajectories as this implies their ongoing movement and thus points to the problems law has in achieving closure in the face of modernization.

2. For the very extensive literature in this area see Firestone (1970); Stanworth (1987); Smart (1987); Zipper (1989); Dewar (1989); Morgan and Douglas (1994); Eekelaar (1994); Millns (1995); Thomson (1997); Chavo (1997); Dewar (1998); Edwards et al. (1999); and Jackson (2001).

3. For the problematic nature of tradition in family value discourses see Nicholson (1997). Also see Mr. Wiltshire, 20 June 1990, HC, vol. 714, col. 1022; Dr. Däubler-Gmelin, 8 December 1989, *PP* 11/183, 14167/14168.

4. See Lord Kennet, 7 December 1989, HL, vol. 513, col. 1025–28; and speaking on the same day and broadly sharing this perspective, Lord Duke of Norfolk, Lord Ashbourne, the Earl of Longford, Baroness Ryder of Warsaw, Lord Harrington, Baroness Elles, Viscount Sidmouth, Viscount Buckmaster, Earl of Perth, Lord Robertson of Oakridge, and the Earl of Cork and Orrery. For Germany, see, for example, Government Bill, 23 February 1988, *BD* 11/1856, 7. Also see the Social Democratic Bill: "The creation of human life for any other purpose than the development into a human person, especially for research purposes, is irreconcilable with the legal and ethical qualities of human life" (16 November 1989, *BD* 11/5710, 13). See also Dr. Seesing, 8 December 1989, *PP* 11/1471.

5. See Beck-Gernsheim, 9 March 1990, *Bundesrechtsausschuss* 11/73, 6; Dr. Seesing, ibid., 71; the Duke of Norfolk, 8 February 1990, HL, vol. 515, col. 996.

6. See the Government Bill (25 October 1989, *BD* 11/5460, 7) and the Green Bill (19 October 1990, *BD* 11/8179, 1/2). For the UK, see Alan Amos, 23

April 1990, HC, vol. 171, col. 106. See Augst (2001) for why these risks feature far more prominently in the German debates.

7. See for example Dr. Däubler-Gmelin, 8 December 1989, *PP* 11 /183, 14168; Green Bill, 19 October 1990, *BD* 11/8179, 3; Ruehmkorf, 9 November 1990, *BP* 624, 639.

8. See Government Statement, 23 February 1988, *BD* 11/1856, 2; Clarke, 2 April 1990, HC, vol. 170, col. 917; Lord Mackay of Clashfern, 7 December 1989, HL, vol. 513, col. 1004.

9. See Lord Ennals, 7 December, HL, vol. 513, col. 1014. The issue of disability is far more controversial in the German debates. See Ms. Schmidt, 24 October 1990, *PP* 11/230, 18213/18214 and Green Bill, 19 October 1990, *BD* 11/8179; Ms. Schmidt, 8 December 1989, *PP* 11/183, 14173; See Dr. Däubler-Gmelin, 24 October 1990, *PP* 11/230, 18211; Amendments of the SPD to the Government Bill, 24 October 1990, *BD* 11/8191, 31.

10. See Ms. Fyfe, 23 April 1990, HC, vol. 171, col. 64; 25 May 1990, Standing Committee B, vol. 1, 147/148; Ms. Richardson, ibid., 150.

11. Also see Giddens: "Science has thus long maintained an image of reliable knowledge which spills over into an attitude of respect for most forms of technical specialism. However, at the same time, lay attitudes to science and to technical knowledge generally are typically ambivalent" (1990, 89). Lee and Morgan (2001, 2) argue that only in the late twentieth century have we learned to view scientific progress "with profound scepticism." Wynne (1996) maintains that this might not be a new phenomenon and wars against assuming epochal changes in the nature of "lay" or "expert" knowledge.

12. Also see Dr. Seesing, 24 October 1990, *PP* 11/239, 18209; Baroness Ryder, 7 December 1989, HL, vol. 513, col. 1067.

13. See Dr. Seesing, 8 December 1989, *PP 11/183*, 14171; Government Paper, 23 February 1989, *BD* 11/1856, 5; Dr. Berghofer-Weichner, 22 September 1989, BP 604, 350; Walter Remmers, ibid., 352; Engelhard, ibid., 357.

14. A few Social Democrats also favor a ban on IVF (*BP* 604, 376/377). On the Green Party's philosophy, see Bause (1999) and Brockmann (1992).

15. Also see Walter Renner, 22 September 1989, *BP* 604, 352.

16. Also see Dr Peter (22 September 1989, *BP* 604, 357); Dr Seesing, 8 December 1989, *PP 11/183*, 14171.

17. See Dr. Buchborn, 9 March 1990, *Bundesrechtsausschuss* 11/73, 142.

18. See Schmidt, 8 December 1989, *PP 11/183*, 14172.

19. Many (radical) feminist writers and critics share this perspective. See, for example, Raymond (1993); Steinberg (1997); Waldschmidt (1993); Corea and Ince (1987); and Mies (1992).

20. See Lord Jakobovitz, 7 December 1989, HL, vol. 513, col. 1075; Duke of Norfolk, ibid., col. 1030; Lord Ennals, ibid., col. 1013; Lord Walton of Detchant, ibid., 1052.

21. Also see Lord Glenarthur, ibid., col. 1042.

22. See Engholm, 25 November 1988, *BP* 595, 435; and Einert, ibid., 440, who speaks of "medical-industrial interests."

23. See: Lord Walton of Detchant, 7 December 1989, HL, vol. 513, col. 1052; Lord Glenarthur, ibid., col. 1042; Lord Ennals, ibid., col. 1013; and Ms. Richardson, 2 April 1990, HC, vol. 170, col. 925.

24. See Lord Meston, 7 December 1989, HL, vol. 513, col.1102 and Lord Ennals, ibid., col. 1015.

25. Earl Jellicoe, ibid., col. 1038. Also see Lord Mc Gregor of Dunnis, ibid., col. 1016–19; Lord Hailsham of Saint Marylebone, ibid., col. 1022; Baroness White, ibid., col. 1045; Lord Winstanley, 20 March 1990, HL, vol. 517, col. 243; and Ms. Richardson, 2 April 1990, HC, vol. 170, col. 927.

26. Lord Hailsham, 8 February 1990, HL, vol. 515, col. 968; Baroness Warnock, 7 December 1989, HL, vol. 513, col. 1036; Lord Flowers (also referring to Darwin), ibid., col. 1061; and Lord Henderson, 8 March 1989, HL, vol. 504, col. 1578.

27. Baroness Faithful, 7 December 1989, HL, vol. 513, col. 1099.

28. Lord Sherfield, ibid., col. 1100.

29. Baroness Warnock's perspective remains central for the majority discourse throughout the debates. Dr. Goodson-Wickes, for example, states that "there are few if any absolutes in this contentious issue. If reassurance is in the air, Warnock provides it" (2 April 1990, HC, vol. 170, col. 961). Also see Ms. Richardson, ibid., col. 924; the Earl of Halsbury, 7 December 1989, HL, vol. 513, col. 1046; Lord Henderson of Brompton, ibid., col. 1097. Morgan and Lee (1991, 4) state that "the [HFE] Act is a Warnock Act."

30. Also see Lord Ennals, 7 December 1989, HL, vol. 513, col. 1046; Lord Glenarthur, ibid., col. 1042; Baroness Nicol, ibid., col. 1062; and Mr. Dafydd Wigly, 2 April 1990, HC, vol. 170, col. 948.

## REFERENCES

Adorno, T. W., and M. Horkheimer. 1999. *Dialectic of Enlightenment.* London: Verso.

Arendt, H. 1993. *Between Past and Future: Eight Exercises in Political Thought.* New York: Penguin.

Augst, C. 2001. Verantwortung für das Denken: Feministischer Umgang mit neuen Reproduktionstechnologien in Großbritannien und der Bundesrepublik. *Jahrbuch für Kritische Medizin: Krankheitsursachen im Deutungswandel* 34: 135–56.

Bauman, Z. 2001 Leben—oder bloß Überleben? *Die Zeit* 2001 (1): 41.

———. 1991. *Modernity and Ambivalence.* Cambridge: Polity Press.

———. 1998. Postmodern adventures of life and death. In *Modernity, Medicine and Health: Medical sociology towards 2000,* ed. G. Scambler and P. Higgs. London: Routledge.

Bause, M. 1999. Natur als Grenze?—Modernes und Gegenmodernes im grünen Diskurs. In *Der unscharfe Ort der Politik—Empirische Fallstudien zur Theorie reflexiver Modernisierung,* ed. U. Beck, M. A Hajer, and S. Kesselring. Opladen: Leske und Burich.

Beck, U. 1986. *Risikogesellschaft: Auf dem Weg in eine andere Moderne.* Frankfurt a.M.: Suhrkamp.

Betta, M. 1995. *Embryonenforschung und Familie: Zur Politik der Reproduktion in Großbritannien, Italien und der Bundesrepublik.* Frankfurt a. M.: Peter Lang.

Brockmann, S. 1992. After nature: Postmodernism and the Greens. *Technology in Society* 14: 299–315.

Bundesärztekammer. 1988. Richtlinien zur Durchführung der In-Vitro-Fertilisation mit Embryotransfer und des intratubaren Gameten- und Embryotransfers als Behandlungsmethoden der menschlichen Unfruchtbarkeit. In *Kommentar zum Embryonenschutzgesetz,* ed. R. Keller, H.-L. Günther, and P. Kaiser. 1992. Stuttgart: Verlag W. Kohlhammer.

———. 1998. Richtlinien zur Durchführung der assistierten Reproduktion. Novellierte Fassung 1998. *Deutsches Ärzteblatt:* A 3166–71.

Chavo, R. A. 1997. The interaction between family planning policies and the introduction of the NRTs. In *Intersections: Women on Law, Medicine and Technology,* ed. K. Petersen. Aldershot/Hants.: Dartmouth.

Connolly, W. 1987. *Politics and Ambiguity.* Madison: Univ. of Wisconsin Press.

Cooke, P. 1990. *Back to the Future.* London: Unwyn Hyman.

Corea, G., and S. Ince. 1987. Report of a survey of IVF clinics in the U.S. In *Made To Order: The Myth of Reproductive and Genetic Progress,* ed. P. Spallone and D. L. Steinberg. Oxford: Pergamon.

Dewar, J. 1989. Fathers in law? The case of AID. In *Birthrights: Law and Ethics at the Beginning of Life,* ed. R. Lee and D. Morgan. London: Routledge.

Dewar, J. 1998. The normal chaos of family law. *Modern Law Review* 61: 467–85.

Douzinas, C., and R. Warrington, with S. McVeigh. 1991. *Postmodern Jurisprudence: The Law of Text in the Text of Law.* London: Routledge.

Durant, J. 1998. Once the men in the white coats held the promise of a better future. . . . In *The Politics of Risk Society,* ed. J. Franklin. Cambridge: Polity Press in association with the IPPR.

Firestone, S. 1970. *The Dialectic of Sex.* New York: William Morrow & Co.

Fitzpatrick, P. 2001. *Modernism and the Grounds of Law.* Cambridge: Cambridge Univ. Press.

———. 1992. *The Mythology of Modern Law.* London: Routledge.

Foucault, M. 1984. What is enlightenment? In *The Foucault Reader,* ed. P. Rabinow. London: Penguin.

Franck, T. 2000. *The Empowered Self: Law and Society in the Age of Individualism.* Oxford: Oxford Univ. Press.

Freud, S. 1985. *Civilization and Its Discontents.* In *The Pelican Freud Library— Civilization, Society and Religion,* vol. 12. Harmondsworth: Penguin.

Giddens, A. 1990. *The Consequences of Modernity.* Cambridge: Polity Press.

———. 1994. Living in a post-traditional society. In *Reflexive Modernization: Politics, Tradition and Aesthetics in the Modern Social Order,* ed. U. Beck, A. Giddens, and S. Lash. Cambridge: Polity Press.

Gieryn, T. F. 1995. Boundaries of science. In *Handbook of Science and Technology Studies,* ed. S. Jasanoff, G. E Markle, J. C. Petersen, T. Pinch. London: Sage Publications.

———. 1999. *Cultural Boundaries of Science: Credibility on the Line.* Chicago: Univ. of Chicago Press.

Jackson, E. 2001. *Regulating Reproduction: Law, Technology and Autonomy.* Oxford: Hart Publishing.

Lee, R., and D. Morgan. 2001. *Human Fertilisation and Embryology: Regulating the Reproductive Revolution.* London: Blackstone.

Mies, M. 1992. *Wider die Industrialisierung des Lebens: Eine feministische Kritik der Gen- und Reproduktionstechnik.* Pfaffenweiler: Centaurus-Verlags-Gesellschaft.

Millns, S. 1995. Making "Social judgements that go beyond the purely medical": The reproductive revolution and access to fertility treatment services. In *Law and Body Politics: Regulating the Female Body,* ed. J. Bridgeman and S. Millns. Aldershot, Hants: Dartmouth.

Morgan, D., and G. Douglas. 1994. The constitution of the family: Three waves for Plato. In *Archiv für Rechts- und Sozialphilosophie 57: Constituting Families,* ed. D. Morgan and G. Douglas. Stuttgart: Franz Steiner Verlag.

Morgan, D., and R. Lee. 1991. *Blackstone's Guide to the Human Fertilisation and Embryology Act 1990: Abortion and Embryo Research, the New Law.* London: Blackstone.

Mulkay, M. 1997. *The Embryo Research Debate: Science and the Politics of Reproduction.* Cambridge: Cambridge Univ. Press.

Nicholson, L. 1997. The myth of the traditional family. In *Feminism and Families,* ed. H. Lindemann Nelson. London: Routledge.

Raymond, J. 1993. *Women as Wombs: Reproductive Technologies and the Battle over Women's Freedom.* San Francisco: HarperCollins.

Santos, B. de Sousa. 1995. *Toward a New Common Sense: Law, Science and Politics in the Paradigmatic Transition.* New York: Routledge.

Smart, C. 1987. "There is of course the distinction dictated by nature": Law and the problem of paternity. In *Reproductive Technologies: Gender, Motherhood and Medicines,* ed. M. Stanworth. Cambridge: Polity Press.

———. 1989. *Feminism and the Power of Law.* London: Routledge.

Stanworth, M. 1987. Reproductive technologies and the deconstruction of motherhood. In *Reproductive Technologies: Gender, Motherhood and Medicine*, ed. M. Stanworth. Cambridge: Polity Press.

Steinberg, D. L. 1997. *Bodies in Glass: Genetics, Eugenics and Embryo Ethics*. Manchester: Manchester Univ. Press.

Thomson, M. 1997. Legislating for the monstrous: Access to reproductive services and the monstrous feminine. *Journal of Social and Legal Studies* 6: 401–24.

Waldschmidt, A. 1993. Halbherzige Verbote, große Regelungslücken: Deutsche Gesetze zur Fortpflanzungsmedizin und Embryonenforschung. In *Die Kontrollierte Fruchtbarkeit: Neue Beiträge gegen die Reproduktionsmedizin*, ed. E. Fleischer and U. Winkler. Vienna: Verlag für Gesellschaftskritik.

Warnock, M. 1985. *The Warnock Report on Human Fertilisation and Embryology: A Question of Life*. London: Blackwell.

Wynne, B. 1996. May the sheep safely graze? A reflexive view of the expert–lay knowledge divide. In *Risk, Environment and Modernity: Towards a New Ecology*, ed. S. Lash, B. Szerszinski, and B. Wynne. London: Sage.

Zipper, J. 1989. What else is new? Reproductive technologies and custody politics. In *Child Custody and Politics of Gender*, ed. C. Smart and S. Sevenhuijsen. London: Routledge.

# 16

# Reconceptualizing Technology Transfer

*The Challenge of Shaping an International System of Genetic Testing for Breast Cancer*

SHOBITA PARTHASARATHY

One of the most compelling signs of globalization is the transnational flow of technology. The world seems to be shrinking, with innovators moving computers, cell phones, medical devices, and drugs across national borders and establishing linkages across countries. But how seamless are these processes? How is transnational technology transfer taking place? How do the specificities of national context, such as different approaches to technology, property, and health care, figure in these processes of globalization?

Many scholars and observers of globalization argue that the uniformity of science and technology provide the bedrock for contemporary processes of globalization (Drori et al. 2003). Other analysts, however, disagree, arguing that the success or failure of technology transfer depends not simply on scientific objectivity but on regulatory frameworks and the organizational cultures of firms (Segerstrom, Anant, and Dinopoulos 1990; Keller and Chinta 1990). Unfortunately, neither of these perspectives helps us understand contemporary controversies over the transnational movement of new technologies such as genetically

modified foods or anti-AIDS medications. In this chapter, I explore
the attempt by an American genomics company to transfer its biotechnology, genetic testing for breast cancer, to suggest that we can understand these challenges to globalization by reconceptualizing the process
of technology transfer. Building upon approaches from the field of science and technology studies (e.g., Winner 1986), I argue that technologies are simultaneously technical and moral objects, not just novel tools
to test blood or grow crops but embodying specific values toward health
care, intellectual property, or environmental conservation.[1] Challenges
to technology transfer and even controversies over globalization occur
when conflicts arise over either the technical approach or the "moral
order" prescribed by the technological object.

When discoveries of two genes linked to inherited susceptibility for
breast cancer (the BRCA genes), were announced in the mid-1990s, attention turned almost immediately to the development of related diagnostics and therapeutics (Davies and White 1995). In both the United
States and Britain, for example, groups began to develop technologies
to test for mutations in the BRCA genes that predicted an inherited
susceptibility to breast or ovarian cancer. By 1998, very different BRCA
testing systems dominated the biomedical landscapes of the two countries. In the United States, Myriad Genetics (Myriad), a start-up biotechnology company that had been credited with finding the first BRCA
gene, used its relative financial strength and legal power, gained through
acquisition of patents and licenses covering both BRCA genes, to become the sole provider of BRCA testing. It offered BRCA testing like
any ordinary medical test: individuals could use its DNA analysis service
through any physician. By allowing access to any individual who received a referral from any physician, Myriad ensured that the potential
market for its service was quite large—it was available to anyone who
could afford it. In Britain, BRCA testing services were provided on a regional basis through the state-run National Health Service (NHS). Its
shape was reminiscent of other specialist services offered by the NHS,
involving both risk assessment and triage. Individuals interested in testing would first provide their family history of breast and ovarian cancer
to a primary- or secondary-care physician in their region. Then, using a
standard that had been developed in consultation with geneticists across
the country, these physicians would classify individuals into low, moder-
ate, and high risk categories and offer services accordingly. Only individuals classified as high risk would be allowed to visit a regional

genetics clinic and access both counseling and laboratory analysis of the BRCA genes.

After systems of genetic testing for breast cancer had been integrated into health care in the United States and Britain, the U.S. provider Myriad attempted to expand its testing service to Europe and Asia by trying to convince health-care professionals of the value of its technology and threatening legal action on the grounds of patent infringement for those who were already providing testing. The firm began its efforts in Britain, hoping to shut down the NHS's national BRCA testing system and have blood samples that were collected in Britain analyzed by its laboratory in Salt Lake City. What happened when Myriad tried to expand its version of BRCA testing services to Britain and the rest of Europe? Would Myriad's technology, together with its way of structuring the identities of patients and health care professionals, be neatly transferable to the British context? And how would the British government, scientists, clinicians, and patients respond to Myriad's attempt to insinuate its approach to BRCA testing into its institutions?

In this chapter, I explore Myriad's attempt to shape a global service of genetic testing for breast cancer, and how British scientists, health-care professionals, and activists responded to these efforts.[2] As we will see, ongoing tensions emerged: in working to extend its patent rights, the company was not simply trying to introduce a single entity of narrow scope into a new geographic region; it was trying to introduce an entire system, encompassing not only the clinical and technical dimensions of the test but also particular roles for that system's participants— patients and health-care professionals—into a cultural context that differed greatly from the one where it took shape.

The chapter begins by describing how Myriad attempted to expand its testing service to Britain. It then explores how British scientists, health-care professionals, and activists challenged the company in three ways: they questioned Myriad's use of patent rights as a justification for expansion of its testing service; they challenged the validity of a BRCA testing system based simply on laboratory services; and they argued that Myriad's system prescribed roles for health care professionals and individuals that were inappropriate in the British context. Next, the chapter describes the negotiations between Myriad and the NHS and the eventual resolution of Myriad's attempted technology transfer. Finally, I conclude with a discussion of the implications of this story for our understanding of the relationship between technology and globalization.

## Myriad Tries to Transfer Its Technology

When Myriad turned its attention to the international market, it adopted a similar approach to its expansion efforts in the United States—to use its legal and economic position to eliminate all BRCA testing providers. Indeed, Myriad had already applied for patents from the European Patent Office (EPO), which covers most countries in Europe, when it applied for American patents in 1994 and 1995 (Shattuck-Eidens et al. 1996; Rommens et al. 2002).[3] By 1998, expecting that the EPO would soon grant its patent applications covering both BRCA genes, it began to market its BRCA testing regime directly to European health-care professionals. Its first strategy was to emphasize that it could provide an accurate laboratory service that would be widely available.

The company began its European expansion efforts in earnest by inviting representatives from the European Familial Breast Cancer Demonstration Project, an initiative designed to investigate methods of management of women at high risk for breast cancer, to tour its laboratories and facilities. It hoped to convince project members, which included delegates from the United Kingdom, France, Italy, Germany, Norway, and the Netherlands, that it could provide services that were more technically accurate than those already available there. Myriad's BRCA testing services involved full sequencing of both BRCA genes, which was considered 99 percent sensitive, while most European services used a variety of methods that ranged in sensitivity from 80 to 95 percent.[4] Company officials argued that cooperation between Myriad and the European Project would allow for an expansion of the "currently limited availability of breast cancer genetic testing in Europe," placing emphasis on the technical superiority of the DNA sequencing services that they could provide (Myriad Genetics 1998). At the time, Dr. Neva Haites, head of the European project, welcomed the visit. "This meeting offers us the possibility of understanding the potential for future collaborations between European centers and Myriad Genetics Laboratories, to ensure that high-risk individuals can be identified and hence are offered optimal screening and management" (Myriad Genetics 1998). Despite Haites's enthusiasm, however, few European health-care professionals seemed interested in using Myriad's services. Most seemed to prefer to continue with their existing national systems of BRCA testing.

By the end of 1998, Myriad had focused its efforts on Britain and taken an aggressive approach, not only explaining the benefits of the company's testing services but also threatening legal action on the grounds of patent infringement if British regional genetics clinics did not begin sending their samples to Myriad's U.S. laboratories.[5] The company's CEO and lawyers presented their case to a biennial meeting of the UK Cancer Family Study Group, which included medical geneticists, molecular geneticists, oncologists, genetic nurses, and genetic counselors involved in providing services or conducting research in the area of inherited cancer risk. Myriad used the same tactic as it had in the United States, arguing that by providing BRCA testing, the British NHS would be in violation of its European patents as soon as they were issued. British health-care professionals, however, were unmoved either by Myriad's promise of a better testing service or its threats to file suit and shut down NHS BRCA testing services. They continued to provide testing as a national system that included risk assessment, triage, and a package of counseling and laboratory analysis services.

Myriad then tried another approach and directly contacted the UK Department of Health (DoH), which was in charge of NHS services. It demanded that the UK DoH pay a licensing fee to continue testing and send samples to Myriad's U.S. laboratories, or risk suits for patent infringement.[6] Meanwhile, the company explored other options; for example, they contacted private laboratories in Britain to see if they were willing to serve as satellite laboratories that would send mutation information and a share of the revenues back to Myriad's BRCA gene databases in Salt Lake City.

## Responding to Myriad

Myriad's concerted effort to pressure the DoH and health care professionals to adopt its testing service led British scientists, activists, clinicians, and government officials to begin organizing targeted responses to the company. Most major scientific and professional organizations, such as the British Society of Human Genetics (BSHG) and the Clinical Molecular Genetics Society (CMGS), a branch of the BSHG devoted to molecular genetics, wrote position papers and official statements questioning the patentability of genes and predicting negative consequences for the ownership of human gene sequences. Patient activist Wendy Watson gave interviews to the media expressing her concern over gene patenting and the commercialization of genetic testing and helped mobilize opposition to Myriad among patient groups. Meanwhile, the UK DoH developed a consultation committee to aid in its discussions with Myriad, which included, along with Watson, the chairperson of CMGS; physicians, counselors, and nurses from regional genetics clinics; and NHS officials involved in purchasing regional services.[7]

As we shall see, opposition to Myriad's effort to expand its BRCA testing monopoly in the UK challenged the validity of the firm's testing system in the British context in two ways. One line of attack questioned the use of patent rights as a justification for Myriad's testing service and expansion efforts, arguing that the patenting of genes was unethical and inappropriate. Another questioned the accuracy of Myriad's testing system in a British context that emphasized the importance of clinical care and had defined specific roles for health-care professionals and individuals.

## The Legitimacy of Patent Rights

In the United States, assignment and acquisition of patent rights over the BRCA1 and 2 genes to Myriad Genetics had not only identified the company as an inventor of the isolated and purified genes but also helped justify Myriad's efforts to become the sole provider of BRCA testing and control how the testing system would be built. While scientists, health-care professionals, and activists in the United States questioned the architecture of Myriad's testing system and the roles it prescribed for health-care professionals and individuals, they largely did not challenge assignment of inventorship of the BRCA genes to Myriad, or the ownership that these patents represented (Parthasarathy 2003; Lewin 1996, A14; National Breast Cancer Coalition 1997).[8] This lack of organized opposition within the scientific community might be better understood by considering the regulatory and industrial environment in the United States; not only were linkages between the university and industrial sectors not uncommon, but technology transfer offices at American universities actively encouraged their scientists to patent their inventions, and some scientists even left academia to start companies and commercialize their own research findings (96th Cong. 1980).[9] But would patents, and the BRCA gene patents in particular, have the same meaning in Britain? In contrast to their American counterparts, European universities did not so actively encourage their scientists to patent their work and had not had such a historically close relationship with the industrial sector. Moreover, very few European scientists left academia to "start-up" their own companies. Recent scholarship underscores these observations, demonstrating that attributions of inventorship and ownership of scientific and technological objects are contingent upon social context (Hilgartner 2002; Biagioli 1998; Boyle 1996; Cambrosio and Keating 1995; Bowker 2001; de Laet 2000).[10] The following discussion emphasizes not only that understandings of ownership and property are nationally specific but also shows how understandings of the patentability of the BRCA genes in the American context were explicitly challenged and rejected in Britain.

## RESISTING THE EU DIRECTIVE ON THE LEGAL PROTECTION OF BIOTECHNOLOGICAL INVENTION

British scientists, health-care professionals, and activists had begun to mobilize against the patenting of genes long before Myriad expanded its testing service, in response to EU legislation designed to strengthen the European biotechnology industry by harmonizing patent law across member countries. The EU Directive on the Legal Protection of Biotechnological Invention (called the Biotech Patent Directive), first introduced in 1988, aimed to make human gene sequences, as well as genetically engineered plants and animals, patentable across the European Union. Opponents of the directive, which included many European governments, companies, health-care professionals, and activist groups, argued that human gene sequences were not patentable because they were discoveries of things already existing in nature rather than inventions of novel things could be subjected to rules of intellectual property (Hawkes 1995; Green 1995; Bremmer 1998; Butler and Arthur 1998). This opposition to gene patenting at the EU level would have a very important influence on Myriad's expansion attempts in the UK. It provided an opportunity for British scientists, health-care professionals, and activists to mobilize and develop their political strategies and for them to develop and publicize arguments against gene patenting that would later prove effective against the American genomics company.

In September 1997, the British Society for Human Genetics (BSHG) issued a statement on "Patenting of Human Gene Sequences and the EU Draft Directive," arguing that the EU directive should not be passed because genes did not fulfill the first criteria of patenting—novelty— and were therefore unpatentable. "[Novelty] cannot reside in the mere description of a nucleotide sequence. It must rest in either novel

methodologies for discovering the sequence or a novel use or application of the sequence" (BSHG 1997). Simply identifying an existing nucleotide sequence, BSHG argued, did not require ingenuity on the part
of the researcher. A number of British geneticists also signed a separate
letter to the EU Parliament that articulated these same concerns. The
letter stated, "As researchers or clinical scientists we urge you to exempt
genes and their elements from patenting" (Andrews et al. 1997). It was
the first time that British scientists had made this type of concerted
grassroots effort to influence policymaking. While American geneticists
had voiced little organized opposition to gene patenting, British scientists had come together to question the attribution of ownership that
had certified Myriad's legitimacy in the United States (American Medical Association 2000).[11]

It should be stated that organizations like BSHG and CMGS
were not uniformly against the practice of patenting inventions. In fact,
BSHG's 1997 statement on the subject noted, "Patenting is a valuable
means of protecting intellectual property and promoting investment in
developing products for the diagnosis and treatment of genetic disease"
(BSHG 1997). These organizations were concerned that patenting genes,
in particular, would be detrimental to the cultures of both research and
health care in Britain. They worried that assigning ownership for gene
sequences themselves would allow patent holders to control all research
on a particular gene for the life of the patent, potentially limiting research opportunities and preventing scientists from working on the
most lucrative and complex biomedical problems. They also wondered
if a focus on intellectual property and commercialization of medical
biotechnology would conflict with European approaches to health care
as a public good.

Wendy Watson, who also campaigned actively against the EU
Directive, had a similar perspective to the BSHG. She asserted, "You
can't patent a gene . . . ! It's not an invention, it's a discovery!"[12] Both
activists like Wendy Watson and mainstream organizations like BSHG
questioned the very idea that genes were patentable inventions, and
thus the basis of Myriad's effort to expand its testing service to the UK
and shut down the NHS's BRCA testing service.

These critics, however, were unsuccessful in stopping the EU legislation. After years of vigorous debate, the EU Parliament and Commission finally passed the Biotech Patent Directive, which allowed the patenting of genes that had been isolated from the body, in May 1998

(European Parliament and Council 1998). However, despite this law that seemed to allow patenting of human genes, controversy continued over whether or not human genes should be patentable and what kind of ramifications patenting would create for European health-care systems. In fact, a number of countries immediately challenged the directive in the European courts (Dickson 2000).

QUESTIONING THE PATENTABILITY OF THE BRCA GENES

The controversy over the Biotech Patent Directive at the EU level had important consequences for the British debate regarding Myriad's expansion campaign. The directive seemed to justify Myriad's patents, but the new law had also sparked tremendous resistance among critics in Britain and had galvanized them to articulate a clear position against the patenting of genes.

It should come as no surprise, then, that British opposition to gene patenting continued and grew even after the directive was passed. As the EU directive was going into its final negotiations, the BSHG continued to assert that genes should not be patentable. Its 1998 statement on "Patenting and Clinical Genetics" reiterated this point. "A natural human gene sequence is part of the human body, and as such should not be patentable. The suggestion that such a sequence might be patentable if it is 'isolated in a pure form' or 'isolated outside of the body' seems to us a sophistry, and should not be allowed" (British Society for Human Genetics 1998)."

These scientific organizations also suggested that Myriad's attempts at EU expansion exemplified the dangers of patenting. The BSHG, for example, argued that testing services controlled by gene patent holders would interfere with downstream innovation. "If the sequence as such is patentable, it will not be possible for anyone at any time to devise a better or different way of genetic diagnosis; this is inequitable."[13] The CMGS, which published a detailed paper on "Gene Patents and Clinical Molecular Genetic Testing in the UK" in 1999, echoed the BSHG's concerns. The paper predicted that allowing gene patent holders to determine the provision of genetic testing services would make genetics services prohibitively expensive, reduce testing services available through the NHS, and jeopardize clinical and laboratory expertise in the NHS by allowing private concerns to provide state-of-the-art services. These groups, which had mobilized in opposition to the EU directive, now used the same arguments to fight Myriad.

COLLECTIVE INVENTORSHIP

British scientists, health-care professionals, and activists asserted further
that even if genes were inventions that could be patented and owned,
Myriad could not claim to be the sole "inventor" of the BRCA genes
(Meek 2000). The discovery was a collective effort, they argued, involving researchers, women, and funding bodies in Britain as well as the
United States. Myriad executives responded to these criticisms in the
British media, arguing that Myriad deserved the title of inventor and
the accompanying benefit because of the time and money it had spent.
"We've invested an enormous amount of man-years in making this discovery and making it applicable. It's only right that we should be protected." While Myriad based its claim to ownership on the resources it
spent mapping and sequencing the BRCA genes, British critics wanted
to adopt a broader definition of resources that included donated blood
from families with histories of breast and ovarian cancer, studies that
contributed to the overall body of knowledge about breast cancer genetics, and money from groups who funded research that led to the
discoveries.

Many British researchers, for example, argued that if authorship
of the genes could be claimed, they deserved some ownership because
they, too, had contributed to the gene discoveries. Sir Walter Bodmer,
one scientist involved in early research on the BRCA1 gene, said,
"Myriad is claiming it contributed far more than it actually achieved. As
a result . . . there is a lot of feeling of unfairness among British scientists"
(Ross 2000). Other scientists simply argued that the BRCA gene discoveries were the result of a protracted collective effort, and the final mapping and sequencing was more a matter of luck than inventiveness.
Andrew Read, chairman of the BSHG, explained, "The whole area of
gene patenting is controversial because it gives the prize to the person
who put the last brick in the wall" (Ross 2000). Scientists frequently used
this type of metaphor to explain their opposition to gene patenting,
tapping into an age-old image of science as both disinterested and collective (Merton 1973; Mulkay 1976; Mitroff 1974). Oncologist Bruce
Ponder noted, "We are uneasy about the principle of patenting genes.
Finding a gene is just the final step in a pyramid of knowledge and the
question is whether it is justifiable for one company to own the patent"
(Connor 1994, 3). Patient activists agreed. Wendy Watson noted, "I do
know that when it got to this stage, it was pure spade work, there was

nothing inventive about it, it was pure spade work."[14] This sense of out-rage, of course, contrasted starkly with the silence of American geneti-cists and activists with regard to Myriad's claims to inventorship and rightful ownership of BRCA testing in the United States.

In addition, other geneticists pointed out that Myriad's claims to sole ownership were particularly offensive because most Britons (as well as most other Europeans and Americans) credited Mike Stratton, a genet-icist at the Institute for Cancer Research in London, not Myriad, with finding the BRCA2 gene. Establishing priority in the BRCA gene dis-coveries was very controversial. The public excitement and potential scientific, medical, and industrial rewards had led a number of scientists to search for the genes and many even referred to research to find the breast cancer genes as a "race" (Davies and White 1995). Researchers throughout the world participated in this race, but Myriad was able to complete the mapping and sequencing of the BRCA1 gene first. The race continued, however, as researchers looked for BRCA2, another gene also thought to be a major cause of hereditary breast cancer. This time, however, the "winner" was much more difficult to determine (Dalpe et al. 2003). The day before Mike Stratton's group published the BRCA2 gene sequence in *Nature* magazine, Myriad announced that it had found the gene and submitted its sequence to GenBank, an inter-national depository of gene sequence information. Both Myriad and Mike Stratton's group filed for American and European patents on the BRCA2 gene, each claiming that they had first mapped and sequenced the gene. This BRCA2 controversy led many of the British scientists and health-care professionals who were part of Britain's small cancer genetics community to feel personally aggrieved by Myriad's proposed expansion. One scientist said that she would rather continue testing and go to jail on the grounds of patent infringement than accept Myriad's patent claims over the breast cancer genes. "At the end of the day, I hope I am locked up, because I'll make such a big deal about it. I mean they say they'll try and enforce this patent but I just hope the NHS doesn't just cave in and pay them money. The other thing is that my mate found BRCA2 at Sutton [Institute for Cancer Research]. So you can imagine how galling that is."[15]

Some commentators protested Myriad's sole authorship by noting that both inventorship and ownership of the genes should be expanded to include all of the contributors to the discovery, such as the families who donated the blood samples that were mined for the genes and the charities who funded the research. One scientist involved in the BRCA gene research noted that it would be unfortunate if the women who helped to find the BRCA genes by donating blood samples would later have to pay Myriad to receive access to testing.[16] Wendy Watson echoed this sentiment, noting, "Nobody has the right to patent this kind of information, which was only found with the help of the many families who had suffered a case of hereditary cancer. . . . It is morally wrong that any company should benefit commercially from that kind of research" (Dobson 1999). Watson also took this position further as she argued that it was not simply Myriad's money that contributed to finding the BRCA genes, but money from UK medical charities as well. "It was charity money that was looking for the gene," she said, "academic money, not private enterprise money that was looking for the gene."[17] Unlike in the United States, where both scientists and other testing providers simply accepted that Myriad's patent rights gave it control over the provision of testing, these commentators argued that the BRCA genes were the result of multiple contributions from a variety of sources and that Myriad had no right to claim sole ownership or control.

Finally, some scientists argued that not only was the BRCA gene discovery itself the product of multiple inventors but also that the process of actually finding the gene was identical to the way hundreds of other genes had been found. From their perspective, there was not even anything novel about the process of finding the BRCA genes as the process of gene discovery was a well-understood, widely used, and fairly uniform process. Scientists engaged in the process of looking for any gene would have followed a process similar to Myriad's. The BSHG noted simply, "The discovery of gene sequence has for some little time been a well-understood process. There is nothing novel or inventive about this in principle, and as such new gene sequences should not be patentable, even where a straightforward utility e.g. diagnostic testing has been specified, unless there has been real progress towards the design of a specific commercial product" (BSHG 1997). One scientist who was involved in breast cancer genetic research in Britain noted, "Most of us are pretty uncomfortable about it [patenting]. That finding the BRCA1 gene, in our view, didn't involve anything really novel. It's novel in the sense that they didn't know it was the BRCA1 gene until it was found, but it was a totally predictable consequence of the work that everybody was doing and there wasn't any particular reason why Myriad should scoop that particular pool, whether they were going to make a lot of money or not. It just didn't seem, to us, that they fulfilled the criteria of originality and so forth that you need for a patent."[18] British scientists were simply not prepared to accept that there was anything novel or unique in finding the BRCA genes that deserved the attribution of sole inventorship and ownership to Myriad.

## Opposing Myriad's Testing System

British critics also responded to Myriad's proposed expansion by challenging the appropriateness of the system for a context where health care was provided by the state, and clinical care was traditionally integrated with laboratory services. They attacked the architecture of the system itself, arguing that Myriad's focus on DNA sequencing missed the importance of counseling about the uncertainty of genetic risk and linking the technical results to preventive care. They also disagreed with the roles of health-care professionals and patients prescribed by Myriad's system, arguing that patients should not be consumers whose health-care decisions were simply facilitated by clinicians.

### TESTING WITHOUT COUNSELING?

In contrast with Myriad's system, the British NHS had built a BRCA testing system based on risk assessment and triage that integrated laboratory analysis and clinical care, demonstrated a commitment to genetic counseling, and was well integrated into the health system. One molecular geneticist suggested directly that the testing-counseling combination was part of the British ethos.

> We have quite a strong ethic, in this country, that suggests that many types of genetic tests should be coupled to access to genetic counseling, in fact it should be a package. . . . pre-test counseling, the test, and then post-test counseling, and then maybe, long term follow-up after that for some individuals. Or at least the possibility that they can come back if they're still worried. So it's part of a continuing care package, the genetic test, the technical test is only in some ways, the easiest part of it. And we wouldn't like to see genetic testing decoupled from access to counseling. And in fact, it may be that genetic testing became discredited if it were decoupled from access to counseling so that's something that we worry about and are quite keen to preserve.[19]

If British physicians were forced to send their samples to Myriad's laboratories in the United States or to Myriad-approved laboratories in

Britain, patients would no longer have to go through regional genetics services to access laboratory analysis, and there would be no guarantee that they would receive appropriate counseling from NHS-trained
health care professionals. Myriad's system would jeopardize the NHS
commitment to providing a package of genetics services. The CMGS
report stated, "Focusing genetic testing in multidisciplinary Regional
Genetic Centres assures the link between diagnosis and counseling that
is the hallmark and assurance of quality in this area of Medicine. . . .
The 'testing-within-counseling' culture may be lost if the laboratories
are divorced from the counselors. At worst, a group of patients and families for whom a genetic diagnosis is made will be at risk from the consequences of weak counseling and may be lost to key follow-up systems"
(Clinical Molecular Genetics Society 1999).

THE ROLE OF LABORATORY ANALYSIS

Myriad's critics also argued that while the company claimed to have a
state-of-the-art testing system by fully sequencing both genes, it was not
accurate for their purposes (Pinch 1993; Mackenzie 1993).[20] For them,
finding a deleterious mutation that could facilitate managing the health
care of women at risk for breast or ovarian cancer was much more important than producing genomic information by fully sequencing both
BRCA genes. Most British geneticists did not apologize for this approach. "We're not as effective as Myriad, because we don't sequence
the gene in quite the same way, but we do it in a more intellectual fashion. . . . If we came to a mutation, we'd stop, whereas they sequence the
whole gene" (Meek 2000). In addition, British opponents also criticized
the technical accuracy of Myriad's methods of DNA analysis by noting
that the "gold standard" method of DNA sequencing that Myriad used
would not necessarily pick up all of the alterations in the BRCA genes.[21]
If Myriad's method of laboratory analysis wasn't even 100 percent sensitive in picking up all mutations, critics argued, it certainly didn't war-
rant relinquishing an approach that enhanced identification of inherited risk through assessment of family history.

THE AUTHORITY OF THE HEALTH-CARE PROFESSIONAL

Not only did the components of Myriad's testing system come under
fire, but British health-care professionals also questioned whether the
roles of the system participants prescribed by Myriad's testing system

were appropriate in the British context. They argued that Myriad's system would remove the gatekeeping authority of health-care professionals and possibly jeopardize the future of genetic medicine in Britain. The NHS BRCA testing system provided genetics clinics with the authority to direct care while demonstrating to administrators that the NHS could provide genetics services for common diseases within the existing NHS culture. By creating an independent laboratory that would allow individuals to circumvent the risk assessment and triage system, Myriad's system would diminish the authority of health-care professionals and enhance the individual's right to demand access to services if implemented. Scientists and health-care professionals suggested that such a system was not appropriate in the British context. One molecular geneticist described the problem in the following manner. "There has been quite a strong emphasis on this in this country on trying to develop, I don't think it's in place yet, but trying to develop a system that gives equitable access to these services, but gives also some kind of gateway function. And the gateway can operate both ways, really, it can operate as a funnel into access to something, but it's also a controlling function. And I do think that if you had completely open access, it wouldn't be a good use of either public or private resources."[22] By allowing individuals to demand access to laboratory analysis, Myriad's service would limit the health-care professional's opportunity to control access to BRCA testing and thereby diminish her authority.

Opponents also worried that Myriad's service would hurt the practice of genetic medicine in the NHS. They predicted, for example, that both molecular and clinical geneticists were likely to be rapidly deskilled as they were prevented in engaging in their own laboratory analysis. "At best, UK centres would be deskilled to the level of sub-contractors of Myriad Genetics for routine work. . . . A feature of Clinical Molecular Genetics in the last ten years has been the rapidity of transferring research funding to tests of clear benefit to patients. Unless Regional Centres and the research groups with whom they collaborate are exposed to the problems of applying leading edge technologies to diagnostics, development will be increasingly confined to commercial companies" (Clinical Molecular Genetics Society, 1999). The potential loss of both expertise and gatekeeping authority particularly concerned specialists in genetics because NHS administrators could interpret BRCA testing as a model and possibly reduce funding for genetics services in the future.

## THE RIGHTS OF THE INDIVIDUAL

Would Myriad's expensive and demand-based testing system interfere with the British commitment to provide all individuals equal access to health care? Some critics argued that if testing were available on demand, it would be provided in an uneven manner across the country. Access would be based on initiative and financial opportunity, rather than a family history of breast and ovarian cancer. The CMGS report stated that such a demand-based system could have very damaging consequences for the overall NHS. "On the one hand it threatens . . . spiralling costs and on the other hand geographic inequalities of access to diagnosis" (Clinical Molecular Genetics Society 1999).

Indiscriminate BRCA testing, many health-care professionals argued, could devastate the NHS mission to provide individuals with the health care that they needed by drastically limiting the number of individuals who could be tested. If the NHS had to pay the high costs of Myriad's system ($2500 per test) within its limited and relatively stable budget, they would be able to offer the test to far fewer individuals than the current system allowed. Neva Haites, who had initially been more open to a relationship between Myriad and European health-care providers, noted, "In a way I'd rather offer a 70 percent service to the whole of the UK rather than a 100 percent service to a tenth of the country" (Connor 1994, 3). Rationing plans would have to become more strict, and many individuals with extensive family histories of breast and/or ovarian cancer might not qualify for BRCA testing services. Wendy Watson questioned, "It's, will genetic testing become more rationed than it should be because of the extra expense. And if that happens, I shall fight it. That's where I am coming from. I'm not particularly bothered if somebody's patented a part of my gene or whatever. That's not the issue. The issue is that it might reduce the number of people who are able to have genetic testing, who may well die because they haven't had genetic testing. And that is wrong."[23] While this statement seems to contradict Watson's strong opposition to gene patenting quoted earlier in this paper, it actually points us to the main concern of Myriad's critics. Scientists, health-care professionals, and even Watson herself were less concerned with the patenting of genes themselves than on the implications that such a practice would have on the provision of health care.

# Resolution

The UK Department of Health, British scientists, health-care professionals, and patient advocates negotiated with Myriad for over a year, trying to reach an agreement that would be acceptable to all parties involved. By late 1999, however, it had become clear that opposition to Myriad was nationwide—neither British health-care professionals nor patients would be likely to welcome the company. Some health-care professionals even threatened to bring Myriad to court if the company tried to enforce its patents. In addition, there were indications that Mike Stratton might sue Myriad for illegally acquiring a license to his patent. Myriad, however, remained persistent. Britain could be the gateway to a potential gold mine of European patients.

During this period, Myriad negotiated with private laboratories to offer BRCA testing to the UK population, and in March of 2000, the company announced that it had issued a license to Rosgen Ltd., an Edinburgh-based private genetics laboratory, which would offer laboratory analysis of the BRCA genes—within a context of pre- and post-test counseling—on a fee-for-service basis (Myriad Genetics 2000). Patients with private health insurance or who could afford to pay out of pocket could utilize the faster and arguably more technically sensitive services. At the time, however, this agreement did not affect Myriad's ongoing negotiations with the NHS.

The testing system created by the Myriad-Rosgen agreement is fascinating for three reasons. First, by not interfering with the NHS's testing services, Myriad lost the majority of revenue from BRCA testing services in Britain. Second, Myriad appeared to learn from its battle and opted to change its testing system to include counseling—thereby creating a service that was more in keeping with the priorities of British health-care professionals and the NHS. Finally, while Rosgen seemed to accept the British approach to counseling, however, it adopted Myriad's definition of the individual's right to choose to access testing. The company implicitly argued that it would provide testing to those individuals who were not able to access NHS services due to lack of risk factors, when it stated in its promotional materials that it would offer the widest possible access. Although British health-care professionals emphasized the ideal of providing everyone with equal access to the system, Rosgen adopted Myriad's approach to provide the "best possible testing" to the widest audience possible.

Despite Myriad's concessions, British health-care professionals were still reluctant to use the Myriad-Rosgen system and chose instead to continue to use the NHS BRCA testing service. When Rosgen sent letters to general practitioners across Britain announcing its service in June 2000, for example, staff of the Southwest Thames Regional Genetics Service issued a vehement response. The service was reluctant to use Myriad's system because it seemed to doubly exploit those who had given blood samples used in isolating the gene. "Much of the original work in mapping the genes for BRCA1 and BRCA2 was done on families in the South West London and Surrey area as part of the charitably funded work by the Cancer Research Campaign at the Institute of Cancer Research and The Royal Marsden Hospital. This work was put in the public domain and Myriad Genetics has claimed a patent for BRCA2 on the basis of sequencing the remainder of the gene. It seems ironic therefore that relatives of these individuals who gave samples to improve medical science for all should potentially have testing prejudiced by this commercial interest" (Southwest Thames Regional Genetics Service 2000).

Finally, in November 2000, Myriad and Rosgen reached an agreement with the UK Department of Health. The settlement allowed the NHS to continue testing without paying royalties or licensing fees to Myriad. In what was hailed as an "unprecedented deal," Myriad and Rosgen agreed to waive royalties on all breast cancer genetic tests that had or would be provided by the National Health Service while Rosgen agreed to provide the NHS with data about the mutations it collected in order to improve the NHS's clinical services. Rosgen would continue to provide testing privately in the UK, to those individuals who could afford the £179 to £2600 fee (depending on which test was performed).

The fate of the deal, however, was threatened in January of 2001 when Rosgen filed for voluntary liquidation for reasons unrelated to their agreement with the NHS. Not only did Rosgen's collapse mean that Myriad no longer had a presence in Britain, but because its deal with the NHS had been based on its license with Rosgen, Myriad could choose to renegotiate, though it had not done so as of fall 2004. It is, however, quite unlikely that Myriad might try to expand its service to Britain again. Since 2000, Myriad has run into similar battles across Europe, and in May 2004, a coalition of European scientific and patient advocacy groups were successful in getting the European Patent Office to revoke Myriad's European BRCA1 gene patent (Pollack 2004, C3).

# Conclusion

When Myriad attempted to expand its BRCA testing service by exerting its patent rights in Britain, it encountered a very different terrain than the one it successfully dominated in the United States. There, a few "cease and desist" letters to competitors and lawsuits sufficed to establish a monopoly. In light of this easy domination at home, and because the British testing system that was available did not involve full-sequence BRCA testing, one might expect that transferring Myriad's technology would be easy. But Myriad's efforts in Britain were unsuccessful because actors throughout the UK health-care system objected to the approaches to biomedicine that were embedded in Myriad's testing system.

Using patent rights, the company tried to transfer not just a full-sequence test but an entire social and economic system based on an independent diagnostic laboratory and a particular moral order in which the laboratory limited an individual's access to testing only by ability to pay, did not require counseling, and imposed few restrictions on the physician who ordered the test. While Myriad's testing system identified the same phenomenon—mutations to the BRCA genes—as did the NHS system, the approach to biomedicine embedded in the company's system could not work in Britain as it had in the United States. In Britain, patent rights were not acceptable as a justification for control over a testing service, and the organizational form of the NHS regional genetics clinic—to provide clinical care and laboratory analysis—emphasized priorities that conflicted with Myriad's system. Embedded in the components of a technology are not only technical hardware and mechanisms to execute certain functions but also social implications that add additional challenges to attempts to transfer technologies across national boundaries.

This story provides us with some insight into the relationship between technology and globalization. First, it demonstrates the importance of national context in the shaping of technology. While scholars of comparative politics and comparative health care systems have often pointed to the importance of regulatory systems, government institutions, and intangible national norms and values in the provision and use of technology, few have opened the "black box" of the technologies to investigate how the technical details themselves are shaped by national context. Opening this "black box" of genetic testing for breast cancer shows us how test providers in the United States and Britain built the

clinical and technical aspects of their new technologies in very different ways. Moreover, these differences in the methods of analyzing the
BRCA genes and providing clinical care were significant. They inspired
British scientists, health-care professionals, and patients to launch their
opposition to Myriad.

Second, the chapter shows that technology transfer, even between
two similar nations, is not simply a matter of moving equipment to a
new environment and explaining the benefits of the new technology to
the recipients. Technologies are deeply embedded in cultures, and their
technical details prescribe particular norms and values. In the Myriad
case, DNA sequencing machines and family medical history information forms were coupled with particular ways of understanding intellectual property and specific roles for the health-care professional and patient. In rejecting Myriad's technology, the British were responding not
only to the testing service but the social system that it prescribed as well.

Third, many scholars argue that technology transfer often serves as
a bridge to strengthen and promote globalization. In this case, however,
the proposed technology transfer became an opportunity to resist American domination and cultural homogenization and thus assert national
identity. The debate transformed from a technical one about the patentability of genes to a nationalist one about the British commitment to
approaching health care as a public good. This rich case provokes us to
ask additional questions about the tales of successful technology transfer
and globalization. Are these technologies really transferred so seam-
lessly and without controversy? When a technology is transferred, is it
really perceived and used as it was in its originating context?

Finally, the prospects of globalization can lead to the formation of
unlikely alliances. In this case, Myriad's BRCA testing service no longer
simply concerned the small groups of scientists who did research on the
genes, the health-care professionals who offered testing services, and the
fraction of individuals who might have an inherited risk of breast or
ovarian cancer. It became an issue that mobilized patients, scientists,
and health-care professionals—experts and laypeople—to consider the
fate of their health-care system.

NOTES

1. Winner has argued that the very design of technologies can prescribe
their social and political implications.

2. This paper is based on fieldwork conducted from 1998 to 2001. I conducted in-depth, semistructured interviews with approximately one hundred individuals involved in the development of genetic testing for breast cancer in the United States and Britain. I used a snowball sampling methodology to identify potential interviewees, first interviewing individuals most visibly involved in the development of the new technology in the two countries, and then relying on their referrals for subsequent interviewees. Interviewees included scientists involved in breast cancer genetics research, geneticists, counselors, and nurses who offered BRCA testing services to the public, innovators who developed BRCA testing, representatives of patient advocacy groups, government officials interested in regulating genetic testing, and scholars interested in the provision of genetic testing. I also analyzed documents provided by interviewees and reports produced by deliberative bodies concerned with the provision of BRCA testing. Finally, I attended meetings of the U.S. Department of Health and Human Services Secretary's Advisory Committee on Genetic Testing, the American Society of Human Genetics, the World Conference on Breast Cancer Advocacy, and the UK Genetics and Insurance Committee.

3. These patents were later granted.

4. UK Geneticist #1. Personal interview. August 10, 1999.

5. UK Genetics nurse. Personal interview. September 1999. British healthcare professionals and even government officials suspected that Myriad wanted to shut down NHS testing services before those of any other countries because Britain likely had the most advanced testing services as well as high incidences of breast cancer.

6. DoH official. Personal interview. October 1999.

7. DoH official. Interview.

8. Jeremy Rifkin, an American critic of biotechnology, did start a petition opposing the patenting of the BRCA genes. This petition was signed by some patient activist groups, such as the National Breast Cancer Coalition and National Ovarian Cancer Coalition, but did not result in any major policy change. In fact, the NBCC later wrote an article in its newsletter announcing an equivocal position on the patenting of human genes. It argued that patenting could be beneficial to research, whereas a ban on patenting genes might have unforeseen consequences.

9. In 1980, the U.S. Congress passed the Bayh-Dole Act, which allowed for private ownership of inventions funded by the taxpayer. Small businesses, nonprofit organizations, and universities could now patent and commercialize inventions that resulted from government-funded research. See Sampat (chapter 3 in this volume).

10. Hilgartner (2002), for example, describes how laboratories participating in the project to map and sequence the human genome defined property

rights and the distinctions between public and private domains in very specific ways. Locally contingent definitions of inventorship and property are also evident in how patent rights are attributed and used. Marianne de Laet (2000) argues that patents "are different things in different places," noting that a single patent can be, simultaneously, a recognition of achievement at a laboratory in the Netherlands, a mechanism to protect innovation at the World Intellectual Property Organization in Switzerland, and a source of information at a government ministry in Africa.

11. While American scientists and physicians voiced little opposition to gene patenting in the 1980s and early 1990s, some groups began to criticize the practice when researchers attempted to patent expressed sequence tags (ESTs), gene fragments with unknown function. These criticisms, however, did not usually cover the patenting of genes with known function (e.g., disease genes).

12. Wendy Watson. Personal interview. June 1998.

13. Ibid.

14. Wendy Watson. Personal interview. October 1999.

15. UK Geneticist #1. Personal interview. August 10, 1999.

16. UK Geneticist #3. Personal interview. August 13, 1998.

17. Wendy Watson. Personal interview. June 1998.

18. UK Geneticist #2. Personal interview. July 27, 1998.

19. UK Molecular Geneticist. Personal interview. December 23, 1999.

20. The idea that accuracy is a social achievement, which is locally contingent and requires negotiation over standards and margins of error, has been widely discussed by scholars of science and technology studies.

21. One molecular geneticist noted, "We do not question the validity of the idea that a sequencing test is currently a technical 'gold standard.' Sequencing is probably not 100 percent sensitive though and the Myriad test does not claim to pick up all mutations—deletions for instance and cannot pick up mutations in other breast cancer predisposition genes apart from BRCA1/2." UK Oncologist #1. Personal interview. September 20, 1999.

22. UK Molecular Geneticist. Personal interview. December 23, 1999.

23. Wendy Watson. Personal interview. June 1998.

### REFERENCES

96th Cong. 1980. *Public Law 96-517* (12 December). Washington, DC.

Adam, S. 2000. *Future Provision of BRCA-Testing Services.* London: Department of Health.

American Medical Association. 2000. Patenting of genes and their mutations. *Report 9 of the Council on Scientific Affairs.*

Andrews, T., et al. 1997. "As researchers or clinical scientists . . ." Letter (14 July).

Biagioli, M. 1998. The instability of authorship: Credit and responsibility in contemporary biomedicine. *FASEB Journal* 12 (January).

Bowker, G. 2001. The new knowledge economy and science and technology policy. *Encyclopedia of Life Support System*. Paris: UNESCO.

Boyle, J. 1996. *Shamans, Software, and Spleens : Law and the Construction of the Information Society*. Cambridge, MA: Harvard Univ. Press.

Bremmer, C. 1998. Euro-MPs clear way for genetic patents. *Times* (London), 13 May.

British Society for Human Genetics. 1998. *BSHG Statement on Patenting and Clinical Genetics*. Available at http://www.bham.ac.uk/BSHG/patent2.htm. 8 December 1999.

———. 1997. *Patenting of Human Gene Sequences and the EU Draft Directive*. Available at http://www.bshg.org.uk/Official%20Docs/patent_eu.htm. 1 April 2001.

Butler, D., and S. Goodman. 2001. French researchers take stand against cancer gene patent. *Nature* 413: 95–96.

Butler, K., and C. Arthur. 1998. Anger as Europe votes to "sell off" genes. *Independent* (London), 13 May.

Cambrosio, A., and P. Keating. 1995. *Exquisite Specificity: The Monoclonal Antibody Revolution*. New York: Oxford Univ. Press.

Clinical Molecular Genetics Society. 1999. *Gene Patents and Clinical Molecular Genetic Testing in the UK*. Executive Committee of the CMGS.

Connor, S. 1994. Concern over Cancer Gene Patent. *Independent* (London), 15 September.

Dalpé, R., L. Bouchard, A.-J. Houle, and L. Bédard. 2003. Watching the race to find the breast cancer genes. *Science, Technology, and Human Values* 28: 187–16.

Davies, K., and M. White. 1995. *Breakthrough: The Race to Find the Breast Cancer Gene*. New York: John Wiley & Sons, Inc.

de Laet, M. 2000. Patents, travel, space: Ethnographic encounters with objects in transit. *Environment and Planning: Society and Space* 18 (2): 152.

Dickson, D. 2000. Politicians seek to block human-gene patents in Europe. *Nature* 404 (6780): 802.

Dobson, R. 1997. Women fight patent on cancer test. *Sunday Times* (London), 20 April.

Drori, G. S., J. W. Meyer, F. O. Ramirez, and E. Schofer. 2003. *Science in the Modern World Polity: Institutionalization and Globalization*. Palo Alto, CA: Stanford Univ. Press.

European Parliament and Council of the European Union. 1998. Directive on the legal protection of biotechnological inventions. *Directive 98/44/EC*.

Green, D. 1995. Parliament scuppers a new patents directive. *Financial Times*, 3 March: 4.

Hawkes, N. 1995. Euro MPs turn down life-form patent law. *Times* (London), 2 March.

Hilgartner, S. 2002. Acceptable intellectual property. *Journal of Molecular Biology* 319 (4): 943–46.

Keller, R. T., and R. R. Chinta. 1990. International technology transfer: Strategies for success. *Executive* 4: 33–43.

Lewin, T. 1996. Move to patent gene is called obstacle to research. *New York Times*, 21 May.

Mackenzie, D. 1993. *Inventing Accuracy: A Historical Sociology of Nuclear Missile Guidance.* Cambridge, MA: MIT Press.

Meek, J. 2000. Money and the meaning of life: Business and science in race to crack the genetic code. *Guardian*, 17 January.

Merton, R. K. 1973. The normative structure of science. Repr. in Merton, *The Sociology of Science*, 267–78. Chicago: Univ. of Chicago Press. (Orig. pub. 1942.)

Mitroff, I. 1974. Norms and counter-norms in a select group of the Apollo moon scientists: A case study of the ambivalence of scientists. *American Sociological Review* 39 (August): 579–95.

Mulkay, M. J. 1976. Norms and ideology in science. *Social Science Information* 15 (4–5): 637–56.

Myriad Genetics. 1998. Myriad Genetics hosting conference of European experts on breast cancer genetic testing. *Press Release*. Salt Lake City, UT.

———. 2001. Myriad Genetics launches predictive medicine business in Brazil. *Press Release* (14 November).

———. 2001. Myriad Genetics launches predictive medicine testing in Germany, Switzerland, and Austria. *Press Release* (27 June).

———. 2000. Myriad Genetics Launches genetic testing in the United Kingdom and Ireland. *Press Release* (March).

National Breast Cancer Coalition. 1997. Gene patenting: Yes or no? *Call to Action! The Quarterly Newsletter of the National Breast Cancer Coalition and the National Breast Cancer Coalition Fund* (Fall/Winter).

Parthasarathy, S. 2003. Knowledge is power: Genetic testing for breast cancer and patient activism in the U.S. and Britain. In *How Users Matter: The Co-Construction of Users and Technology,* ed. N. Oudshoorn and T. Pinch. Cambridge, MA: MIT Press.

Pinch, T. 1993. "Testing—one, two, three . . . testing!": Towards a Sociology of Testing. *Science, Technology & Human Values* 18: 25–41.

Pollack, A. 2004. Patent on test for cancer is revoked by Europe. *New York Times*, 19 May.

Rommens, J. M., J. Simard, F. Couch, A. Kamb, B. L. Wever, and S. Tavtigian, Endorecherche Inc., HSC Research and Development Ltd. Partnership, Myriad Genetics, Inc., Trustees of the Univ. of Pennsylvania. 2002.

*Chromosome 13-linked Breast Cancer Susceptibility Gene.* European Patent 1260520.

Rosgen Ltd. 2000. UK company announces licensing agreement for breast cancer genetic testing. *Press Release* (March).

Ross, E. 2000. Scientists object to gene patent. *Associated Press,* 18 January.

Segerstrom, P. S., T. C. A. Anant, and E. Dinopoulos. 1990. A Schumpeterian model of the product life cycle. *American Economic Review.* 80: 1077–91.

Shattuck-Eidens, D., Y. Miki, D. E. Goldgar, A. Kamb, M. H. Skolnick, J. Swenson, R. W. Wiseman, A. P. Futreal, K. D. Harshman, and S. V. Tavtigian, Myriad Genetics Inc., Univ. of Utah Research Foundation, U.S. Department of Health. 1996. *Method for Diagnosing a Predisposition to Breast or Ovarian Cancer.* European Patent 0699754.

Southwest Thames Regional Genetics Service. 2000. *Stop Press: Myriad Genetics Attempt to Monopolize Breast Cancer Testing.* St. Georges Hospital Medical School. Available at http://www.genetics-swt.org/oldnews.htm. 28 June 2001.

Winner, L. 1986. Do artifacts have politics? *The Whale and the Reactor: A Search for Limits in an Age of High Technology.* Chicago: Univ. of Chicago Press.

# ABOUT THE EDITORS

DAVID H. GUSTON is professor of political science and associate director of the Consortium for Science, Policy and Outcomes at Arizona State University, where he also directs the NSF-funded Center for Nanotechnology in Society. His book *Between Politics and Science: Assuring the Integrity and Productivity of Research* (Cambridge Univ. Press) won the 2002 Don K. Price award from the American Political Science Association for best book in science and technology policy. He is a co-author (with M. Jones and L. M. Branscomb) of *Informed Legislatures* (Univ. Press of America) and co-editor (with K. Keniston) of *The Fragile Contract* (MIT Press). He received his PhD from MIT in political science in 1993. He is a fellow of the American Association for the Advancement of Science.

DANIEL SAREWITZ is professor of science and society and director of the Consortium for Science, Policy and Outcomes at Arizona State University. He is the author of *Frontiers of Illusion: Science, Technology, and the Politics of Progress* (Temple Univ. Press) and coeditor (with A. Lightman and C. Desser) of *Living with the Genie: Essays on Technology and the Quest for Human Mastery* (Island Press) and (with Roger Pielke, Jr. and Radford Byerly, Jr.) *Prediction: Science, Decision-Making, and the Future of Nature* (Island Press). From 1989 to 1993, he worked as a Congressional Science Fellow and then science consultant to the House Science, Space, and Technology Committee. He received his PhD in geological sciences from Cornell University in 1986.

# ABOUT THE CONTRIBUTORS

CHARLOTTE AUGST is the policy manager at the Human Fertilisation and Embryology Authority, the governmental body regulating fertility services and the use of human embryos for treatment and research. Previously, she worked in public health genetics and as a researcher for Dr. Ian Gibson, Member of Parliament and Chair of the Science and Technology Select Committee. She received her PhD from the University of London in 2001.

MICHAEL BARR earned his PhD in philosophy from the University of Durham in England. He currently works at the London School of Economics and Political Science where he is investigating the ethical and social impact of genome-based medicines for depression.

GRANT BLACK is an assistant professor of economics at Indiana University–South Bend and serves as director of the Bureau of Business and Economic Research and the Center for Economic Education. He received his PhD in economics from Georgia State University in 2001.

MARK B. BROWN is an assistant professor in the Department of Government at California State University, Sacramento. He received a PhD in political science from Rutgers University in 2001 and worked for two years as a postdoctoral fellow at the Institute for Science and Technology Studies at Bielefeld University in Germany.

KEVIN ELLIOTT received his PhD in 2004 from the Program in History and Philosophy of Science at the University of Notre Dame. He is currently an assistant professor in the Department of Philosophy at Louisiana State University.

PATRICK FENG is an assistant professor in the faculty of Communication and Culture at the University of Calgary. He received his PhD in science and technology studies from Rensselaer Polytechnic Institute in 2002 and recently completed a two-year postdoctoral fellowship at Simon Fraser University in Vancouver, Canada.

PAMELA M. FRANKLIN received her doctorate in the Energy and Resources Group at the University of California, Berkeley, in May 2002. She served as a Congressional Science and Technology Fellow, sponsored by the American Association for the Advancement of Science. She is currently working in the Climate Change Division at the U.S. Environmental Protection Agency.

CAROLYN GIDEON is assistant professor of international communication and technology policy at the Fletcher School of Law and Diplomacy at Tufts University. She received her PhD in public policy from Harvard University in 2003.

TENÉ HAMILTON FRANKLIN is a genetic counselor at Meharry Medical College, where she provides genetic counseling services to underserved populations. In addition, she provides community-based education about genetics to the African American community.

BRIAN A. JACKSON is a physical scientist on the research staff of the RAND Corporation. He received a PhD from the California Institute of Technology in bio-inorganic chemistry in 1999.

SHOBITA PARTHASARATHY is currently an assistant professor at the Ford School of Public Policy at the University of Michigan. She received her PhD from the Department of Science and Technology Studies at Cornell University in 2003.

JASON PATTON received his PhD in science and technology studies (STS) from Rensselaer Polytechnic Institute. He studies the relationship between social and material changes in urban transportation infrastructure. He works for the City of Oakland, California, on multimodal street design and environmental justice planning.

A. ABIGAIL PAYNE is an associate professor of economics at McMaster University. She holds a Canada Research Chair in Public Economics and is also the director of the Public Economics Data Analysis Laboratory (PEDAL). PEDAL is funded through grants from the Canada Foundation for Innovation, Ontario Innovation Fund, and McMaster University. Her research on higher education has been funded through grants from the National Science Foundation, Social Sciences and Humanities Research Council of Canada, and the Andrew W. Mellon Foundation. She received her PhD in Economics from Princeton University and her JD from Cornell University.

BHAVEN N. SAMPAT is an assistant professor at the School of Public Health and School of International and Public Affairs at Columbia University. He taught at Georgia Institute of Technology from 2001 to 2003 and was a Robert Wood Johnson Foundation Scholar in Health Policy Research at the University of Michigan from 2003 to 2005. He received his doctorate in economics from Columbia in 2001.

CHRISTIAN SANDVIG is an assistant professor of Speech Communication at the University of Illinois at Urbana–Champaign and an Associate Fellow of Socio-Legal Studies at Oxford University.

SHERYL WINSTON SMITH is a lecturer at the Carlson School of Management at the University of Minnesota. She received her PhD in Public Policy from Harvard University in 2004. Her research focuses on innovation and international trade and investment.

# INDEX

Absorbtive capacity of firms, 175
Administrative Procedures Act (APA), 103
African-American community, genetics research, 277–79, 285–86, 288–90
Agglomeration, small-business innovation, 78–79, 85–87, 90–92
Akami Technologies, 249–50
Altruism, genetic donation and, 301–2
Anomalies, scientific: deliberation as response to, 136–41, 142; effects on science, 125–27, 134–38, 141–42; policy and, 124–27, 134–42; public opinion and, 134, 135–36
Arndt, Rudolph, 131
Arndt-Schultz law, 131
Autonomous science, public research and, 11–12
Autonomy, patient, 291, 294–96, 303, 304–7

Bayh-Dole Act, 9, 60–74
Bell Company, 257–59
Beneficial hormesis, 130–31
Benveniste affair, 143n.5
Biobanking: in Great Britain, 292–94; informed consent and, 294–303, 304–7; in U.S., 294
Bioethics, ethics advisory committees, 14
Biotech Patent Directive, 339
Breast cancer (BRCA) gene testing, 334–35
Business services, small business innovation, 84, 86, 89, 91

Bus systems, 219–21; and rapid transit, 222, 224–27

Cable industry, open access debate, 247–48
Calebrese, Edward, 127, 134–35, 143n.1
Chemical hormesis, 124–25, 127–34
Chevron v. NRDC, 104
Chloroform drinking water standard, EPA, 104–11
Christian ethics: reproductive technologies and, 318
Citations measuring innovation, 78
Cohen-Boyer technique licensing, 69
Colombia University, Research Corporation at, 58–59
Commercialization of research, 19
Committee on Science, Engineering, and Public Policy (COSEPUP), 43–45
Communications technology policy, historical development of, 257–62
*Communities of Color and Genetics Policy Project*, 276, 279–90
Communities of practice, 219–21
Community consent, 299–300
Competition: international, 184–85; new technology development and, 257–61, 263–69
Competitive local exchange companies, 260
Computer equipment industry, U.S., 173–74, 180–87

365